121 **Topics in Current Chemistry**

Fortschritte der Chemischen Forschung

Managing Editor: F. L. Boschke

Host Guest Complex Chemistry III

Editors: F. Vögtle, E. Weber

With Contributions by
J. E. Baldwin, Th. A. Kaden, O. Manabe, P. Perlmutter,
S. Shinkai, R. Sinta, J. Smid, M. Takagi,
Y. Takeda, K. Ueno

With 72 Figures and 29 Tables

Springer-Verlag
Berlin Heidelberg GmbH
1984

This series presents critical reviews of the present position and future trends in modern chemical research. It is addressed to all research and industrial chemists who wish to keep abreast of advances in their subject.

As a rule, contributions are specially commissioned. The editors and publishers will, however, always be pleased to receive suggestions and supplementary information. Papers are accepted for "Topics in Current Chemistry" in English.

ISBN 978-3-662-15290-4 ISBN 978-3-540-38750-3 (eBook)
DOI 10.1007/978-3-540-38750-3

Library of Congress Cataloging in Publication Data. (Revised for vol. 121) Main entry under title:
Host guest complex chemistry.
(Topics in current chemistry = Fortschritte der chemischen Forschung; 98, 101)
Includes bibliographical references and index.
1. Electron donor-acceptor complexes -- Addresses, essays, lectures. I. Vögtle, F. (Fritz), 1939 — II. Blasius, Ewald. III. Series: Topics in current chemistry; 98, etc.
QD1.F58 vol. 98 [QD474] 540s 541.2'242 81-8864

Originally published by Springer-Verlag Berlin Heidelberg New York in 1984
Softcover reprint of the hardcover 1st edition 1984

2152/3020-543210

Table of Contents

Preface IX

The Solvent Extraction of Metal Ions by Crown Compounds
Y. Takeda 1

Crown Compounds as Alkali and Alkaline Earth Metal Ion Selective Chromogenic Reagents
M. Takagi, K. Ueno 39

Photocontrol of Ion Extraction and Ion Transport by Photofunctional Crown Ethers
S. Shinkai, O. Manabe 67

Macroheterocyclic Ligands on Polymers
J. Smid, R. Sinta 105

Synthesis and Metal Complexes of Aza-Macrocycles with Pendant Arms having Additional Ligating Groups
Th. A. Kaden 157

Bridged, Capped and Fenced Porphyrins
J. E. Baldwin, P. Perlmutter 181

Author Index Volumes 101–121 221

Tables of Contents of the Volumes 98 and 101

Host Guest Complex Chemistry I

Crown-Type Compounds — An Introductary Overview
E. Weber, F. Vögtle

Concept, Structure, and Binding in Complexation
D. J. Cram, K. N. Trueblood

Complexation of Uncharged Molecules and Anions by Crown-Type Host Molecules
F. Vögtle, H. Sieger, W. M. Müller

Analytical Applications of Crown Compounds and Cryptands
E. Blasius, K.-P. Janzen

Host Guest Complex Chemistry II

Structural Chemistry of Natural and Synthetic Ionophores and their Complexes with Cations
R. Hilgenfeld, W. Saenger

Dynamic Aspects of Ionophore Mediated Membrane Transport
G. R. Painter, B. C. Pressman

Bioorganic Modelling — Stereoselective Reactions with Chiral Neutral Ligand Complexes as Model Systems for Enzyme Catalysis
R. M. Kellogg

Phase-Transfer Catalyzed Reactions
F. Montanari, D. Landini, F. Rolla

Closing Remarks

Preface

Since C. J. Pedersen reported the synthesis of the first crown ethers in 1967, interest in organic multidentate ligands has continually increased. The research concerning these compounds is broad, ranging from synthesis of new ligand topologies to investigations of selective binding properties and their application in the catalysis of synthetic organic reactions to analytical separation and determination problems. This volume is the third in a series, presenting a selection of contributions in the interesting field of modern coordination chemistry with organic ligands. At present, the functional facet of neutral ligands is perhaps the most exciting one. Therefore the overall contents of the various chapters is "to let the crown compounds work".

Chapter 1, by Y. Takeda, presents a detailed introduction of the solvent extraction of metal ions by crown compounds. Fundamental physical equilibria, which are helpful in understanding extraction processes are explained, and the various factors affecting the extent of ion extraction are discussed. Examples of analytical applications (separation analysis, colorimetric determination) of this technique round up the picture.

Chapter 2, by M. Takagi and K. Ueno, shows crown compounds functioning as alkali and alkaline earth metal ion selective chromogenic reagents. This report deals with recent developments in the synthesis of crown derivatives, which have chromophoric functional groups within its molecular skeleton (crown ether dyes). These compounds are designed to bring about specific colour changes on the interaction with certain alkali/alkaline earth metal ions, thus being able to serve as probes of photometric reagents for these metal ions. The utility of modified types of indicator systems in the fluorescence analysis of ion concentration is also demonstrated, showing the idea of chromoionophores to be realized in a broad sense.

Chapter 3, by S. Shinkai and O. Manabe, offers a review over the fascinating field of coupled ion transport. Artificial membrane systems affected by an energy switch of light, a pH-antenna, a redox-couple and other types of responsive functionalities introduced into a crown carrier are discussed "at work". Studies on liposomal membranes are also included. The scope of the principle of responsive functionality control by modified crown compounds shows a new entry into enzyme mimickry in respect to certain coupled ion/electron transport processes.

Chapter 4, by J. Smid and R. Sinta, deals with the synthesis and the binding properties of polymeric crown/cryptand ligands (linear polymeric and those anchored to insoluble supports). Extensive information on the complexation behaviour in aqueous and nonaqueous media is given. The effect of variables in polymer structure is analyzed and applications in ion transport, exchange resin chromatography and polycrown catalysis are demonstrated.

Chapter 5, by T. A. Kaden, describes an interesting new class of ligands which combine structural characteristics of a macrocyclic oligoaza ligand with specific properties of coordinating flexible side chains. Insight into the relations between reactivity and structure of the pendant functional groups and of the bound metal ion is given. The different complex geometries in respect to the functional group appendages (axial coordination) and possible applications regarding enzyme analogy are also discussed.

The concluding *chapter 6*, by J. E. Baldwin and P. Perlmutter, reviews the chemistry of diverse metalloporphyrin derivatives, which over the last decade have become known as "bridged, capped and fenced porphyrins". The article is divided into sections dealing with synthetic models for haemoglobin, myoglobin, cytochrome P-450, cytochrome *c* oxidase. Each section includes a brief description of the natural system with the properties, which should be considered in a reliable model.

As a general conclusion these contributions demonstrate that the crown chemistry has reached a level beyond merely synthesizing ligands and the formation of ordinary crown complexes, but is concerned with the sophisticated functionalization of crown compounds in terms of styling new controlling processes and regulation systems in response to a functionality couple. The spreading of these principles into different fields of application (dye stuffs, drugs, micelles, enzyme models, membrane ionophores etc.) are coming off broadly.

This book therefore aims at researchers and students in organic, physical, analytical, and inorganic chemistry, but should also be of interest for many from the biological, physiological and physiochemical fields.

The editors wish to acknowledge their indebtedness to all who have helped in bringing this book to completion.

Bonn, September 1983 F. Vögtle, E. Weber

List of constitutional formulae of crown compounds and of their notations/abbreviations frequently used in this volume[a]

Formula	Notation	Abbreviation
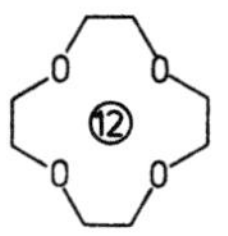	[12] crown-4	12C4
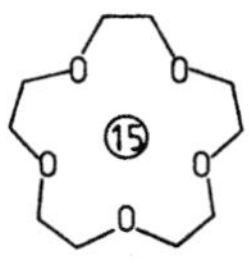	[15] crown-5	15C5
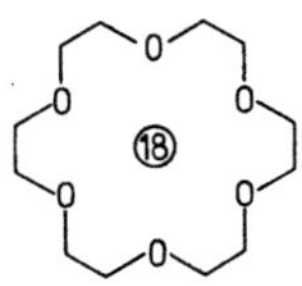	[18] crown-6	18C6
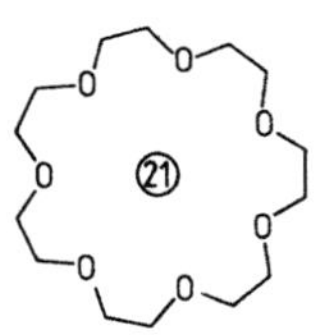	[21] crown-7	21C7
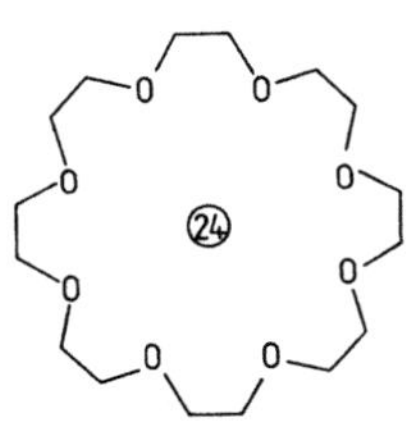	[24] crown-8	24C8
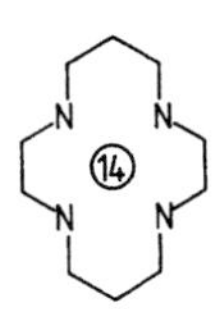	Tetraaza [14] coronand-4	TA14C4
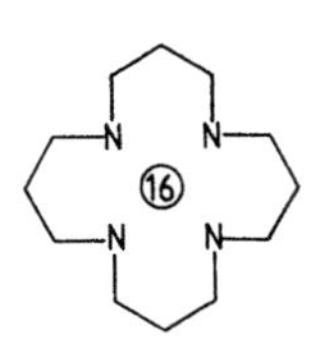	Tetraaza [16] coronand-4	TA16C4

Formula	Notation.	Abbreviation
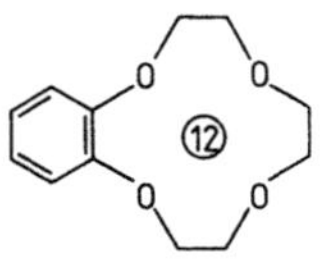	Benzo [12]crown - 4	B12C4
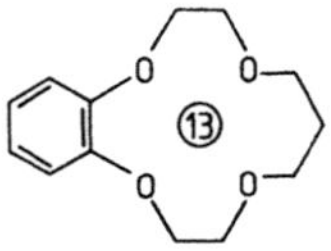	Benzo [13] crown - 4	B13C4
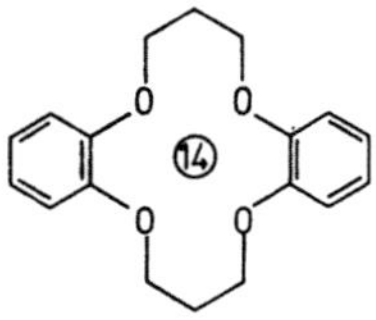	Dibenzo [14] crown - 4	DB14C4
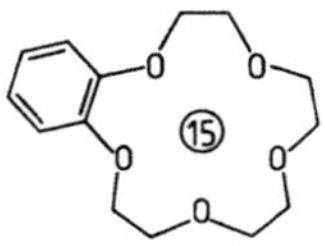	Benzo [15] crown - 5	B15C5
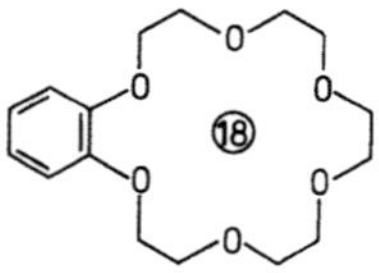	Benzo [18] crown - 6	B18C6
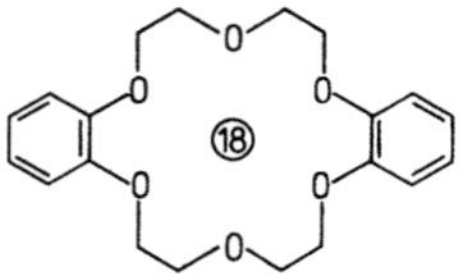	Dibenzo [18] crown - 6	DB18C6
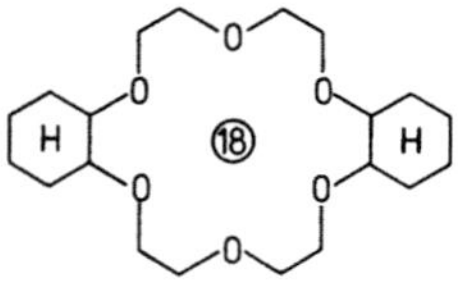	Dicyclohexano [18] crown - 6	DCH18C6
	Dibenzo [21] crown - 7	DB21C7

Formula	Notation	Abbreviation
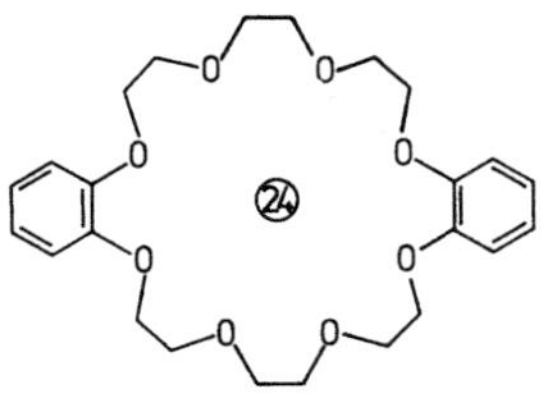	Dibenzo[24]crown-8	DB24C8
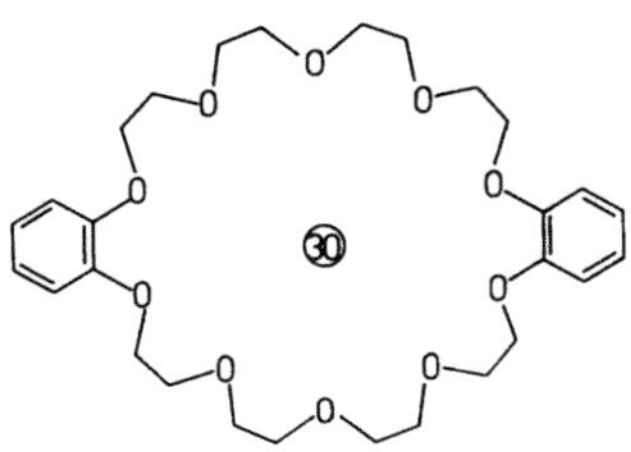	Dibenzo[30]crown-10	DB30C10
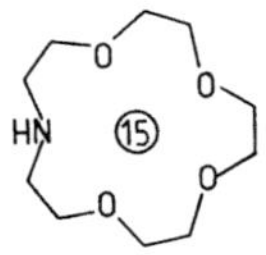	Monoaza[15]coronand-5	MA15C5
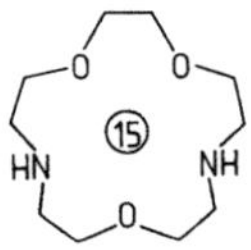	Diaza[15]coronand-5	DA15C5
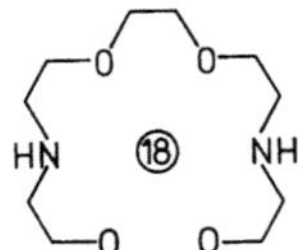	Diaza[18]coronand-6	DA18C6
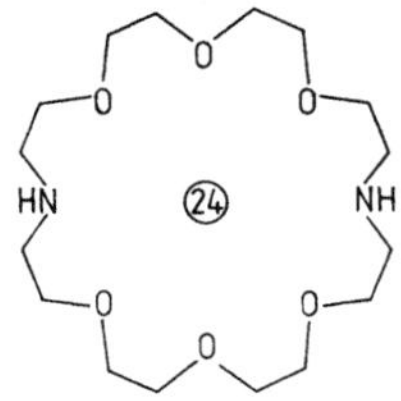	Diaza[24]coronand-8	DA24C8

Formula	Notation	Abbreviation
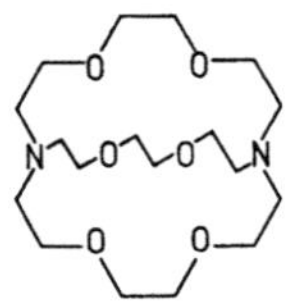	Cryptand [2.2.2]	[2.2.2]

[a] For more detailed information on crown/cryptand nomenclature see ("Crown-Type Compounds — An Introductory Overview") Weber, E., Vögtle, F.: Top. Curr. Chem. *98*, 1 (1981); sect. 2 and 3.1.

The Solvent Extraction of Metal Ions by Crown Compounds

Yasuyuki Takeda

Department of Chemistry, Faculty of Science, Chiba University, Yayoi-chō, Chiba 260, Japan

Table of Contents

1 Introduction 2

2 Equilibrium for Extraction of Metal Ions by Crown Ethers 2

3 Factors Affecting Extraction 10
3.1 Distribution of a Free Crown Ether 10
3.2 Stability of a Crown Ether Complex in Aqueous Solution 11
3.3 Extraction of a Crown Ether Complex 13
3.4 Counter Anion 15
3.5 Diluent 17
3.6 Coexisting Species in the Aqueous Phase 19
3.7 Structure of a Crown Ether 19

4 Extraction of Higher-Order Crown Ether Complexes 26

5 Extraction by Bis- and Poly-Crown Ethers 27

6 Application of Crown Ethers to Extraction Analysis 32
6.1 Separation Analysis 32
6.2 Colorimetric Determination 33

7 Acknowledgements 36

8 References 36

1 Introduction

Solvent extraction is a convenient and useful method in solution and analytical chemistry. It can be employed to concentrate and separate metal ions, and to determine stoichiometries and stabilities of complexes extracted into an organic phase. It thus makes sense to use solvent extraction for studying the complexing ability of crown ethers (including cryptands) toward metal ions. Crown ethers consist of hydrophilic cavities and hydrophobic exteriors, and form stable complexes particularly with alkali and alkaline earth metal ions by holding the ions in the central cavity. To date, numerous papers report on metal ion binding properties of newly synthesized crown ethers determined by solvent extraction.

In general, the selectivity of crown ethers for metal ions in solvent extraction systems largely depends on the relation between the size of the metal ion and the cavity of the crown ether; in other words, the metal ion which fits the crown ether cavity best is the most extractable. Extraction equilibrium constants and distribution ratios are very important for analysis as they are a measure of extractability of a metal ion. Since extraction equilibrium is a result of several constituent equilibria, stability of a crown ether complex with a metal ion in an aqueous solution is not always a dominant factor in determining the magnitude of the extraction equilibrium constant. Therefore, it is important to analyze the extraction equilibrium by means of the constituent equilibria, and furthermore, to investigate the influence of each of the constituent equilibria on the extraction equilibrium. This would give a valuable hint for the molecular design of new crown ethers with high extraction-selectivity for a given metal ion.

In order to compare the relative complexing powers of various crown ethers for alkali metal ions, Pedersen [1)] was first to employ the solvent extraction method. He reported that alkali metal ions which more optimally fit the crown ether cavity are more extractable. Frensdorff [2)] elaborately studied the extraction of sodium and potassium picrates with dicyclohexano[18]crown-6 (DCH18C6)[1] A and B, and dibenzo[18]crown-6 (DB18C6). These experiments established the basic equilibria of the extraction process. In order to clarify the role of crown ethers in the extraction process, he analyzed the overall extraction equilibrium by studying three of the constituent equilibria: complex formation in the aqueous phase and distribution of the uncomplexed and the complexed crown ether.

2 Equilibrium for Extraction of Metal Ions by Crown Ethers

Extraction equilibrium between an aqueous phase of a metal ion (M^{m+}), a counter anion (A^-), and an organic phase of a crown ether (L), may be represented by the following equation:

$$M^{m+} + L_o + mA^- \rightleftharpoons (MLA_m)_o \qquad K_{ex} = \frac{[MLAm]_o}{[M^{m+}]\,[L]_o\,[A^-]^m}, \tag{1}$$

1 A list of the frequently used notations and their explanation are given on page XI

where K_{ex} is the extraction equilibrium constant; MLA_m designates as ion pair between a crown ether complex with the methyl ion (ML^{m+}) and the counter anion; subscript "o" and the lack of subscript denote an organic and an aqueous phase, respectively; square brackets indicate molar concentrations. The extraction equilibrium is considered to consist of the following constituent equilibria:

— distribution of the free crown ether between the two phases:

$$L \rightleftharpoons L_o \qquad D_L = \frac{[L]_o}{[L]}, (D_L = \text{distribution constant of a crown ether}) \tag{2}$$

— complexation reaction of the crown ether with the metal ion in the aqueous phase:

$$M^{m+} + L \rightleftharpoons ML^{m+} \qquad K_{ML} = \frac{[ML^{m+}]}{[M^{m+}][L]}, \tag{3}$$

$(K_{ML}$ = complex-formation constant)

— ion-pair extraction of the crown ether-metal ion complex with the counter anion:

$$ML^{m+} + mA^- \rightleftharpoons (MLA_m)_o \qquad K_{ex'} = \frac{[MLA_m]_o}{[ML^{m+}][A^-]^m} . \tag{4}$$

If a nonpolar solvent is used as an organic phase, dissociation of the ion pair MLA_m in the organic phase will be negligible:

$$K_{ex} = D_L^{-1} K_{ML} K_{ex'} . \tag{5}$$

The all-activity coefficients of the chemical species mentioned here are assumed as unity. However, when the concentrations of the metal ions and/or the counter anions in the aqueous phase are high, it is necessary to take into account the activity coefficients of the ions.

The distribution ratio of the metal ion can be defined as follows:

$$D_M = \frac{[M]_{o,t}}{[M]_t},$$

and calculated from the experimental data where $[M]_{o,t}$ and $[M]_t$ are total metal concentrations in the organic and aqueous phases, respectively. Assuming that every ion-pair formation is negligible in the aqueous phase because of the high dielectric constant of water we arrive at:

$$D_M = \frac{[MLA_m]_o}{[M^{m+}] + [ML^{m+}]} . \tag{6}$$

The $[MLA_m]_o$ value can be obtained experimentally. In the case of $[M^{m+}] \gg [ML^{m+}]$ Eq. 6 becomes:

$$D_M = K_{ex}[A^-]^m [L]_o . \tag{7}$$

The values of $[A^-]$ and $[L]_o$ can be obtained from the mass balance. By way of example, $\lg(D_M/[A^-])$ vs. $\lg[L]_o$, $\lg(D_M/[A^-]^2)$ vs. $\lg[L]_o$ and $\lg(D_M/[L]_o)$ vs. $\lg[A^-]$ plots are illustrated in Figs. 1–3, respectively. Figure 1 corresponds to the [18]crown-6 (18C6)-univalent metal picrate system, and both Figs. 2 and 3 correspond to the 18C6-bivalent metal picrate system. Figure 1 shows a linear relationship with a slope of 1 in every case, indicating that 18C6 forms a 1:1 complex with the univalent metal ion. Figures 2 and 3 always show lines with slopes of 1 and 2,

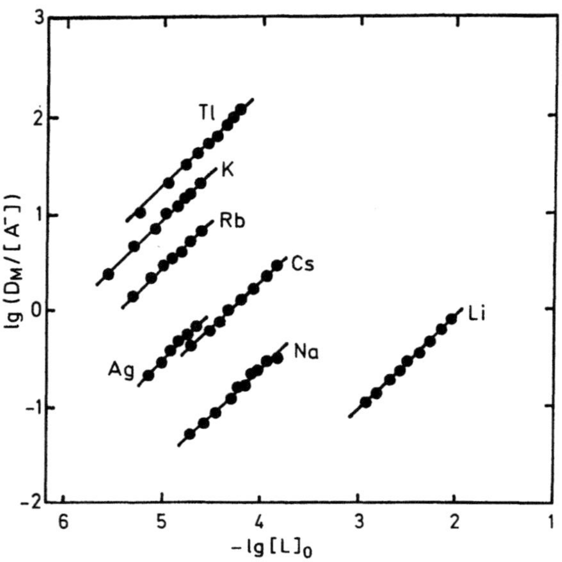

Fig. 1. Plots of $\lg(D_M/[A^-])$ vs. $-\lg[L]_0$ for the 18C6-univalent metal picrate in the benzene-water system at 25 °C [5)]

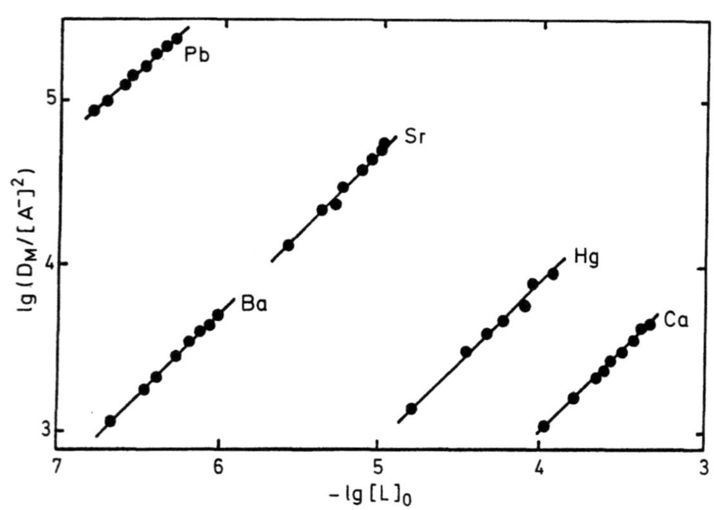

Fig. 2. Plots of $\lg(D_M/[A^-]^2)$ vs. $-\lg[L]_0$ for the 18C6-bivalent metal picrate in the benzene-water system at 25 °C [6)]

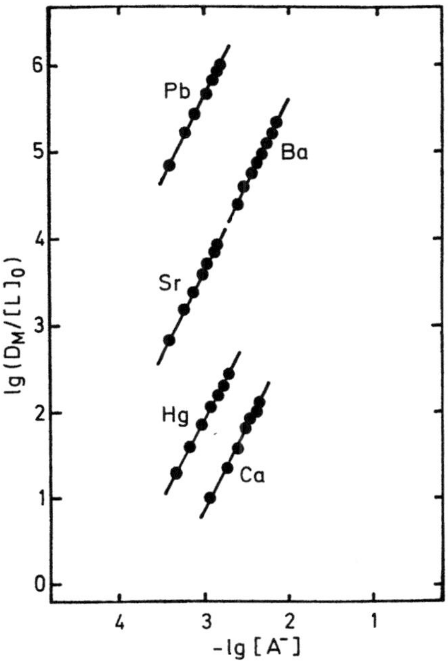

Fig. 3. Plots of lg ($D_M/[L]_0$) vs. $-$lg [A^-] for the 18C6-bivalent metal picrate in the benzene-water system at 25 °C [6]

respectively. This indicates that 18C6 forms a 1:1:2 complex with the bivalent metal ion and the picrate ion.

The picrate anion is widely used as a counter anion in the crown ether extraction system. The reason is as follows: A crown ether complex with a metal picrate is more easily extracted into an organic phase because of the large molar volume of the picrate anion. Furthermore, the extracted picrate complex can be determined by colorimetry owing to the colored anion.

lg K_{ex} values of crown ether complexes with metal picrates at 25 °C for benzene-water system are shown in Table 1. When a polar solvent is used as a diluent, it is more complicated to analyze the extraction equilibrium since dissociation of an ion pair in on organic phase is not negligible. Thus, benzene was chosen as a solvent because of its nonpolarity. Plots of lg K_{ex} vs. crystal ionic radius (r_c) for univalent and bivalent metal systems are given in Figs. 4 and 5, respectively. It can be seen from Fig. 4 and Tables 1 and 2 that the nearer the crystal ionic radius of alkali metal approaches to the cavity radius of crown ether, the more extractable the alkali metal ion is; exceptions being the sequence of Li^+ and Na^+ for the 12C4 system, and Li^+ for both 15C5 and B15C5 systems. The lower extractability of 12C4, 15C5, and B15C5 complexes with lithium picrate than expected on the basis of the size-fit idea may be attributed to the fact that Li^+ and probably crown ether complexed Li^+ [18] are by far the most strongly hydrated of all the alkali metal ions. A possible explanation for the difference in the extractability sequences of Li^+ and Cs^+ for 15C5 and B15C5 is as follows: The extractability of the crown ether complex with the univalent metal picrate primarily depends on the interaction of the univalent metal ion in the crown ether cavity with water molecules [9]. Since among the alkali metal ions included in crown cavities, the interaction of lithium ions with

Table 1. Lg K_{ex} values of crown ether complexes with metal picrates at 25 °C for the benzene-water system

Crown ether	Metal ion											
	Li^+	Na^+	K^+	Rb^+	Cs^+	Ag^+	Tl^+	Ca^{2+}	Sn^{2+}	Ba^{2+}	Hg^{2+}	Pb^{2+}
12C4	0.77 [3]	0.95 [3]	0.57 [3]	0.29 [3]	0.21 [3]	1.96 [4]	2.22 [4]	—	—	—	—	—
15C5	1.29 [5]	3.90 [5]	2.58 [5]	2.14 [5]	1.90 [5]	4.45 [5]	3.80 [5]	3.72 [6]	5.70 [6]	5.20 [6]	—	6.46 [6]
B15C5	1.26 [7]	3.29 [7]	1.93 [7]	1.44 [7]	1.08 [7]	3.33 [4]	3.23 [4]	—	—	—	—	—
18C6	1.92 [5]	3.39 [5]	5.97 [5]	5.43 [5]	4.38 [5]	4.44 [5]	6.30 [5]	7.03 [6]	9.72 [6]	9.72 [6]	7.97 [6]	11.75 [6]
		3.31 [12]	6.00 [12]	5.43 [12]	4.28 [12]							
DB18C6	—	2.1 [13]	4.65 [8]	3.75 [8]	3.07 [8]	3.56 [9]	4.49 [10]	4.01 [6]	5.38 [6]	—	6.22 [6]	7.16 [6]
			4.4 [13]	3.8 [13]	2.9 [13]	3.6 [14]			5.3 [15]			6.9 [15]
DB24C8 [11]	0.67	2.05	2.79	3.00	3.15	3.09	4.06	3.07	3.99	6.86	—	6.32

Table 2. Cavity radii of crown ethers and crystal ionic radii of metals (Å)

Cation	Crystal ionic radius [16]	Crown ether	Cavity radius [17]
Li^+	0.60	12C4	0.6—0.75
Na^+	0.95	15C5	0.85—1.1
K^+	1.33	18C6	1.3—1.6
Rb^+	1.48	24C8	>2.0
Cs^+	1.69		
Ag^+	1.26		
Tl^+	1.40		
Ca^{2+}	0.99		
Sr^{2+}	1.13		
Ba^{2+}	1.35		
Hg^{2+}	1.10		
Pb^{2+}	1.20		

water seems to be strongest, hindrance for hydration of the B15C5 complex by the benzo group attached to 15C5 may be much more effective with Li^+ than with Cs^+. Thus, in the case of B15C5, Li^+ is more extractable than Cs^+. The following data may support the above discussion though they have been obtained in the nitrobenzene-water system. The numbers of water molecules coextracted with the Li^+-15C5, Li^+-B15C5, Cs^+-15C5 and Cs^+-B15C5 complexes into the nitrobenzene phase are 2.0, 1.4, 0.2 and 0.4, respectively [18]. For the lithium complex the number of water molecules decreases remarkably on moving from 15C5 to B15C5, while

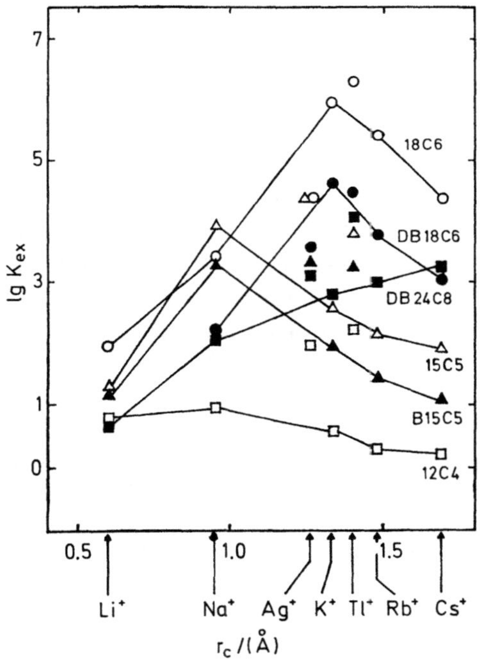

Fig. 4. Plots of lg K_{ex} at 25 °C between benzene and water vs. r_c for univalent metal picrate-crown ether systems
□: 12C4, △: 15C5, ▲: B15C5, ○: 18C6, ●: DB18C6, ■: DB24C8

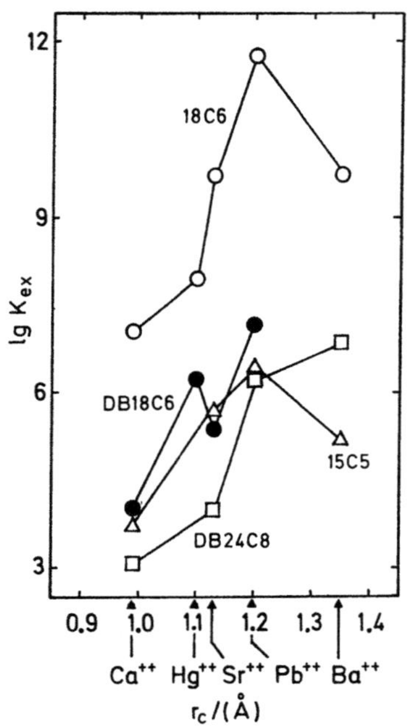

Fig. 5. Plots of lg K_{ex} at 25 °C between benzene and water vs. r_c for bivalent metal picrate-crown ether systems

for the caesium complex it increases only slightly. The univalent metal ions of Ag^+ and Tl^+ have a more suitable size for the crown ether cavity and show greater extractability, except for 12C4. This behavior is similar to that of the alkali metal ions (Fig. 4 and Tables 1 and 2). 15C5, B15C5, 18C6, and DB18C6 show pronounced cation selectivity for alkali metal ions (the former two for Na^+ and the latter two for K^+). On the other hand, 12C4 with a small cavity size compared to alkali metal ions (except of Li^+) and DB24C8 with a large cavity size in relation to alkali metal ions display no remarkable selectivity for any alkali metal ion.

Selectivity tendencies of crown ethers for alkaline earth metal ions coincide with the size-fit idea, except in the case of 15C5-Ca^{2+} (Fig. 5 and Tables 1 and 2). The much lower extractability of the 15C5 complex with calcium picrate than expected by the size-fit idea may be due to the fact that Ca^{2+} and probably 15C5 complexed with Ca^{2+} are the most strongly hydrated of all the alkaline earth metal ions (Ca^{2+}, Sr^{2+}, and Ba^{2+}). Although Pb^{2+} is larger than the cavity of 15C5 and Ba^{2+} fits more closely into the 18C6 cavity than Pb^{2+} (Table 2), both 15C5 and 18C6 show highest extraction efficiency for Pb^{2+} (Fig. 5).

The most remarkable substituent effect of two benzo groups attached to 18C6 on the extractability of Hg^{2+} and Sr^{2+} can be seen in Fig. 5. Here, the extractability sequence for Hg^{2+} and Sr^{2+} with 18C6 is $Hg^{2+} < Sr^{2+}$, while for DB18C6 it is $Hg^{2+} > Sr^{2+}$, although the crystal radii of Hg^{2+} and Sr^{2+} are nearly equal (Table 2).

It is interesting that DB24C8 reveals by far the highest selectivity for Ba^{2+} among all the alkaline earth metal ions (Fig. 5).

Jaber et al. [19] determined K_{ex} of DCH18C6 complexes with alkali and alkaline earth metal picrates between CH_2Cl_2 and water at 23 °C. The K_{ex} sequences for the alkali and alkaline earth metal ions are $K^+ > Rb^+ > Cs^+ > Na^+ > Li^+$ and $Ba^{2+} > Sr^{2+} > Be^{2+} > Ca^{2+} > Mg^{2+}$, respectively.

Judging from hard-soft principle, it can be considered that lanthanides and actinides as well as alkali and alkaline earth metals form very stable complexes with crown ethers. Recently some solvent extraction studies for lanthanides and actinides by means of crown ethers have been reported. Yakshin et al. [20] extracted nitrates of U^{6+}, Pu^{4+}, Pu^{6+}, Np^{4+}, and Np^{6+} by various crown ethers from an aqueous nitric acid solution into 1,2-dichloroethane and determined the stoichiometries of the extracted complexes to be $MO_2L(NO_3)_2$ for U^{6+}, Pu^{6+}, and Np^{6+}; and $ML_2(NO_3)_4$ for Pu^{4+} and Np^{4+}. DCH18C6 and DCH24C8 show the most effective extractability for actinides irrespective of the oxidation numbers, while DB18C6 and DB24C8 show the poorest extractability. Wang et al. [21] studied solvent extraction of lanthanides (Ce^{3+}, Nd^{3+}, Eu^{3+}, Tm^{3+}, and Yb^{3+}) and actinides (Th^4, UO_2^{2+}, and Am^{3+}) with 4′-methyl-B15C5 (4′-MB15C5) from an aqueous picric acid solution into chloroform at 25 ± 1 °C. They found the extracted lanthanide complexes to have a stoichiometry of 1:2:3 (metal/crown ether/picrate) and determined the extraction equilibrium constants of the complexes. The extractability sequences of lanthanides and actinides with 4′-MB15C5 are Nd^{3+} ($r_c = 0.995$ Å) > Eu^{3+} (0.950) > Ce^{3+} (1.034) > Tm^{3+} (0.869) > Yb^{3+} (0.858) and Th^{4+} ($\approx Nd^{3+}$, $r_c = 0.99$ Å) > Am^{3+} ($\approx Ce^{3+}$, $r_c = 1.07$ Å) ≫ UO_2^{2+} (0.89), respectively. In terms of thermodynamics, Kojima et al. [22] investigated solvent extraction of uranyl nitrate by DCH18C6 into various organic solvents in order to elucidate the effect of the diluents on the extraction. DCH18C6 extracted uranyl nitrate into cyclohexane, toluene and benzene by forming $UO_2L_2(NO_3)_2$, and into chlorobenzene and nitrobenzene by forming both $UO_2L(NO_3)_2$ and $UO_2L_2(NO_3)_2$. The extractability of uranyl nitrate by DCH18C6 in organic solvents increases in the order of cyclohexane < toluene < benzene < chlorobenzene < nitrobenzene. The enthalpy ($\Delta H°$) and entropy changes ($\Delta S°$) for the extraction process into benzene and nitrobenzene are all negative.

For further investigation on the role of crown ether in the extraction process, it is very important to obtain $\Delta H°$ and $\Delta S°$ for the extraction reaction of a crown ether with a metal ion. However, there is very little data on $\Delta H°$ and $\Delta S°$ at present. Thermodynamic quantities for the extraction of B15C5 and DB18C6 complexes with alkali metal picrates into benzene are listed in Table 3. Table 3 points out that the extraction reactions are all exothermic and the $\Delta S°$ values are all negative. For every system the $-\Delta H°$ value is larger than the $-T\,\Delta S°$ value and the major part of the

Table 3. Thermodynamic quantities for the extraction of B15C5 and DB18C6 complexes with alkali metal picrates into benzene

		$\Delta H°$/kJ mol^{-1}	T $\Delta S°$/kJ mol^{-1} (at 25 °C)
B15C5 [23]	Na^+	−74	−55
	K^+	−50	−38
DB18C6 [8]	K^+	−68.2	−41.6
	Rb^+	−68.6	−47.1
	Cs^+	−66.9	−49.4

former is canceled by the latter. The overall enthalpy (ΔH°) and entropy changes (ΔS°) consist of constituent enthalpy and entropy changes. Thus, it would be very interesting to analyze ΔH° and ΔS° by the constituent thermodynamic quantities and to discuss the overall extraction process in detail in terms of the constituent thermodynamic quantities.

3 Factors Affecting Extraction

3.1 Distribution of a Free Crown Ether

The distribution coefficient of a crown ether, D_L in Eq. 5, is one of the most important factors in determining the magnitude of K_{ex}. For example, the smaller the D_L value becomes, the larger the K_{ex} value will be. This means that a crown ether, which is more soluble in water, would exhibit higher extractability for a metal

Table 4. Values of D_L at 25 °C

Crown ether	Solvent	D_L
12C4 [3]	C_6H_6	1.5×10^{-1}
15C5	C_6H_6	$1.5_6 \times 10^{-1}$ [6]
	$C_6H_5NO_2$	2.2×10^{-1} [24]
B15C5 [7]	C_6H_6	2.0×10
18C6	n-$C_{16}H_{34}$	3.9×10^{-4} [25]
	C_6H_6	$6.3_4 \times 10^{-2}$ [6]
	C_6H_6	5.72×10^{-2} [12]
	CH_2Cl_2	4.50 [26]
	CH_3OH[a]	2.2×10^{-2} [25], 9.0×10^{-2} [27]
	$C_6H_5NO_2$	1.0×10^{-1} [24]
DCH18C6-A [2]	n-C_6H_{14}	6.3
	$CFCl_2CF_2Cl$	1.2×10
	CH_2Cl_2	4.0×10^3
DCH18C6-B [2]	n-C_6H_{14}	7.1
	$CFCl_2CF_2Cl$	1.7×10
	CH_2Cl_2	3.7×10^3
DB18C6[c]	CCl_4	3.9×10^2 [28]
	C_6H_6	8.0×10^2 [8]
	$CHCl_3$	8.0×10^3 [15]
	CH_2ClCH_2Cl	1.0×10^4 [29]
	CH_3OH[a]	$7._1 \times 10$ [30]
	$C_6H_5NO_2$	2.3×10^4 [24]
	CH_3CN[a]	$3._2 \times 10^3$ [30]
	DMF[a]	$2._8 \times 10^3$ [30]
	DMSO[a]	$2._5 \times 10^3$ [30]
	PC[b]	$8._1 \times 10^2$ [30]
DB24C8	C_6H_6	1.1×10^3 [11]
	$C_6H_5NO_2$	$\sim 10^4$ [24]

[a] Water-miscible solvent; [b] Propylene carbonate; [c] Solubilities of DB18C6 in various solvents at 25.0 °C are presented in ref. [43]

ion. The crown ether with the higher extraction efficiency would be available for extraction determination and separation. However, if a crown ether itself is very soluble in water, the crown ether complex with a metal ion would also be very soluble in water and thus, the high solubility of the complex in water would considerably lower the extractability of the complex. The two opposite effects must be taken into account for molecular design of a new crown ether as an extractant for chemical analysis. Values of D_L at 25 °C are shown in Table 4. As a rule, the larger the molar volume of a chemical species, the more extractable it is from an aqueous phase into an organic phase. In the case of 12C4, 15C5 and 18C6 the D_L value for the same solvent does not increase with an increase in the size of the crown ether; it is of interest that, for both benzene and nitrobenzene, the D_L value of 18C6 is the smallest. 18C6 is much more soluble in water than in organic solvents except CH_2Cl_2, and, on the contrary, the reverse for CH_2Cl_2. When a substituent group, such as a cyclohexano or a benzo group, is attached to a crown ether ring, the D_L value for the same organic solvent increases considerably.

Luo et al.[31] determined the distribution constants of 14-membered saturated macrocyclic ligands, 1,4,8,11-tetraazacyclotetradecane (TA14C4) and the two diastereoisomeric 5,7,7,12,14,14-hexamethyl-TA14C4 and their protonated species in several solvent systems at 25 °C and $\mu = 0.1$ mol dm^{-3} ($NaClO_4$) from the variations of the distribution ratios as a function of hydrogen ion concentration in the range of $5 < pH < 13$. The authors discussed the dilution effect on the distribution constants.

3.2 Stability of a Crown Ether Complex in Aqueous Solution

It can be seen from Eq. 5 that stability of a crown ether complex with a metal ion in an aqueous solution also governs the magnitude of K_{ex}. The more stable the crown ether complex in an aqueous solution, the more extractable it is. Plots of lg K_{ex}, lg $K_{ex'}$, and lg K_{ML} at 25 °C vs. crystal ionic radius of an univalent metal for the 15C5- and 18C6-picrate systems[5] are given in Figs. 6 and 7, respectively. The extractions for both 15C5 and 18C6 systems were carried out between benzene and water. Figs. 6 and 7 indicate that the lg K_{ex} sequences for the 15C5 system do not depend on the lg K_{ML} sequences but depend completely on the lg $K_{ex'}$ sequences. On the other hand, those for the 18C6 system largely depend on the lg K_{ML} sequences. The lg K_{ex} sequences for the DB18C6/Na^+, K^+, Rb^+, Cs^+/picrate anion system (benzene/water, at 25 °C) are governed primarily by the lg $K_{ex'}$ sequences[12].

For bivalent metal ions the lg K_{ex} sequences for the 15C5/Sr^{2+}, Pb^{2+}, Ba^{2+}/picrate anion system do not depend on the lg K_{ML} sequences, but depend entirely on the lg $K_{ex'}$ sequences; on the contrary, those for the 18C6/Ca^{2+}, Hg^{2+}, Sr^{2+}, Pb^{2+}, Ba^{2+}/picrate anion and the DB18C6/Ca^{2+}, Sr^{2+}, Pb^{+2}/picrate anion systems largely depend on the lg K_{ML} sequences[6]. The extractions for the 15C5, 18C6, and DB18C6 systems were conducted at 25 °C between benzene and water. The above data indicate that selectivity tendency of a crown ether for metal ions is not always in agreement with that in an aqueous solution. lg K_{ML} values of 1:1 complexes of 15C5,

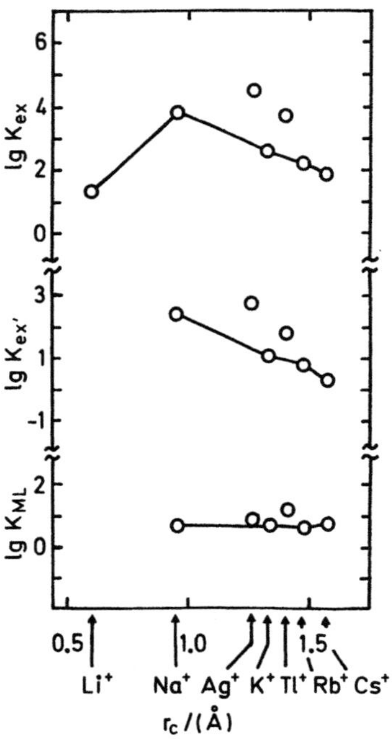

Fig. 6. Plots of lg K_{ex}, lg $K_{ex'}$, and lg K_{ML} at 25 °C vs. r_c of univalent metal ions for the 15C5 system [5)]

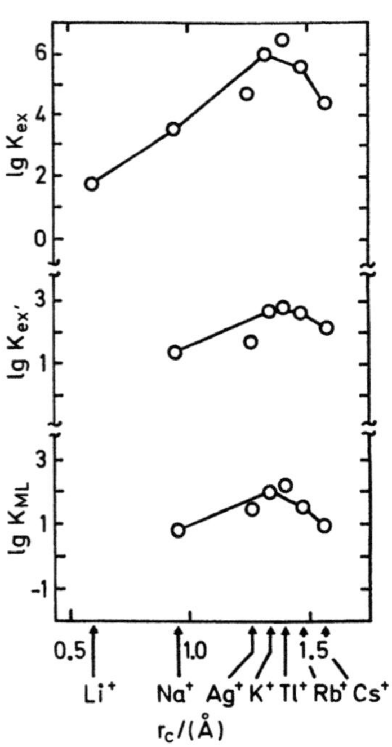

Fig. 7. Plots of lg K_{ex}, lg $K_{ex'}$, and lg K_{ML} at 25 °C vs. r_c of univalent metal ions for the 18C6 system [5)]

18C6, and DB18C6 with metal ions in an aqueous solution at 25 °C are listed in Table 5.

Table 5. lg K_{ML} values of 1:1 complexes of crown ethers with metal ions in an aqueous solution at 25 °C

	Na^+	K^+	Rb^+	Cs^+	Ag^+	Tl^+	Ca^{2+}	Sr^{2+}	Ba^{2+}	Hg^{2+}	Pb^{2+}
15C5 [32]	0.70	0.74	0.62	0.8	0.94	1.23	—	1.95	1.71	1.68	1.85
18C6 [32]	0.80	2.03	1.56	0.99	1.50	2.27	<0.5	2.72	3.87	2.42	4.27
DB18C6 [28]	1.16	1.67	1.08	0.83	1.41	1.50	<0	1.00	ca. 1.95	—	1.89

3.3 Extraction of a Crown Ether Complex

As can be seen from Eq. 5, the magnitude of K_{ex} is determined by $K_{ex'}$ as well as by D_L and K_{ML}. $K_{ex'}$ represents the extractability of a crown ether complex with a metal ion from an aqueous phase into an organic phase. The values of K_{ex}, D_L, and K_{ML} can be determined experimentally, however, that of $K_{ex'}$ cannot. $K_{ex'}$ can be calculated by Eq. 5. lg K_{ex} values of crown ether complexes with metal picrates and free energies of hydration at 25 °C are given in Table 6.

It can be seen from Table 6 that, when one or two benzo groups are attached to a crown ether ring (more hydrophobic), the crown ether complex with a metal ion is much more extractable (15C5 → B15C5 and 18C6 → DB18C6). The size sequences of 15C5, 18C6, and DB18C6 are 15C5 < 18C6 < DB18C6. For benzene, the lg $K_{ex'}$ value of the same metal ion for these three crown ethers increases with an increase in the size of the crown ether, except for Na^+-15C5 and 18C6, and Ag^+-15C5 and 18C6. It is of great interest that, although the D_L value of 18C6 between benzene and water is smaller than that of 15C5, the 18C6 complex with the same metal ion is much more extractable than the 15C5 complex, except for Na^+ and Ag^+. A sodium ion is the most favorable fit for the 15C5 cavity, while Na^+ is smaller than the 18C6 cavity (Table 2). Thus, deviation of the charge in the crown ether complex with Na^+, which enhances hydrophilicity and reduces extractability of the complex, may be much larger for 18C6 than for 15C5. 18C6 may more effectively screen the surface charge of Ag^+ trapped in its cavity than 15C5 because of its larger size (Table 2). However, since the size of Ag^+ is larger than the cavity size of 15C5, Ag^+ is located partly outside the plane of the 15C5 ring. Thus, the exposed part of Ag^+ can interact strongly with the π-electrons of benzene. The reason why, in each case of Na^+ and Ag^+, the lg $K_{ex'}$ value of 15C5 is larger than that of 18C6 may be attributed to the above discussion.

For both alkali and alkaline earth metal ions, using the same solvent, the more closely the metal ion fits into the crown ether cavity, the more extractable the crown ether-metal ion complex is except for the 18C6 complex with Ba^{2+}. The same tendency can be seen with Ag^+ and Tl^+.

Table 6 includes some cases in which, for the same solvent, the lg $K_{ex'}$ values of the same crown ether with both univalent and bivalent metal ions do not differ

Table 6. lg K_{ex}, values of crown ether complexes with metal picrates and free energies of hydration, ΔG_h°, at 25 °C

		Metal ion										
		Na^+	K^+	Rb^+	Cs^+	Ag^+	Tl^+	Ca^{2+}	Sr^{2+}	Ba^{2+}	Hg^{2+}	Pb^{2+}
	$-\Delta G_h^\circ$/kJ mol^{-1} [33]	411	338	320	297	479	343	1593	1447	1318	1825	1497
Crown ether	Solvent											
15C5	C_6H_6	2.4 [5]	1.0 [5]	0.7 [5]	0.3 [5]	2.7 [5]	1.8 [5]	—	2.9 [6]	2.7 [6]	—	3.8 [6]
B15C5 [7]	C_6H_6	—	2.9	—	—	—	—	—	—	—	—	—
18C6	C_6H_6	1.4 [5]	2.7 [5]	2.7 [5]	2.2 [5]	1.7 [5]	2.8 [5]	>5.3 [6]	5.8 [6]	4.7 [6]	4.3 [6]	6.3 [6]
		>1.8 [12]	2.70 [12]	2.7 [12]	2.2 [12]							
	CH_2Cl_2 [a]	3.2	4.8	—	—	—	—	—	—	—	—	—
DB18C6	C_6H_6	3.8 [13]	5.9 [b]	5.6 [b]	5.2 [b]	5.1 [9]	5.9 [10]	>6.9 [6]	7.3 [6]	—	—	8.2 [6]
			5.6 [13]	5.6 [13]	5.0 [13]				7.2 [15]			7.9 [15]
	$CHCl_3$	4.8 [13]	6.8 [13]	7.0 [13]	6.3 [13]	5.8 [14]	7.1 [13]	—	7.8 [15]	—	—	8.6 [15]

[a] Calculated by Eq. 5 from the values of K_{ex} [26], D_L [26], and K_{ML} [32]

[b] Calculated by Eq. 5 from the values of K_{ex} [8], D_L [8], and K_{ML} [28]

as much from one another. This indicates that the charge of the metal ion held in the crown ether cavity is effectively shielded by the crown ether and that the complexed cations are roughly equal in chemical nature.

It is noteworthy that in each crown ether-solvent system the largest lg $K_{ex'}$ value is seen for Pb^{2+}. Although the crystal radii of Hg^{2+} and Sr^{2+} are nearly equal Hg^{2+}. This may be due to the fact that the $-\Delta G_h^\circ$ value of Hg^{2+} is much larger than (Table 2), the lg $K_{ex'}$ value of 18C6 in benzene is much larger for Sr^{2+} than for that of Sr^{2+} (Table 6). As may be seen from Tables 2 and 6, for 18C6 and DB18C6 and the same solvent, the lg $K_{ex'}$ values of K^+ and Tl^+, having very much smaller $-\Delta G_h^\circ$ values and fitting the cavities of 18C6 and DB18C6 much better than Ag^+, are larger than that of Ag^+. The lg $K_{ex'}$ values of Rb^+ and Cs^+, which also have very much smaller $-\Delta G_h^\circ$ values than Ag^+ but do not fit the cavities of 18C6 and DB18C6 as well as Ag^+, are also larger than that of Ag^+; the lg $K_{ex'}$ value of Ag^+, which fits more closely into the 18C6 and DB18C6 cavities than Na^+, is larger than that of Na^+, although the $-\Delta G_h^\circ$ value of Ag^+ is comparably much higher. It was found that the structure of the crown ether complex with the center cation does not require the stripping of the ion's entire solvation shell [18], since solvent contacts are still possible in the direction perpendicular to the plane of the ring. It thus appears that the extractability of metal ion complexes with one and the same crown ether is determined primarily by both the ratio of the ion-to-cavity size and the interaction of the metal ion trapped in the cavity with water molecules [9].

For the same metal ion complexes of 18C6 and DB18C6 are, respectively, more extractable for CH_2Cl_2 and $CHCl_3$ than for benzene (Table 6). This may be attributed to the stronger interaction of the hydrogen atoms of CH_2Cl_2 and $CHCl_3$ with the crown ether oxygen atoms compared to those of benzene [13,15].

Iwachido et al. [18] extracted alkali and alkaline earth metal salts of 2,2′,4,4′,6,6′-hexanitrodiphenylamine into nitrobenzene in the absence or presence of several crown ethers or cryptands with different hole sizes in order to study the quantitative effect of crown ethers and cryptands on the coextraction of water. The NMR and near-infrared spectra show that the coextraction of water is caused by the hydration of the cations. The complexation between these cations and the crown ethers causes a significant decrease of the hydration number; e.g., in the 1:1 metal-crown ether complexes more than half of the water molecules are removed, whereas in 1:1 metal-cryptand complexes less than one water molecule remains unremoved. Complexes with a 1:2 stoichiometry are also found; some are stable and carry virtually no water molecules. The number and type of donor atoms and the presence of fused benzo rings in the crown ether have no significant effect on the coextraction of water.

3.4 Counter Anion

For the same crown ether-metal ion complex, the extractability of the ion pair into the same organic solvent is largely governed by the chemical nature of the counter anion. In general, a counter anion with a larger molar volume, such as picrate, dinitrophenolate, dipicrylaminate, and tetraphenylborate, makes the ion-pair extraction more efficient.

Jawaid and Ingman[34] determined K_{ex} of Na^+, K^+, and Ca^{2+} complexes of DCH18C6 with 2,5- and 2,6-dinitrophenol, picric acid, and dipicrylamine in the CH_2Cl_2/H_2O system at 20 ± 1 °C. These data are as follows:

Counter anion	lg K_{ex}	
	Na^+	K^+
2,5-dinitrophenolate	—	3.98
2,6-dinitrophenolate	—	4.26
picrate	4.20	6.36
dipicrylaminate	7.81	8.16

The figures indicate that the extractability increases with the hydrophobicity of the counter anion; at the same time selectivity decreases. Yakshin et al.[35] extracted DB18C6-alkali metal ion complexes with various inorganic anions into 1,2-dichloroethane. The extractability sequences of the anion with the DB18C6-alkali metal ion complex are

$$ClO_4^- > I^- > NO_3^- > Br^- > OH^- > Cl^- > F^- .$$

Distribution ratios of metal ions between an aqueous phase of a hard anion such as Cl^-, NO_3^-, and SO_4^{2-} and a hydrocarbon or chlorocarbon phase of a crown ether are too low to be useful for a practical application of crown ethers for selectively removing metal ions from natural and industrial water sources which usually contain hard anions. An alternative solution to this problem is the use of macrocyclic complexing reagents for which metal extraction does not involve concomitant transfer of the aqueous phase anion into the organic phase. Thus, Bartsch et al.[36-40] incorporated a carboxylic acid group into a crown ether molecule in order to achieve high selectivity and extractability. The synthesized crown ethers with the combination of size selectivity, by means of the ion-to-cavity size ratio, and an ionizable group act as bifunctional complexing reagents. Solvent extraction of alkali metal ions by *sym*-dibenzo[16]crown-5-oxyacetic acid (*1*) into $CHCl_3$ was studied[36]. Extractability is insensitive to a change of the aqueous phase anion from chloride to sulfate which demonstrates that the anion is not extracted into the organic phase. For competitive ion extractions the selectivity is $K^+ > Na^+ > Rb^+ \geq Cs^+ > Li^+$ at pH 6–7 and $Na^+ > K^+ > Rb^+ \approx Cs^+ > Li^+$ at pH 8–12. Extractabilities and selectivities of *1* surpass those of *sym*-dibenzo[16]crown-5 methyl ether or phenoxyacetic acid. Competitive solvent extractions of alkali metal ions into $CHCl_3$ by 2-(*sym*-dibenzo[16]crown-5-oxy)butanoic acid (*2*), 2-(*sym*-dibenzo-[16]crown-5-oxy)hexanoic acid (*3*), 2-(*sym*-dibenzo[16]crown-5-oxy)octanoic acid (*4*), 2-(*sym*-dibenzo[16]crown-5-oxy)decanoic acid (*5*), and 5-(*sym*-dibenzo[16]crown-5-oxy)pentanoic acid (*6*) were reported[37]. The series of *2–6* possess a common DB16C5 ring system and carboxylic acid portions of varying lipophilicity (*2–5*) or of a different spacial relationship with the crown ether cavity (*6*). Above pH 10, *2–6* show by far the highest selectivity for Na^+ of all the alkali metal ions. The extractabilities of the same reagent for the other alkali metal ions do not differ as much from one another. *3*, *4*, and *5* are sufficiently lipophilic to totally remain in the organic phase when $CHCl_3$ solutions of these are brought in contact with highly alkaline aqueous solutions, which significantly increases the extractability. Competitive solvent extractions of alkali metal ions by *sym*-dibenzo[13]crown-4-oxyacetic acid (*7*), *sym*-dibenzo[14]crown-4-oxyacetic acid (*8*), and *sym*-dibenzo[19]crown-6-oxyacetic acid (*9*) into $CHCl_3$ were conducted [38]. At the pH for maximum metal extractability the selectivity orders vary with increasing crown ether cavity size as follows: $K^+ > Rb^+ > Na^+ \approx Cs^+ > Li^+$ for 7 (pH 6), $K^+ > Na^+ > Rb^+ > Cs^+ \approx Li^+$ for 8 (pH 5.4), and $K^+ > Rb^+ > Na^+ \approx Cs^+ > Li^+$ for 9 (pH 7.2). In terms of selectivity and extractability *9* is the best of the four crown ether carboxylic acids (*1*, *7*, *8*, and *9*) and *7* is the worst. Competitive solvent extractions of alkali metal ions by *sym*-bis[4(5)-*tert*-butylbenzo][16]crown-5-oxyacetic acid (*10*) into $CHCl_3$ were investigated in order to probe the effect of varying the attachment site for the lipophilic group upon cation complexation [39]. Ligand *10* is a structural isomer of *5*. Although the overall extraction behavior of *5* and *10* for alkali metal ions is quite similar (selectivity order for *10* at above pH 9: $Na^+ \gg K^+ > Rb^+ > Cs^+$), *10* exhibits complete exclusion of Li^+ from the $CHCl_3$ phase and slightly enhanced Na^+/K^+ selectivity compared to *5*. Solvent extraction of alkaline earth metal ions by *1*, *7*, and *9* into $CHCl_3$ was reported [40]. For competitive ion extractions, the selectivity orders with *1*, *7*, and *9* for maximum metal ion extractabilities are $Ba^{2+} > Ca^{2+} > Sr^{2+} > Mg^{2+}$ at above pH 6.5, $Ba^{2+} > Ca^{2+} \approx Sr^{2+} \approx Mg^{2+}$ in the acidic pH region and $Ba^{2+} > Ca^{2+} > Sr^{2+} > Mg^{2+}$ at pH 8, respectively. The metal/crown ether carboxylic acid ratio of the extracted complex is 1:2 for each reagent. In terms of extractability and selectivity, the order for *1*, *7*, and *9* is *1* > *9* > *7*. Ligands *1*, *7*, and *9* are superior extraction reagents to closely related neutral crown ethers and to phenoxyacetic acid.

3.5 Diluent

Extractability and selectivity for solvent extraction with crown ethers are greatly affected by organic solvents.

Rais et al. [41] extracted sodium and caesium dipicrylaminates by DB18C6 into $CHCl_3$, $C_6H_5NO_2$, CH_2Cl_2, C_6H_5Cl, propylene carbonate, and CH_3NO_2. The distribution ratios of Na^+ and Cs^+ are the largest for $CHCl_3$. On the contrary, those for CH_3NO_2 and propylene carbonate are the smallest although these two diluents are polar solvents. The difference in the distribution ratios of Na^+ and Cs^+ (i.e., selectivity) varies by changing the solvents. Danesi et al. [42] investigated the extraction of alkali metal picrates by DB18C6 at 22 ± 1 °C into various nitrobenzene-toluene mixtures whose dielectric constant varies from 3.4 to 35. The affinity of DB18C6 for the alkali metal ions has been found to vary in the order $K^+ > Rb^+ > Cs^+ > Na^+ > Li^+$ for all diluent compositions. The extraction equilibrium constants, which were defined by $[ML^+]_o [A^-]_o/[M^+] [L]_o [A^-]$, show that the extractability of the alkali metal ions decreases as the diluent composition is varied from pure nitrobenzene to pure toluene. The extracted ion pair dissociates in the higher dielectric constant mixtures and, on the contrary, does not dissociate in the lower dielectric constant mixtures.

In terms of practical application of crown ethers, in order to enhance selectivity and extractability of the crown ethers for metal salts with inorganic anions, Marcus and Asher [43] used various protic solvents which solvate the anions effectively. Crown ethers used in their work were DCH18C6 and DB18C6, and alkali metal ions were sodium and potassium. Ion pairs and separate ions are extracted into solvents of low and sufficiently high dielectric constants, respectively. Potassium salts with various anions were extracted by DB18C6 into *m*-cresol at 25 °C. The order of extractability is $F^- > CH_3CO_2^- > NO_3^- > I^- > Br^- > Cl^- > SO_4^{2-}$. The order of difficulty to remove them from the aqueous phase is given by the order of their Gibbs free energies of hydration. The positions of NO_3^- and particularly F^- with respect to this series are out of place. This is due to the more negative Gibbs free energies of solvation by the environment that the ions find in the *m*-cresol phase (stronger hydrogen bond), than expected from their hydration properties. As a good solvent for selective extraction of KCl the aforementioned authors recommended a protic solvent which has, a) a low dielectric constant (permitting ion pairing), b) a low Gibbs free energy of transfer of Cl^- from water, c) a high solubility of water, but d) a low solubility in water (to avoid excessive loss of solvent). It has been demonstrated by Asher and Marcus [44] that extraction of KCl by DB18C6 into *m*-cresol at 25 °C is very fast, equilibrium being reached within one minute; K^+ is selectively extracted over Na^+, and both are highly selectively extracted over Mg^{2+} and Ca^{2+}, all as chlorides. Other solvents, besides *m*-cresol, have been tested in order to find some which would show at least as good extraction but lower aqueous solubility. Marcus et al. [45] described the selective extraction of KCl from aqueous brines containing any or all of NaCl, $MgCl_2$, and $CaCl_2$, by crown ethers dissolved in substituted phenols. A system selected for convenient availability, rather than for optimal properties, is 0.1 mol dm^{-3} DB18C6 in 2,4-xylenol (75 volume %) diluted with toluene (25 volume %). With a crude two-stage countercurrent batchwise extraction, 83% removal of potassium from the feed brine and an enrichment factor of 14 for potassium relative to sodium (and $\gg 10^3$ relative to magnesium and calcium) in the aqueous product relative to the feed brine are achieved. No chemicals are consumed in this process, and the solvent was fully recycled. Losses of the extractant and solvent to the aqueous phases by solubility are minimized by proper choice of these reagents and the possible use of a diluent. McDowell and Shoun [46] increased extractability of KCl from $MgCl_2$ aqueous solutions with DCH18C6 by adding 2-ethylhexanol or nonylphenol to a benzene phase. The result for the alcohol system indicate an adduct of one crown ether and one or two alcohol molecules per KCl, while, the extracted species for the phenol systems is DCH18C6 · KCl · nonylphenol.

Iwachido et al. [47] determined D_M values of postassium at 25 °C between an aqueous potassium picrate solution and 57 organic solvents (8 alcohols, 12 ketones, 6 esters, 3 ethers and 28 hydrocarbons among them 24 halo-, 2 nitro-substituted, and 2 unsubstituted ones) in the presence or in the absence of 18C6. The presence of 18C6 markedly enhances the extractability of potassium when the halogenated hydrocarbon solvents are used, but only a slight enhancement can be observed for the oxygen-containing solvents. Among the properties showing good correlations with D_M, Hansen's three-dimensional solubility parameters and Dimroth's E_T values are most satisfactory for application to a wide variety of solvents.

3.6 Coexisting Species in the Aqueous Phase

Effects of ionic strength and salting out on extraction of metal ions by crown ethers were studied [26,46,48,49]. McDowell and Shoun [46] investigated HCl dependence and aqueous ionic strength dependence of KCl extraction by 0.1 mol dm^{-3} DCH18C6 in a 25% benzene-75% 2-ethylhexanol mixture. In the case of ionic strength dependence the $MgCl_2$ concentration was varied from 0.03 to 3.3 mol dm^{-3}. The observed increase in D_M probably reflects the change in chloride concentration or activity. For an $AgNO_3$/DB18C6/$CHCl_3$ extraction system, Gloe et al. [48] found that the lg K_{ex} value for an Ag(DB18C6)NO_3 complex is slightly larger for a 2 mol dm^{-3} HNO_3 aqueous phase than for a 1 mol dm^{-3} HNO_3 aqueous phase. Kolthoff [26] determined D_L values of 18C6 between CH_2Cl_2 and water at various concentrations of LiCl at 25 °C. These data are as follows:

LiCl concn/mol dm^{-3}	0	0.04	0.12	0.20	0.40
D_L	4.5	5.1	5.8	6.4	7.4

Furthermore, the effect of ionic strength on the extraction constant of potassium picrate by 18C6 in CH_2Cl_2 were studied. The salting out effect on D_L greatly affects the concentration of K^+ left in the aqueous phase. Abashkin et al. [49] found maximum alkali metal extractability by DCH18C6 at a given HNO_3 or $HClO_4$ concentration, but did not in the HCl system.

3.7 Structure of a Crown Ether

Tušek et al. [50] introduced four *t*-butyl groups to the benzene rings of DB18C6 in order to enhance its solubility in an organic phase. They compared the extractability of bis(3,5-di-*t*-butylbenzo)[18]crown-6 (D-*t*-B-DB18C6) for caesium picrate into nitrobenzene with that of DB18C6. The results show that the extraction power of D-*t*-B-DB18C6 is slightly lower than that of DB18C6. Various electronic effects of the aromatic substituent of DB18C6 upon extractabilities and selectivities of the crown ethers for metal ions were studied by Pannell et al. [51]. They extracted sodium and potassium 4-[*p*-(dimethylamino)phenylazo]benzenesulfonates into CH_2Cl_2 by DB18C6 and the 8 substituted DB18C6 ethers (Z)- and (E)-dinitro, (Z)- and (E)-diamino, tetrabromo, octachloro, mono- and bis(tricarbonylchromium)). The lg K_{ex} values for MLA (temp.?) are as follows:

Crown ether *	Na^+	K^+
(*E*)-$(NH_2)_2$DB18C6	4.16	4.67
DB18C6	3.99	4.59
$[Cr(CO)_3]$DB18C6	3.97	4.14
$[Cr(CO)_3]_2$DB18C6	3.81	3.51
(*E*)-$(NO_2)_2$DB18C6	3.43	3.19
(*Z*)-$(NO_2)_2$DB18C6	3.46	3.13
Br_4DB18C6	3.42	3.38
Cl_8DB18C6	3.34	2.81

The sodium salt extraction is considerably less susceptible to the substituent effect compared to potassium salt. The crown ethers with cumulatively large electron-withdrawing substituents exhibit a reversal of the normal ion extraction selectivity of 18C6 ethers ($K^+ > Na^+$).

Gokel et al. [52] reported extractabilities and selectivities of a group of lariat crown ethers (*11–22*) for Na^+ and K^+ in the CH_2Cl_2/water system (temp. and anion (picrate)

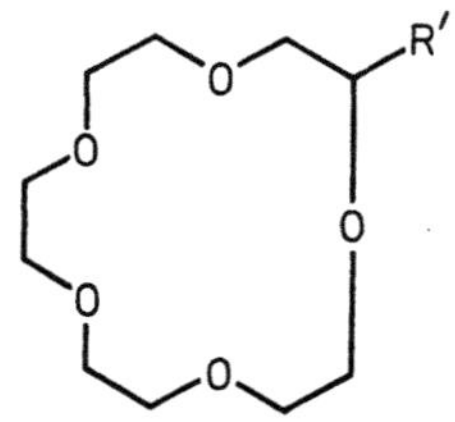

11 R′ = H
12 R′ = CH_2OH
13 R′ = CH_2OMe
14 R′ = CH_2OCOMe
15 R′ = CH_2OPh
16 R′ = $CH_2OC_6H_4OMe$-*o*
17 R′ = $CH_2OC_6H_4OMe$-*p*
18 R′ = CH_2OCH_2Ph
19 R′ = CH_2OCMe_3
20 R′ = $CH_2O(CH_2)_2OMe$
21 R′ = $CH_2O(CH_2)_2O(CH_2)_3Me$
22 R′ = $CH_2O(CH_2)_2O(CH_2)_2OMe$

?). The Na^+ extraction constant for *16* is more than twice that for the isomer *17*. Enhanced extractability is observed for compounds which have donor groups correctly aligned for an axial interaction with the complexed metal ions (Na^+: *16*, *20*, *21*, *22*, and K^+: *16*, *17*, *20*, *21*, *22*). Cho and Chang [53] extracted alkali metal (Li, Na, K) picrates into CH_2Cl_2 at 25 °C by a series of N,N′-disubstituted derivatives of DA18C6 (*23–26*). All the N,N′-disubstituted derivatives of *23a* exhibit enhanced extractabilities compared to *23a* itself, although the selectivities are somewhat less

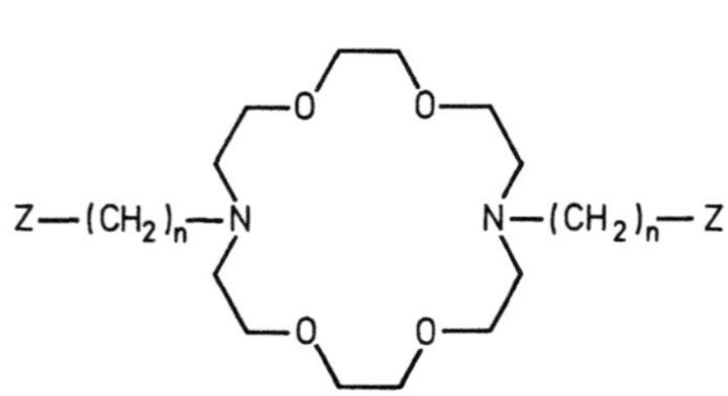

a n = 0
b n = 3

23 Z = H
24 Z = N-Phthalimido
25 Z = NH_2
26 Z = OH

pronounced. An extremely high but non-selective extractability is seen for *25b*. Nakatsuji et al. [54] studied complexation properties for Na^+, K^+, Ca^{2+}, and Ba^{2+} by several crown ethers having one amide group (*27–29*). They extracted the metal

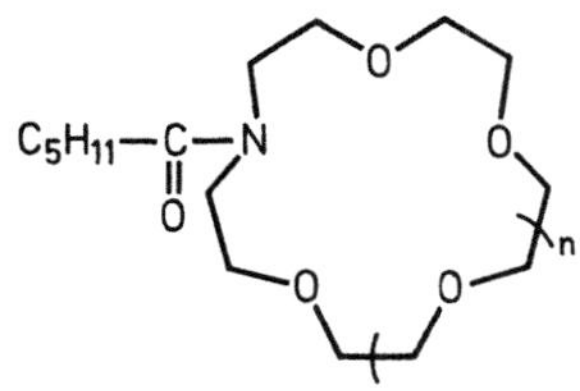

27

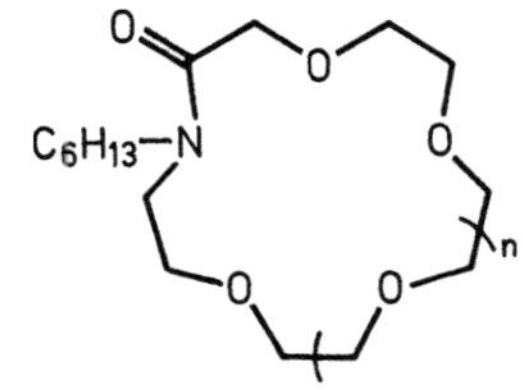

28

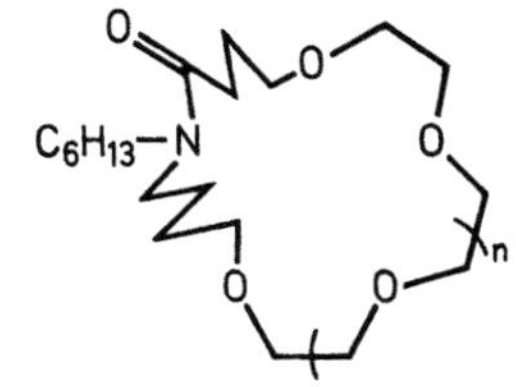

29

27a - *29a* : n = 2
27b - *29b* : n = 1

picrates at 25 °C into CH_2Cl_2. The extractability of the crown ethers having an amide group outside the crown ring (*27*) is relatively low compared with *28* and *29*. Extraction properties of *29* are rather similar to the corresponding *28*; by introducing spacers an increased extractability of *29a* for Ba^{2+} and *29b* for Ca^{2+} is observed. It was found that the preference of Ca^{2+} to Na^+ is specific for crown ethers such as *28* and *29*. The authors concluded that the orientation of the amide group is one of the most important factors determining the extraction properties.

Inoue et al. [55] extracted alkali metal (Na—Cs) picrates into $CHCl_3$ at 25 ± 0.1 °C by distilbeno-, tetraphenyl- and tetracyclohexano-crown ethers (*30–32*). Ligands

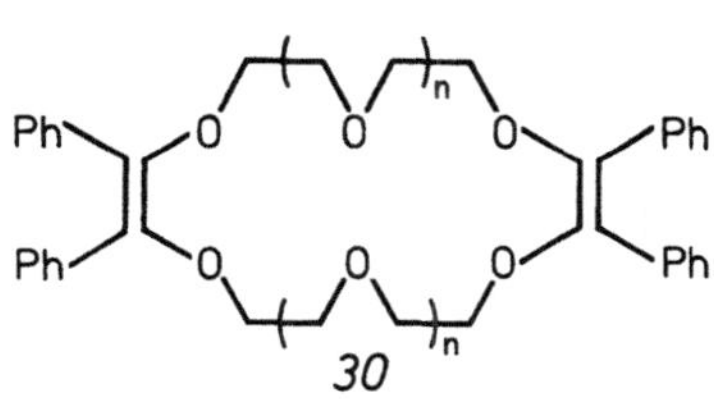

30

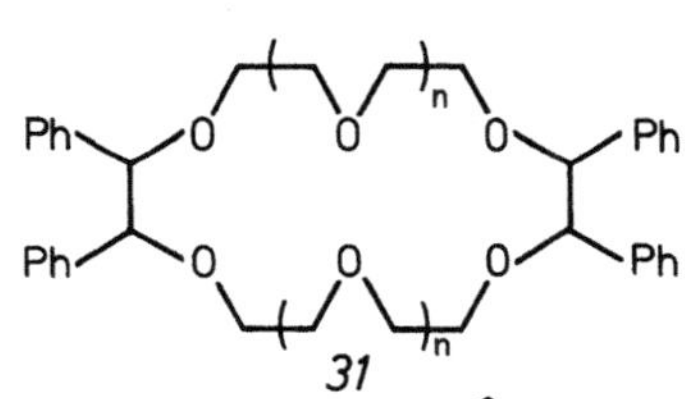

31

a n = 1
b n = 2

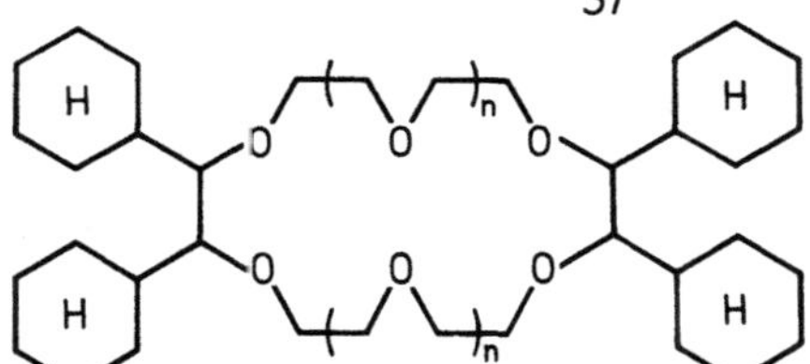

32

30a, *31a*, and *32a* show the highest selectivity for K^+ of all the alkali metal ions. No pronounced selectivity is seen for *30b*; *31b* exhibits the highest selectivity for Rb^+, *32b* for Na^+. The extractabilities of *30a*, *31a*, and *32a* for the same alkali metal ion are very much lower than with 18C6, but higher than with DB18C6. The extractabilities of *30b*, *31b*, and *32b* are almost comparable to those of *30a*, *31a*, and *32a*. Akabori et al. [56] reported extractabilities of polyoxa[n]ferrocenophanes (*33* and *34*)

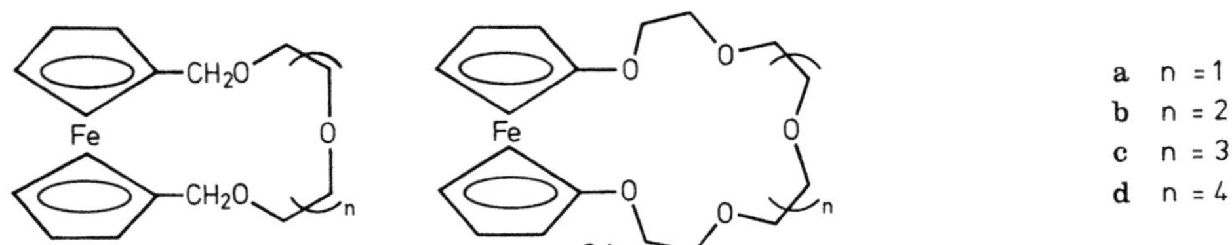

33 *34*

for alkali metal (Li—Cs), barium, thallium, and silver picrates in the CH_2Cl_2/water system (temp. ?). Ligands *33b–d* and *34b* show no selectivity and a little or no extractability for the alkali metal ions, while *34a* has a moderate extractability. *33b–d* exhibit high extractability for Ag^+, and both *33d* and *34a* for Tl^+, however, *34a–b* decompose rapidly with Ag^+. The results indicate that the interaction of the iron atom with a metal ion in the crown ether cavity has an influence on the stability of the complex. Olsher and Jagur-Grodzinski [57] found a crown ether which has high selectivity and extractability for Li^+. They determined K_{ex} values of lithium picrate with B13C4, B12C4, DB14C4, B15C5 and DB18C6 at 25 °C in the tricresyl phosphate-ethylene chloride (1:1) mixture/water system. B13C4 shows the largest extractability of all the crown ethers. The benzene and the CH_2Cl_2 solutions of B13C4 selectively extract LiSCN. The $[Li^+]_o/[M^{m+}]_o$ values of Na^+, K^+, Mg^{2+}, and Ca^{2+} are 2.4, 43, 202, 354 and 2.5, 44, 216, 355 for benzene and CH_2Cl_2, respectively. The highest selectivity and extractability of B13C4 for Li^+ were explained in terms of the size-fit concept; the ratio of the size of Li^+ to that of the B13C4 cavity is ca. 1.0, being the most favorable fit. Gloe et al. [58] extracted various metal nitrates into $CHCl_3$ at 22 ± 1 °C by B15C5, hexanoyl-B15C5, *t*-butyl-B15C5, BMA15C5, DCH18C6 and DB18C6. The extractability series for DCH18C6 and DB18C6 are:

$$Hg^{2+} > Sr^{2+} > Ag^+ > K^+ > Na^+ \quad \text{and}$$
$$Hg^{2+} > K^+ > Tl^+ > Ag^+ > Rb^+ > Cs^+ > Na^+ > Sr^{2+} ,$$

respectively. The unsubstituted B15C5 shows the highest extractability for each of Na^+, K^+, Ag^+, and Hg^{2+} of all the 15C5 derivatives. K^+ and Li^+ were extracted into $CHCl_3$ by DB18C6 using $AuCl_4^-$ as a counter anion. Moreover, the effects of counter anions and diluents on the extractability of crown ethers for metal ions were investigated.

Metal extraction properties of a series of photoresponsive crown ethers were studied (for further details see contribution . . . (Shinkai)). The crown ethers which contain photoisomerizable functional groups (e.g., —N=N—) undergo conformational change of the crown ether ring in response to photoirradiation. Thus, photocontrolled selective metal ion-extraction can be achieved by using crown ethers. Photo-isomerization of the *E*- form of the following compounds to the *Z*- form is performed by photo-(uv) irradiation. Shinkai et al. [59] reported the extraction of alkali metal salts with methyl orange into benzene by an azobenzene-bridged azacrown ether (*35*) at 30 °C. The order of extractability of (*E*)-*35* is $K^+ > Na^+ > Li^+ > Rb^+ \approx Cs^+$ and that of (*Z*)-*35*, $K^+ > Na^+ > Rb^+ > Li^+ \approx Cs^+$. (*E*)-*35*

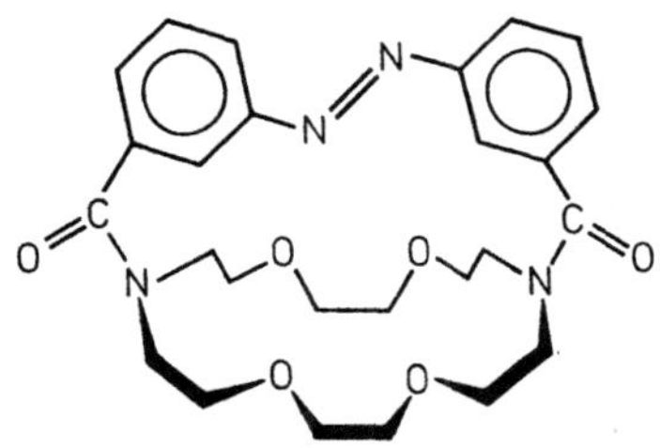

35

shows higher extractability for Li^+ and Na^+ than (*Z*)-*35* and the reverse for K^+. This was explained by the fact that the size of the azacrown ether moiety of (*Z*)-*35* is more expanded than (*E*)-*35* due to the steric distortion of the azobenzene bridge. Extraction behavior of a photoresponsive cryptand, containing an azopyridine-bridge (*36*), for heavy metal ions (Cu^{2+}, Ni^{2+}, Co^{2+}, Hg^{2+}, and Pb^{2+}) was compared

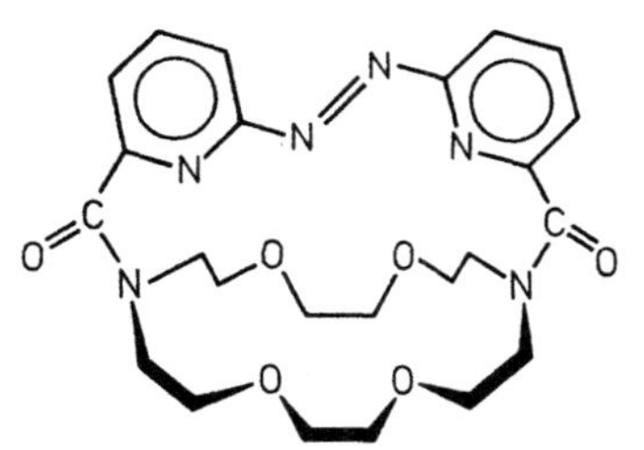

36

with that of *35* [60]. The extraction was conducted at 30 °C in the *o*-dichlorobenzene *n*-butyl alcohol (80:20 by volume) mixture/water system by using a picrate anion at pH 4–5. The extraction trends of (*E*)- and (*Z*)-*36* for Na^+ and K^+ are similar to those of *35*. For Cu^{2+}, Ni^{2+}, Co^{2+}, and Hg^{2+}, a distinct difference in the extractability is not observed between (*E*)- and (*Z*)-*35*. However, (*E*)-*36* very effectively extracts the metal ions, whereas (*Z*)-*36* scarcely does. The results suggest that the enhanced extractability of (*E*)-*36* compared with (*Z*)-*36* is due to the coordination of the azopyridine-bridge to the metal ion trapped in the cavity of *36*. Shiga et al. [61] extracted alkali metal ions into benzene using azocrown ethers (*37* and *38*) and tetrabromophenolphthalein ethyl ester as a counter anion at pH 9 (temp.?). Na^+ is more extractable than K^+ for *37*, however, the contrary applies for *38*. An extraction study of 20% photo-isomerized *37* and *38* indicated that while the selectivity between Na^+ and K^+ remained the same, the amount of extracted metal ions decreased to 80% of the non-irradiated controls. Inspection of the CPK model of the *Z*-structure indicates that the conformation of the crown ether ring is highly strained and unfavorable for the interaction with metal ions. Shinkai et al. [62]

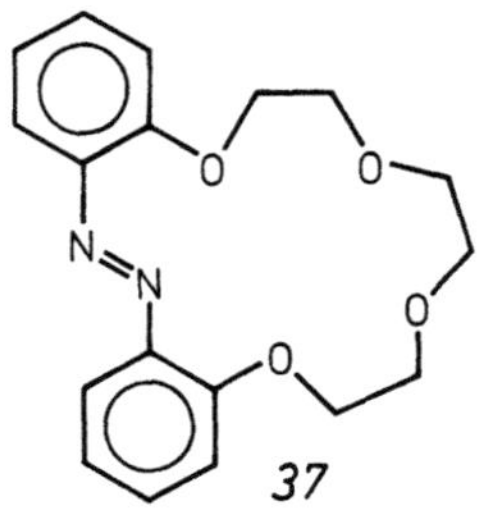

37

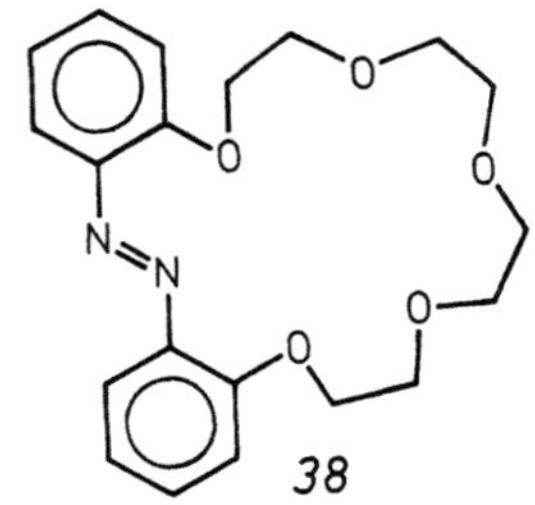

38

reported the extraction alkali metal (Na—Cs) picrates into an *o*-dichlorobenzene *n*-butyl alcohol (80:20 by volume) mixture by *39* at 30 °C. (*E*)-*39* completely lacks

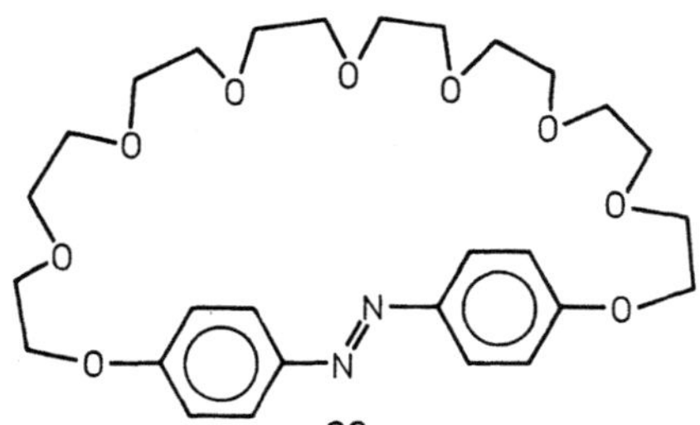

39

extractability for metal ions, whereas (*Z*)-*39* is capable of extracting considerable amounts of metal ions, maximum extractability being observed for Rb^+. The inspection of the CPK models of *39* suggests that the poly(oxyethylene) chain of the (*E*)-*39* is extended linearly, while the *Z*-*39* behaves as an analogue of [21]crown-7. Extraction of alkali metal (Li—Cs) salts with methyl orange into *o*-dichlorobenzene by a bis(crown ether) with azo-linkage (*40*) (30 °C?) was conducted [63]. The

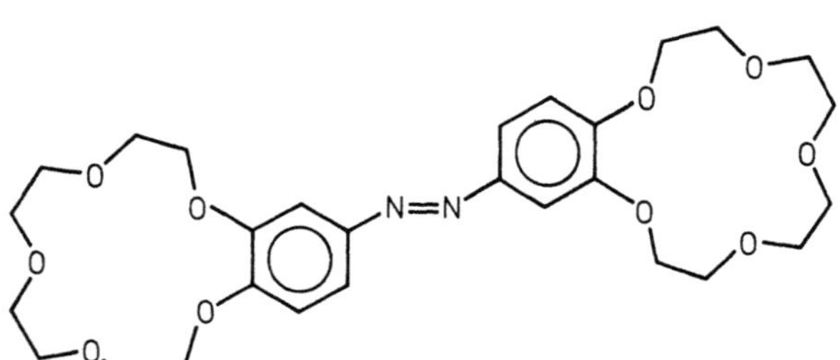

40

order of the extractability for (*E*)-*40* is $Na^+ \approx Rb^+ > Cs^+ > Li^+ \gg K^+$ and that for (*Z*)-*40* $Rb^+ > K^+ > Cs^+ > Li^+ \gg Na^+$. K^+ is efficiently extracted by (*Z*)-*40* but not by (*E*)-*40*. A considerably large selectivity of (*E*)-*40*/(*Z*)-*40* is observed for Na^+/K^+ (238 fold). (*Z*)-*40* extracts Li^+ very efficiently. Extraction of alkali metal salts of methyl orange with a photoresponsive crown ether polymer (*41*) into *o*-dichlorobenzene at 30 °C was reported [59]. The extractability of (*E*)-*41* is $Na^+ > K^+ > Cs^+ > Li^+ \approx Rb^+$. The order is similar to that of the cyclic

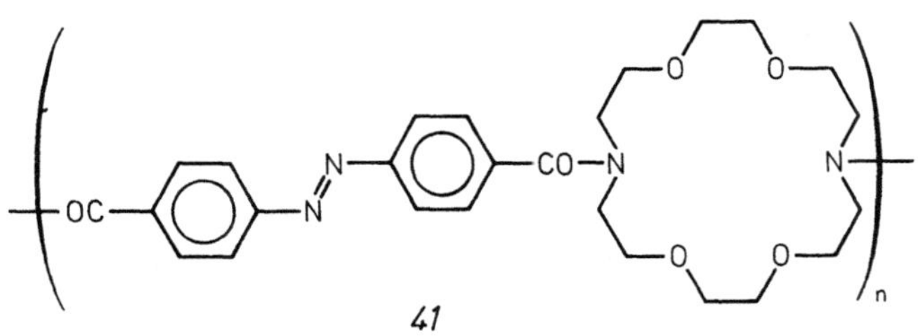

41

analogue *35*. The extractability of K^+ is somewhat greater for (*Z*)-*41* than for (*E*)-*41*. The extractability of Na^+ for (*E*)-*41* is markedly reduced under photo-irradiation. However, a clear co-operative effect of adjacent crown ether rings of *41* on the metal ion cannot be observed. Shinkai et al. [64] studied extraction of alkali metal ions (Na^+—Cs^+) and Ca^{2+} by a crown ether with a photoresponsive carboxylate cap (*42*) into *o*-dichlorobenzene and *n*-butyl alcohol at 30 °C and 20 °C,

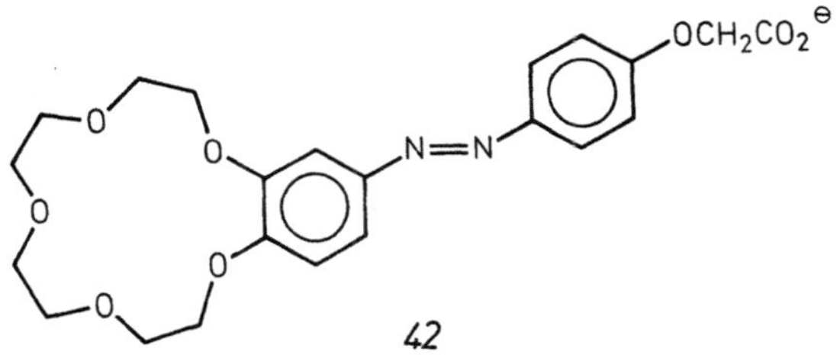

42

respectively. No metal ion was extracted into *o*-dichlorobenzene by *E-42*, while, when extraction was conducted after photo-irradiation, about 5% K^+ were extracted. However, other metal ions were not extracted. For *n*-butyl alcohol, in most cases, (*E*)-*42* extracted metal ions more efficiently than (*Z*)-*42*; only for K^+ (*Z*)-*42* acts as a better extractant than (*E*)-*42*. This phenomenon for K^+ should be rationalized in terms of some co-operative behavior of the carboxylate group and the crown ether.

In addition, effects of substituent groups on extraction efficiencies of crown ethers for alkali metal ions have been reported [65,66].

Thiacrown ethers show higher affinity for metal ions belonging to the soft acids compared to crown ethers containing only oxygen atoms. Solvent extraction with thiacrown ethers was investigated [67–69]. Sevdić and Meider [67,68] extracted Ag^+ and Hg^{2+} from chloride and perchlorate aqueous solutions into nitrobenzene at 24 ± 1 °C by 1,4,8,11-tetrathiacyclotetradecane (TTP) and 1,4,8,11,15,18,22,25-octathiacyclooctacosane (OTO). AgL^+, AgL_2^+ (only in the presence of ClO_4^-), HgL^{2+} (only in the presence of ClO_4^-), and $HgLCl_2$ (in the absence of ClO_4^-) complexes form in the organic phase (L = TTP or OTO). The formation of AgL_2^+ is attributed to the large size of ClO_4^-. The extractability of the Ag^+ complexes is higher for OTO than for TTP. This is due to the cavity size of the thiacrown ether and to the number of sulfur atoms (OTO: 8, TTP: 4). The extractability of the Hg^{2+} complexes is larger for OTO than for TTP. The difference in the extractabilities between OTO and TTP is somewhat greater for $HgLCl_2$ than for HgL^{2+}.

1,4,7,10,13,16-hexathiacyclooctadecane

(HTO) and TTP were used for mercury (II) and silver (I) picrate extraction into nitrobenzene at 24 ± 1 °C [69]. The extractability of the silver picrate complex markedly increases with an increase in the number of sulfur atoms in the thiacrown ether from 4 to 6. The extractability of the silver picrate complex with TTP is much higher than that of the mercury picrate complex.

4 Extraction of Higher-Order Crown Ether Complexes

When MLA undergoes further complex formation with another ligand (L′) in the organic phase, the equilibrium can be described as in Eq. 8:

$$\begin{aligned} MLA_o + L'_o &\rightleftharpoons MLL'A_o \\ K_{(MLL'A)_o} &= K_{ex}(MLL'A)/K_{ex}(MLA) \\ &= [MLL'A]_o/[MLA]_o\,[L']_o \end{aligned} \qquad (8)$$

where $K_{ex}(MLL'A) = [MLL'A]_o/[M^+]\,[L]_o\,[L']_o\,[A^-]$, $K_{ex}(MLA) = [MLA]_o/[M^+]\,[L]_o\,[A^-]$, and $K_{(MLL'A)_o}$ is the formation constant for the MLL′A complex in the organic solvent. lg $K_{(MLL')_o}$ values (A^- is a picrate anion) in a benzene solution at 25 °C are as follows:

L	L′	M		
		K	Rb	Cs
12C4 [70]	12C4	—	1.81	1.54
15C5 [70]	TBP	—	0.81	0.68
15C5 [70]	15C5	—	2.40	1.97
B15C5 [7]	TBP	—	1.32	1.16
B15C5 [7]	B15C5	2.37	2.17	1.72
DB18C6 [8]	DB18C6	—	2.7	2.5

where TBP is a tributyl phosphate. These data indicate that, for the same alkali metal ion, the larger the number of the donor oxygen atoms of another ligand (L′) and the more closely the metal ion fits into the crown ether cavity, the greater is the complex-formation constant for the *synergistic* reaction. When TBP is present in the extraction system, relatively enhanced extractability and selectivity of the crown ethers for the alkali metal ions are observed. Extraction of Ag^+ with DB18C6 as ion pairs with a picrate ion into $CHCl_3$ was found to be synergistically enhanced by trioctylphosphine oxide (TOPO)[71]. This is due to an adduct formation of the complex, $Ag(DB18C6)(TOPO)_2^+$, in the $CHCl_3$ phase. lg $\{[ML(TOPO)_2A]_o/[MLA]_o [TOPO]_o^2\}$ value at 25 °C is 2.1. Synergistic increase was found on rubidium ion extraction into $CHCl_3$ by mixed DB18C6 and *n*-caprylic acid, bis(2-ethylhexyl)phosphoric acid (HDEHP), or TBP[58]. Kinard and McDowell[72] reported on a study of alkali (Li—Cs) and alkaline earth (Ca—Ba) metal ion extraction into benzene by HDEHP mixed with 12C4, CH15C5, DCH18C6, DCH21C7, or DCH24C8 (temp.?). These systems were found to produce a synergistic effect in the extraction of the metal ions. The synergistic effect is size selective; that is, it tends to be largest for the metal ions that best fit the crown ether cavity. The synergistic effects by the alkaline earth metal ions are smaller than those of the alkali metal ions. The stoichiometry of the alkali metal ion complex [probably ML(DEHP)(HDEHP)] for the DCH18C6-HDEHP system was discussed.

5 Extraction by Bis- and Poly-Crown Ethers

Metal ion extractability and selectivity of poly- and bis(crown ether)s containing pendant crown ether moieties are different from those of their corresponding monomeric analogues.

Kopolow et al.[73] reported that poly(vinyl B15C5) (PVB15C5) and poly(vinylbenzo[18]crown-6) (PVB18C6) extract alkali metal picrates into CH_2Cl_2 more efficiently than the corresponding monomeric crown ethers (cf. contribution of Smid . . .). This is partially caused by differences in distribution coefficients between the respective crown species. However, the extractability of the polymers for the metal ions is considerably enhanced in metal ion-crown ether systems which form stable 2:1 crown to metal ion complexes. This is usually the case when the size of the metal ion exceeds that of the crown ether cavity. Judging from the K_{ex} values (temp.?), the selectivity sequences of PVB15C5 for the alkali metal ions (based on Na^+ equal to unity) are $Li^+:Na^+:K^+:Rb^+:Cs^+ = 0.3:1:29:28:3.9$, while for PVB18C6 the ratio is 0.24:1:23:21:43. Gramain and Frere extracted various metal picrates into $CHCl_3$ by *43*[74], *44*[74], *45*[75], and *46*[75] at 25 °C. With all of the metal ions a stoichiometry of 1:1 is observed. Stability constants of the polymers with alkali and alkaline earth metal picrates in $CHCl_3$ are comparable to those of the corresponding monomeric analogues. At best, a slight decrease in stability is observed when the ring is included in the polymeric backbone. The selectivities for the polymers and for the corresponding monomeric analogue are almost the same; those for *43* and *44* are $K^+ > Sr^{2+} > Ba^{2+} > Ca^{2+} > Na^+ > Rb^+ > Li^+ \approx Cs^+$ and $Sr^{2+} > Ba^{2+} > Ca^{2+} > K^+ > Na^+ > Li^+ \approx Cs^+ > Rb^+$, respectively. With Cu^{2+}, Cd^{2+} and Pb^{2+}, *43* shows higher stabilities than *44* independent of the metal ion size[74]. Selectivities are $Pb^{2+} \gg Cu^{2+} \gg Cd^{2+}$ for *43* and $Cu^{2+} = Pb^{2+} > Cd^{2+}$ for *44*.

Kimura et al.[76] studied extraction of alkali metal picrates into $CHCl_3$ at 25 ± 0.1 °C by tactic and atactic poly(methacryloyloxymethyl B15C5) (*47*). For all the polymers the extractability is in the order: $K^+ > Rb^+ > Na^+ \approx Cs^+$. The iso-*47* is a little more selective for K^+ and Rb^+ than the other polymers. Alkali metal picrate extraction properties of tactic and radical poly(methacryloyloxymethyl

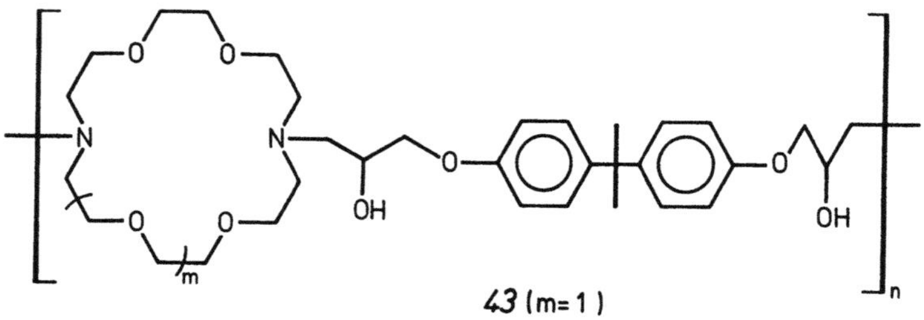

43 (m = 1)
44 (m = 0)

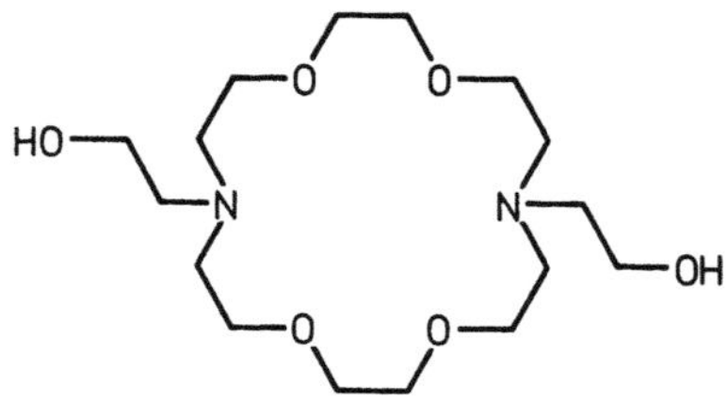

45

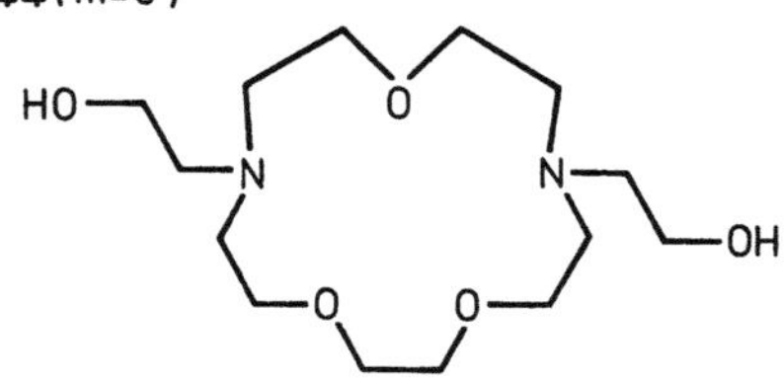

46

B18C6) (*48*), and *d,l*- and *meso*-bis(crown ether)s, which respectively are the dimer models of the isotactic and syndiotactic poly(crown ether)s *47* and *48*, were investigated in the $CHCl_3$-water system at 25 ± 0.1 °C [77]. The extractability of *48* decreases in the order: $Cs^+ > K^+ > Rb^+ > Na^+$. The extractability of iso-*48* for Cs^+ and Rb^+ is larger than that of syndio- and rad-*48*. The results indicate that isotactic *47* and *48* extract particular metal ions, which can be easily sandwiched by two neighboring crown ether units, a little more effectively than syndiotactic ones. No difference in extractability is observed between the corresponding *d,l*- and *meso*-bis(crown ether)s. This suggests that the rigidity of the polymers, in combination with the tactic configurations, causes the difference in the metal ion extractability.

Yagi et al. [78] extracted alkali metal (Li—Cs) picrates into CH_2Cl_2 at about 25 °C by methacrylamide B15C5 (*49*), methacrylamide B18C6 (*50*), their homopolymers (poly-*49* and *50*), and copolymers of *49* or *50* with methyl methacrylate containing various amounts of crown ether moieties. Poly-*49* and *50* show higher extractability than the monomers. In comparison with *49* and *50* the selectivities of poly-*49* and *50* increase for K and Rb and for Cs, respectively. It was demonstrated that poly-*49* and *50* act more effectively as extracting reagents than PVB15C5 and PVB18C6 [73]. High selectivity of poly-*49* for K and Rb, and of poly-*50* for Cs disappears with copolymers.

In order to evaluate the effects of chain rigidity and/or helicity on the metal ion extractability, Anzai et al. [79] investigated extraction of alkali metal picrates into CH_2Cl_2 at 25 °C by α-helical poly(glutamate) containing 4′-MB15C5 side chains. The polymer exhibits higher extractability than the small molecule analogue. The selectivity orders for the polymer and monomer are $K^+ > Rb^+ > Cs^+ > Na^+ > Li^+$ and $K^+ > Na^+ > Rb^+ > Cs^+ > Li^+$, respectively. They concluded that the flexibility of the polymer chain is not an essential requirement for a co-operative metal ion binding. Extraction of alkali metal picrates into CH_2Cl_2 by *51*–*54* was conducted at 25 °C [80]. Polymers show higher extractability than the monomers. The selectivity sequence for *53* and *54* is $K^+ > Rb^+ \gg Cs^+$

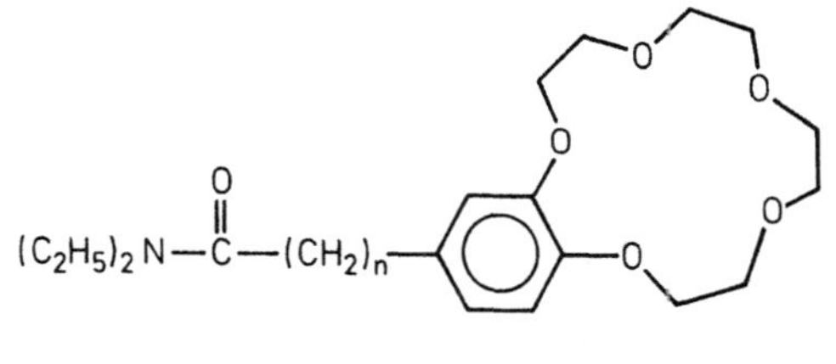

51 n = 0
52 n = 2

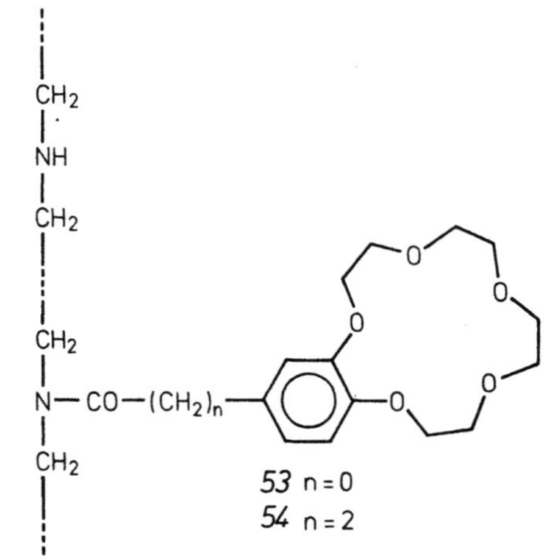

53 n = 0
54 n = 2

> Na^+ > Li^+, while for *51* and *52* it is $K^+ > Na^+ > Rb^+ > Cs^+ \approx Li^+$. The extractability of *53* and *54* is influenced by the electronic nature of the substituents, their chain length and their flexibility at 4′-position of B15C5. The difference in the extractabilities between *53* and *54* may partly be ascribable to the difference in content of polymers of the crown ether.

Warshawsky et al. [81] reported that polymeric pseudocrown ethers, incorporating oxyethylene and oxypropylene units, extract $FeCl_4^-$ into toluene from mixed hydrochloric and phosphoric acids, and that extractability depends on HCl concentration, extraction becoming efficient at HCl concentrations exceeding 4 mol dm^{-3}.

To date, various kinds of polymers such as polystyrenes, polymethacrylates, polypeptides and polyethyleneimines containing crown ethers in their side chain have been used in extraction studies.

Extractions of metal picrates into $CHCl_3$ were conducted at 25 ± 0.1 °C using a series of poly- and bis(crown ether)s and their corresponding monomeric analogues (*55*–*59*) [82–87]. Extractability sequences of *56* (n = 0) and *57* (n = 0) for alkali metal ions are $Na^+ > K^+ > Rb^+ \geq Cs^+$ and $K^+ \approx Rb^+ \approx Cs^+ > Na^+$, respectively. Ligand *56* (n = 0) improves the Na^+ selectivity and extractability of *55* (n = 0), while *57* (n = 0) does not [82]; *56* (n = 1 and 2) and *57* (n = 1 and 2) extract alkali metal (Na—Cs) picrates more effectively than the corresponding monocyclic crown ethers. In particular, *56* (n = 1) and *57* (n = 1) and *57* (n = 2) are remarkably effective extracting reagents for K^+ and Rb^+, and for Cs^+, respectively [83]. *56* (n = 1 and 2) and *57* (n = 1 and 2) show higher extractability for Ag^+ and Tl^+ than the corresponding monocyclic crown ethers. Especially *56* (n = 1) and *57* (n = 1) are quite effective extracting reagents for Tl^+ [84]. Ligands *56* (n = 1 and 2) and *57* (n = 1 and 2) exhibit larger extractability for alkaline earth metal picrates than the corresponding monocyclic analogues. The extractability sequence of *55*, *56*, *57* (n = 1), and *56* (n = 2) is $Ba^{2+} > Sr^{2+} > Ca^{2+} > Mg^{2+}$, while that of *55* and *57* (n = 2) is $Sr^{2+} > Ba^{2+} > Ca^{2+} > Mg^{2+}$. Particularly, *56* (n = 1), *57* (n = 1), and *57* (n = 2) have relatively high extractability and selectivity for Ba^{2+} and Sr^{2+}, respectively [85]. The extractability of *55*, *56*, and *57* (n = 3) for alkali metal ions decreases in the order $Cs^+ > Rb^+ > K^+ > Na^+$. The extractability

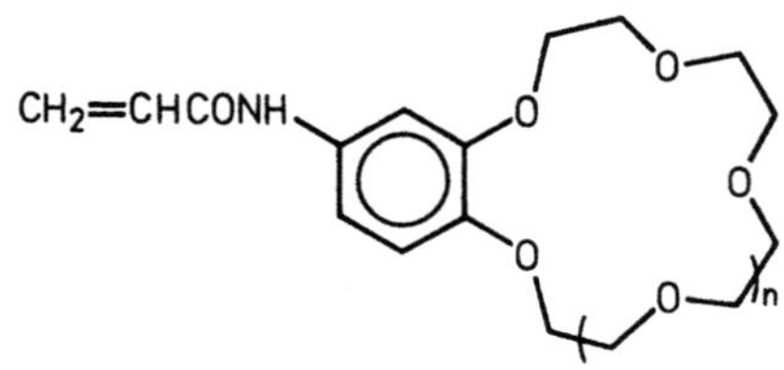

55 (n = 0,1,2,3)

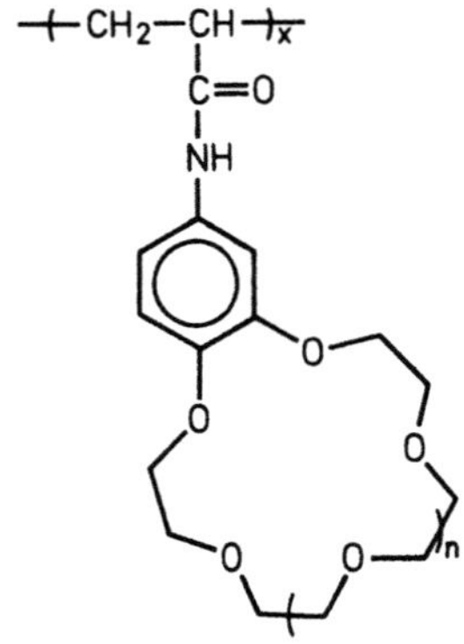

56 (n = 0,1,2,3)

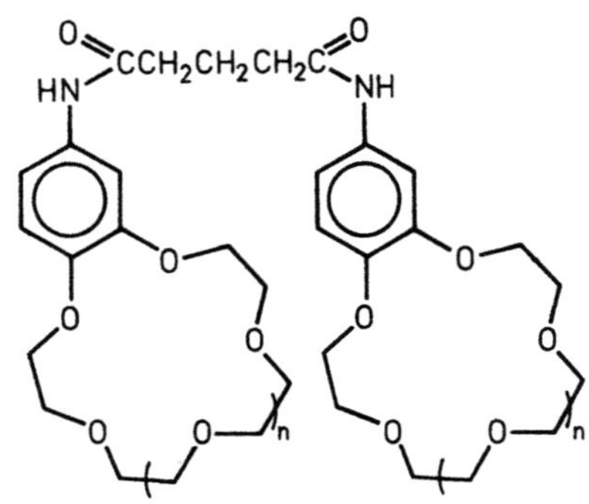

57 (n = 0,1,2,3)

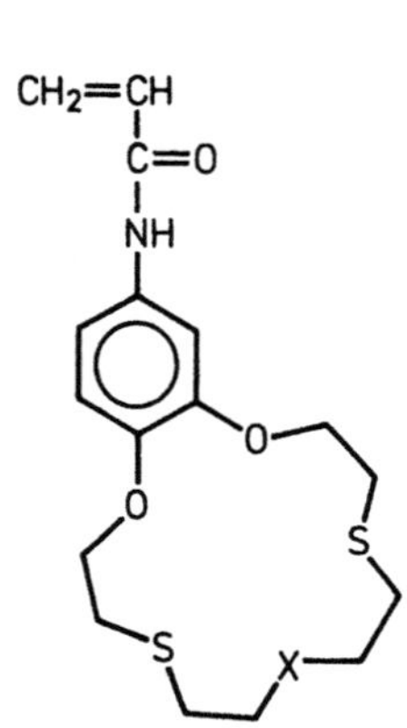

58

a X = —S—
b X = —S(CH$_2$)$_3$S—

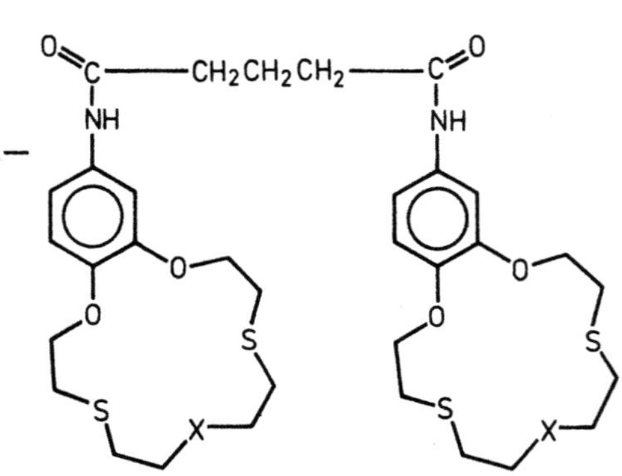

59

of the crown ether derivatives for the same alkali metal ion decreases in the order *56* > *57* > *55* (n = 3). The alkali metal ions do not form 2:1 crown ether unit to cation complexes with *55*, *56*, and *57* (n = 3)[86]. The extractability sequence of *58* and *59* for silver picrate is *59b* > *59a* > *58b* ≫ *58a*. The bis (thiacrown ether)s extract Ag^+ more effectively than the monomeric analogues. As compared with D_M values for silver picrate of PB15C5 and PB18C6, those of the thiacrown ethers are extremely larger[87]. Maeda et al.[88] extracted alkali metal (Na—Cs) picrates into CH_2Cl_2 by 12C4, *60*, and *61* at 25 °C. Ligand *61* (n = 1) has larger extractability and

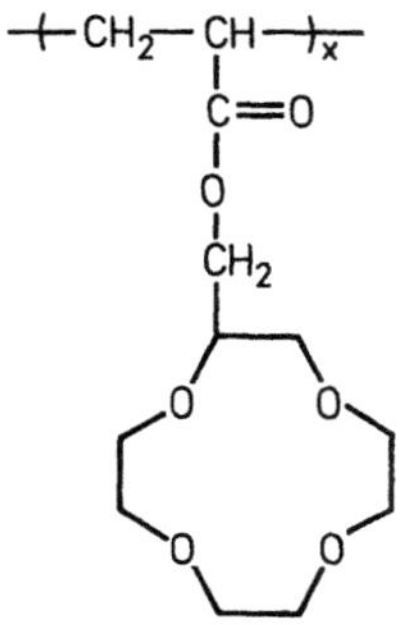

60

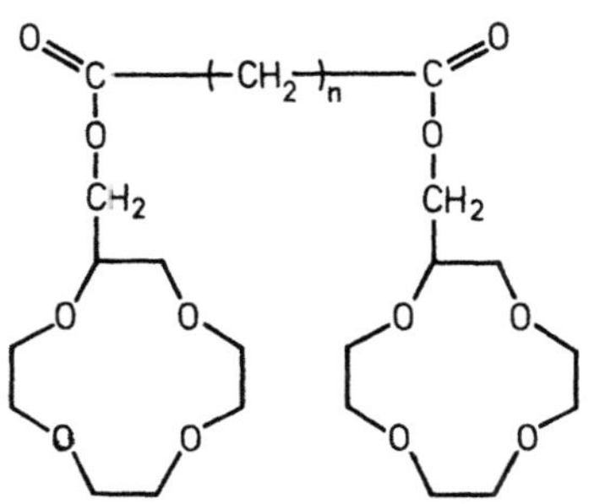

61 (n=1,3)

selectivity for Na^+ than *61* (n = 3) and 12C4. Moreover, *60* has much larger extractability for Na^+ than *61* and 12C4. Kimura et al.[89] studied effects of stereochemical structure of bis(crown ether)s (*62* and *63*) on their extractability for alkali metal

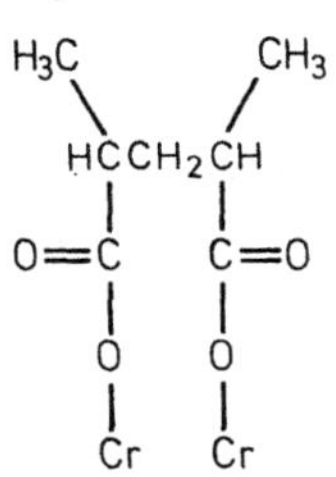

62

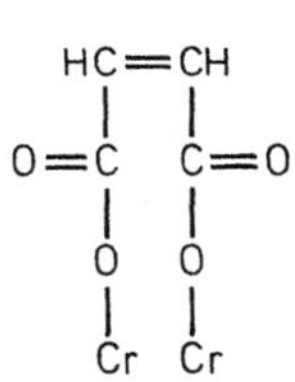

63

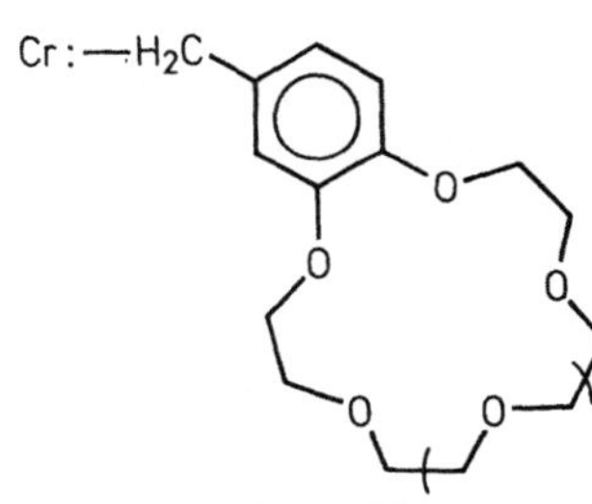

(n = 1, 2)

(Na—Cs) picrates in a $CHCl_3$-water system at 25 °C. *Meso*- and *d,l*-*62* (n = 1) exert a nearly similar extractability. The extraction power of (*Z*)-*63* (n = 1 and 2) is very different from that of the corresponding (*E*)-*63*. The ΔH (<0 for all systems) and ΔS (<0 for all systems) values for the overall extraction reaction were determined and the extraction properties of the (*Z*)- and (*E*)-*63* were discussed in detail. Wong and Ng[90] extracted alkali metal picrates into $CHCl_3$ at 20 °C by bis(crown ether)s with B15C5 and B18C6 moieties bridged by polyethylene glycol ($-CH_2(OCH_2CH_2)_n-OCH_2-$; n = 0–3). For the metal ions which form sandwich complexes, the extractabilities by bis(crown ether)s are much greater than those by the corresponding monomeric analogues, while for those forming non-sandwich complexes, the extractabilities by bis(crown ether)s are slightly lower than those by the corresponding monomeric analogues. The length of the bridge linking the two crown moieties affects extraction efficiency. The selectivity orders are $K^+ > Rb^+ > Na^+ > Cs^+$ for bis-B15C5 and $K^+ \approx Cs^+ > Rb^+ > Na^+$ for bis-B18C6. Particularly, bis-B18C6 shows the most effective extraction power for K^+ and Cs^+.

Poly- and bis(crown ether)s exhibit remarkable extractability and selectivity for metal ions forming 2:1 crown ether unit-cation complexes because of co-operative effects of adjacent crown ether units on the metal ions.

6 Application of Crown Ethers to Extraction Analysis

Numerous fundamental data for extraction of metal ions by crown ethers have been reported to date, while examples for application of crown ethers to extraction analysis are not so frequent.

6.1 Separation Analysis

Mitchell and Shanks [91] accurately determined traces of Na (in a synthetic standard solution and silicon dioxide samples) after neutron activation by selectively extracting the Na into $CHCl_3$ as DCH18C6 complex with tetraphenylborate. Jepson and Dewitt [92] studied the enrichment of heavy isotopes of Ca by the exchange reaction: $^{40}Ca^{2+} + {}^{44}CaL^{2+}{}_o \rightleftharpoons {}^{44}Ca^{2+} + {}^{40}CaL^{2+}{}_o$. For the extraction of $CaCl_2$ into $CHCl_3$ or CH_2Cl_2 by DCH18C6 the single stage separation factor for the isotopes differing by one mass unit is 1.0010 ± 0.0002.

Kimura et al. [93] separated a small amount of Sr^{2+} (0.7–20 mg) from a large amount of Ca^{2+} (1.5 g) by an extraction method using a $CHCl_3$ solution of DCH18C6. A picrate ion was used as a counter anion for the extraction. The average recovery of Sr^{2+} was 97%, and the Ca^{2+} content in the final strontium fraction was decreased to $10^{-5} - 10^{-4}$ parts of the initial amount. A small amount of Ba^{2+} ($2.0 \times 10^{-5} - 5.00 \times 10^{-4}$ mol dm^{-3}) in the presence of a large amount of Ca^{2+} (5.00×10^{-2} mol dm^{-3}) was precisely determined with a combined extraction procedure using a nitrobenzene solution of 18C6 with atomic absorption spectrophotometry employing the N_2O-acetylene flame [94]. A picrate ion was used as a counter anion. The average recovery of Ba was 100 ± 3%. Ba^{2+} (3.00×10^{-4} mol dm^{-3}) was separated from Sr^{2+} ($3.00 \times 10^{-4} - 1.70 \times 10^{-2}$ mol dm^{-3}) by means of extracting with a nitrobenzene solution of DB24C8 using a picrate ion as a counter anion [95]. The average recovery of Ba^{2+} was 97% and only 10^{-3} parts of the Sr^{2+} in the initial sample solution was recovered. Gerow and Davis [96] tried to remove Sr^{2+} and Cs^+ from an aqueous HNO_3 solution by an extraction method using a variety of 24C8 compounds in CCl_4 or kerosene mixed with tributyl phosphate. The highest D_m values obtained were 1.45 ± 0.05 for Cs^+ and 200 for Sr^{2+} (25 ± 2 °C).

Reyes et al.[97] studied the extraction of Na^+, K^+, and Cs^+ by some cryptands with various organic solvents using ClO_4^- as a counter anion. In a mixture of traces of Na^+, K^+, and Cs^+, the two smaller cations were extracted by a nitrobenzene solution of cryptand [2.2.2] and the Na^+ was separated from the K^+ in a few cycles. Cryptand [2.2.2] and [2.2.1]/nitrobenzene systems can be used for quantitative determination of Na^+ and K^+ by the substoichiometric method.

Sevdić et al.[98] studied the extraction of Ag^+ and Hg^{2+} into nitrobenzene by 1,4,8,11-tetrathiacyclotetradecane (TTP) from HCl, $HClO_4$, picric acid, and HNO_3 solutions, and found a separation method for micro-amounts of Ag^+ (10^{-8} — 10^{-7} mol dm^{-3}) from Hg^{2+} ($\leqq 10^{-4}$ mol dm^{-3}). There is no interference of Co^{2+}, Zn^{2+}, and Cd^{2+}.

6.2 Colorimetric Determination

Crown ethers are used for the spectrophotometric determination of traces of alkali and alkaline earth metal ions by means of a solvent extraction method. These studies are classified into two groups: a) ion-pair extraction of crown ether-metal ion complex with an anionic dye, and b) chromogenic crown ether reagents.

a) Sumiyoshi et al.[99] reported colorimetric determination of K^+ in blood serum by *ion-pair extraction* using a benzene solution of 18C6 and bromocresol green. The extraction must be conducted at pH < 4. 500 ppm Na^+ does not interfere with the determination of 5 ppm K^+. Kina et al.[29] reported that, for the extraction in a 1,2-dichloroethane-water system, univalent cations, Sr^{2+}, Ba^{2+}, and Pb^{2+} drastically interfere with the determination of K^+ with DB18C6 and anilinonaphthalene sulfonate. Kolthoff[26] presented examples of the use of extraction analysis for determining traces of K^+ in lithium and sodium salts using 18C6 and picrate anions in a CH_2Cl_2-water system. Sanz-Medel et al.[100] fluorimetrically determined K^+ in blood serum by an extraction method using a 1,2-dichloroethane solution of 18C6 and eosine. Takagi et al.[101] described a photometric extraction procedure for the determination of Na^+ in blood serum using a toluene solution of cryptand [2.1.1] and a picrate anion. A 350-fold amount of K^+ (by weight) does not interfere with the determination of a μg ml^{-1} level of Na^+, nor do other alkali metal ions. Interferences by multivalent cations are masked by EDTA. Szczepaniak and Juskowiak[102] reported the procedure of spectrophotometric determination of traces of Pb^{2+} by extraction with a chlorobenzene solution of [2.2.2] and eosine. High molar absorptivity was obtained for the ion-pair complex of Pb^{2+}. There is no interference from Bi^{3+}, Sn^{2+}, Hg^{2+}, Zn^{2+}, Cd^{2+}, Cu^{2+}, Ca^{2+}, Pd^{2+}, Ag^+, and Tl^+.

In contrast to the monocyclic crown ethers, the ion-pair extraction of the metal ion-cryptand complex depends on the pH and the basic nature of the cryptand. The extraction becomes pH-independent at high pH-levels where the protonation of the cryptand can be neglected. Lower pH causes some extraction of the protonated cryptand, which makes the extraction pattern of the metal ion complicated.

Yoshio et al.[103] extracted $Co(CNS)_4^{2-}$ into 1,2-dichloroethane by the ammonium ion complex with 18C6 or DCH18C6. There is no interference from Ni^{2+}, Cu^{2+}, Zn^{2+}, Pb^{2+}, and Sn^{2+}, while that of iron does interfere. The 18C6 complex with K^+

selectivity extracted the Cu-zincon anionic complex into $CHCl_3$ in the presence of the Zn-zincon chelate anion [104]. By measuring the absorbance of the $CHCl_3$ phase, Cu was determined spectrophotometrically in the presence of Zn.

b) Light-absorbing anions were incorporated into crown ether molecules in order to use the crown ethers as metal ion-selective *photometric reagents*. Nakamura et al. [105] studied extraction of alkali metal ions into $CHCl_3$ by 4′-picrylamino-B15C5 derivatives (HL). The reagents extract alkali metal ions as colored complexes as ML · HL or ML. The extractability decreases in the order: $K^+ > Rb^+ > Cs^+ > Na^+ \gg Li^+$. 10–800 ppm K^+ is determined spectrophotometrically in the presence of other alkali and alkaline earth metal ions. Alkali metal ions are extracted into $CHCl_3$ by forming colored complexes with 4′-picrylamino-B18C6 derivatives (HL) (*64*) as ML [106]. The extractability sequence is:

$$K^+ > Rb^+ > Cs^+ > Na^+ \gg Li^+$$

The colorimetric determination of 4–40 ppm K^+ is possible in the presence of other alkali and alkaline earth metal ions. Pacey and Bubnis [107] extracted K^+ into $CHCl_3$ as a colored complex with 4″-cyano-2″,6″-dinitro-4′-amino-B15C5. 10–700 ppm K^+ is determined spectrophotometrically in the presence of excess Na^+. Nakamura et al. [108] extracted alkali metal ions into

1,2-dichloroethane by N-(2-hydroxy-5-nitrobenzyl)-aza-15C5

(*65*) and

N-(2-hydroxy-5-nitrobenzyl)-aza-18C6

(*66*). The selectivity sequences are

$$Li^+ > Na^+ > K^+ \gg Rb^+ \approx Cs^+$$

for *65*, and

$$K^+ > Na^+ > Rb^+ > Li^+ > Cs^+$$

for *66*. Chromoionophore *65* is useful for photometrically determining 10–100 ppm Li^+ as well as for concentrating Li^+ from a dilute solution. The selectivity of *66* for K^+ against Na^+ is less than that of *64*, however, the extractability for K^+ is greater because of the phenolate anion which allows the coordination to the metal ion in the crown ether cavity.

2-hydroxy-5-(4-nitrophenylazo)phenoxymethyl-15C5

(*67*) and -18C6 (*68*) were successfully used for the extraction-spectrophotometric determination of ppm-levels of Na^+ and K^+, respectively, in synthetic samples [109]. In this work 1,2-dichloroethane was employed as a diluent. Extraction efficiencies for

67 and *68* are generally much higher than those of *64* and *65*. The charge-separated structure of the metal complex derived from *64* is not favorable for the extraction. A low proton dissociation constant of *65* inevitably leads to a low extraction efficiency. Na^+ in human blood serum was spectrophotometrically determined by an extraction method using a 1,2-dichloroethane solution of

(2-hydroxy-3,5-dinitrophenyl)oxymethyl-15C5

(*69*) [110]. The molar absorptivity of the Na^+ complex is 1.4×10^4 l mol^{-1} cm^{-1} at 423 nm. *69* shows a lower reagent blank compared with other reagents (e.g., *64–68*) described in literature [105, 106, 108, 109]. The determination of the metal ion by *69* is much less sensitive to pH variation in the extraction procedure. From the practical point of view these are the most important advantages of *69* over the others. Nishida et al. [111] studied the extraction of alkaline earth metal ions into 1,2-dichloroethane by

N,N′-bis(2-hydroxy-5-nitrobenzyl)-1,10-diaza-4,7,13,16-tetraoxacyclooctadecane

(*70*) and

N,N′-bis(2-hydroxy-5-nitrobenzyl)-1,10-diaza-4,7,13-trioxacyclopentadecane

(*71*). The relative ease of extraction is $Ca^{2+} > Sr^{2+} > Ba^{2+} \gg Mg^{2+}$ for both reagents. The extraction is more effective with *70* than with *71*, selectivity being more pronounced with *71* than with *70*. The extraction-photometric determination of Ca^{2+} in blood serum is possible by using *70*. Nishida et al. [112] reported that fluorescent crown ether reagents,

N-(4-methylumbelliferone-8-methylene)-monoaza-15C5

(*72*) and -18C6 most effectively extract Li^+ and K^+ into 1,2-dichloroethane, respectively. The similar fluorescent crown ether derived from diaza-18C6 (*73*) is selective for Ca^{2+}. The fluorimetric determination of Li^+ and Ca^{2+} is feasible by using *72* and *73*, respectively.

Finally, the transfer rate of K^+ from aqueous picrate solutions into 1,2-dichloroethane solutions of DB18C6 at 25 °C was reported by Danesi et al. [113]. The rate of extraction of K^+ was explained in terms of chemical reactions taking place at the water-organic interphase. The reaction order for both K^+ and picrate ion is equal to one. The reaction order for DB18C6 was found to vary with the DB18C6 concentration from an initial value of one down to zero.

In conclusion, a large number of papers which deal with extraction properties of various crown ethers for metal ions have been reported. However, the data particularly for lanthanides and actinides are few. From the hard/soft principle, crown ethers seem to be promising analytical reagents for extraction separation, -enrichment and -determination of lanthanides and actinides.

7 Acknowledgements

The author would like to thank Dr. Naoaki Fukada, Department of Chemistry, Faculty of Science, Chiba University, for his helpful discussions and assistance.

8 References

1. Pedersen, C. J.: Fed. Proc., Fed. Am. Soc. Exp. Biol. *27*, 1305 (1968)
2. Frensdorff, H. K.: J. Am. Chem. Soc. *93*, 4684 (1971)
3. Takeda, Y.: Bull. Chem. Soc. Jpn. *53*, 2393 (1980)
4. Takeda, Y. et al.: ibid. *55*, 3438 (1982)
5. Takeda, Y., Gotō, H.: ibid. *52*, 1920 (1979)
6. Takeda, Y., Katō, H.: ibid. *52*, 1027 (1979)
7. Takeda, Y. et al.: ibid. *54*, 3727 (1981)
8. Sadakane, A. et al.: ibid. *48*, 60 (1975)
9. Takeda, Y., Takahashi, F.: ibid. *53*, 1167 (1980)
10. Sekine, T. et al.: ibid. *51*, 645 (1978)
11. Takeda, Y.: ibid. *52*, 2501 (1979)
12. Iwachido, T. et al.: ibid. *51*, 629 (1978)
13. Hasegawa, Y. et al.: ibid. *54*, 2427 (1981)
14. Hasegawa, Y. et al.: J. Inorg. Nucl. Chem. *43*, 633 (1981)
15. Sekine, T. et al.: ibid. *41*, 571 (1979)
16. Pauling, L. (Ed.): The Nature of the Chemical Bond, Cornell University Press, 1960[3]
17. Izatt, R. M., Christensen, J. J. (Eds.): Synthetic Multidentate Macrocyclic Compounds, Academic Press, New York 1978
18. Iwachido, T. et al.: Bull. Chem. Soc. Jpn. *53*, 703 (1980)
19. Jaber, A. M. Y. et al.: J. Inorg. Nucl. Chem. *39*, 1689 (1977)
20. Yakshin, V. V. et al.: Dokl. Akad. Nauk SSSR *241*, 159 (1978)
21. Wang, W. et al.: Chem. J. Chinese Univs. *2*, 431 (1981)
22. Kojima, T. et al.: Bull. Res. Lab. Nucl. React (Tokyo Inst. Technol.) *6*, 23 (1981)
23. Takeda, Y.: Bull. Chem. Soc. Jpn. *56*, 931 (1983)
24. Iwachido, T.: Chem. Lett. 1511 (1977)
25. Chantooni, Jr., M. K., Kolthoff, I. M.: Proc. Natl. Acad. Sci. (USA) *78*, 7245 (1981)
26. Kolthoff, I. M.: Can. J. Chem. *59*, 1548 (1981)
27. Abraham, M. H., Ling, H. C.: Tetrahedron Lett. *23*, 469 (1982)
28. Shchori, E. et al.: J. Chem. Soc., Dalton 2381 (1975)
29. Kina, K. et al.: Bunseki Kagaku *27*, 291 (1978)
30. Kolthoff, I. M., Chantooni, Jr., M. K.: Anal. Chem. *52*, 1039 (1980)
31. Luo, Chuen-Lu et al.: J. Chinese Chem. Soc. *26*, 61 (1979)
32. Izatt, R. M. et al.: J. Am. Chem. Soc. *98*, 7620 (1976)
33. Friedman, H. L., Krishnan, C. V.: Water, Franks, F. (Ed.), Vol. *3*, p. 55, Plenum, New York 1973
34. Jawaid, M., Ingman, F.: Talanta *25*, 91 (1978)
35. Yakshin, V. V. et al.: Dokl. Akad. Nauk SSSR *252*, 373 (1980)
36. Strzelbicki, J., Bartsch, R. A.: Anal. Chem. *53*, 1894 (1981)
37. Strzelbicki, J., Bartsch, R. A.: ibid. *53*, 2251 (1981)
38. Strzelbicki, J. et al.: Sep. Sci. Technol. *17*, 635 (1982)
39. Charewicz, W. A. et al.: Anal. Chem. *54*, 2094 (1982)
40. Strzelbicki, J., Bartsch, R. A.: ibid. *53*, 2247 (1981)
41. Rais, J. et al.: Proc. Int. Solv. Ext. Conf. *2*, 1705 (1974)
42. Danesi, P. R. et al.: J. Inorg. Nucl. Chem. *37*, 1479 (1975)
43. Marcus, Y., Asher, L. E.: J. Phy. Chem. *82*, 1246 (1978)
44. Asher, L. E., Marcus, Y.: Proc. Int. Solv. Extr. Conf., Toronto 1977, CIM Special Vol. *21*, 130 (1979)
45. Marcus, Y. et al.: Hydrometallurgy *7*, 27 (1981)

46. McDowell, W. J., Shoun, R. R.: Proc. Int. Solv. Extr. Conf., Toronto 1977, CIM Special Vol. *21*, 95 (1979)
47. Iwachido, T. et al.: Bull. Chem. Soc. Jpn. *55*, 2378 (1982)
48. Gloe, K. et al.: Z. Chem. *19*, 382 (1979)
49. Abashkin, V. M. et al.: Dokl. Akad. Nauk SSSR *257*, 1374 (1981)
50. Tušek, Lj. et al.: J. Inorg. Nucl. Chem. *37*, 1538 (1975)
51. Pannell, K. H. et al.: J. Am. Chem. Soc. *99*, 1457 (1977)
52. Gokel, G. W. et al.: J. Chem. Soc., Chem. Commun. 1053 (1980)
53. Cho, I., Chang, S.: Bull. Korean Chem. Soc. *1*, 145 (1980)
54. Nakatsuji, Y. et al.: Chem. Lett. 1571 (1982)
55. Inoue, Y. et al.: ibid. 781 (1982)
56. Akabori, S. et al.: ibid. 1393 (1982)
57. Olsher, U., Jagur-Grodzinski, J.: J. Chem. Soc., Dalton Trans. 501 (1981)
58. Gloe, K. et al.: Isotopenpraxis *18*, 170 (1982)
59. Shinkai, S. et al.: J. Am. Chem. Soc. *102*, 5860 (1980)
60. Shinkai, S. et al.: Chem. Lett. 499 (1982)
61. Shiga, M. et al.: ibid. 1021 (1980)
62. Shinkai, S. et al.: Tetrahedron Lett. *23*, 2581 (1982)
63. Shinkai, S. et al.: Chem. Lett. 283 (1980)
64. Shinkai, S. et al.: Tetrahedron Lett. *21*, 4463 (1980)
65. Rebek, J. Jr., Wattley, R. V.: J. Heterocycl. Chem. *17*, 749 (1980)
66. Bogatskii, A. V. et al.: Dokl. Akad. Nauk SSSR *250*, 1389 (1980)
67. Sevdić, D., Meider, H.: J. Inorg. Nucl. Chem. *39*, 1403 (1977)
68. Sevdić, D., Meider, H.: ibid. *39*, 1409 (1977)
69. Sevdić, D. et al.: ibid. *42*, 885 (1980)
70. Takeda, Y.: Bull. Chem. Soc. Jpn. *54*, 526 (1981)
71. Hasegawa, Y. et al.: Chem. Lett. 1075 (1981)
72. Kinard, W. F., McDowell, W. J.: J. Inorg. Nucl. Chem. *43*, 2947 (1981);
Kinard, W. F. et al.: Sep. Sci. Technol. *15*, 1013 (1980);
McDowell, W. J. et al.: Int. Solvent Extr. Conf., Proc., Vol. *1*, 80 (1980)
73. Kopolow, S. et al.: Macromolecules *6*, 133 (1973)
74. Gramain, Ph., Frere, Y.: ibid. *12*, 1038 (1979)
75. Gramain, Ph., Frere, Y.: J. Chim. *3*, 53 (1979)
76. Kimura, K. et al.: Polym. Bull. *1*, 403 (1979)
77. Kimura, K. et al.: Makromol. Chem. *182*, 1579 (1981)
78. Yagi, K. et al.: Makromol. Chem., Rapid Commun. *1*, 263 (1980)
79. Anzai, J. et al.: ibid. *3*, 55 (1982)
80. Anzai, J. et al.: ibid. *3*, 399 (1982)
81. Warshawsky, A. et al.: Hydrometallurgy *4*, 93 (1979)
82. Maeda, T. et al.: Bull. Chem. Soc. Jpn. *55*, 3506 (1982)
83. Kimura, K. et al.: Talanta *26*, 945 (1979)
84. Maeda, T. et al.: Fresenius' Z. Anal. Chem. *298*, 363 (1979)
85. Maeda, T. et al.: ibid. *313*, 407 (1982)
86. Kimura, K. et al.: ibid. *313*, 132 (1982)
87. Ōue, M. et al.: Chem. Lett. 275 (1982)
88. Maeda, T. et al.: ibid. 1573 (1981)
89. Kimura, K. et al.: ibid. 611 (1979)
Kimura, K. et al.: Talanta *27*, 801 (1980)
90. Wong, K. H., Ng, H. L.: J. Coord. Chem. *11*, 49 (1981)
91. Mitchell, J. W., Shanks, D. L.: Anal. Chem. *47*, 642 (1975)
92. Jepson, B. E., Dewitt, R.: J. Inorg. Nucl. Chem. *38*, 1175 (1976)
93. Kimura, T. et al.: Chem. Lett. 563 (1977)
94. Takeda, Y. et al.: ibid. 1377 (1978)
95. Takeda, Y. et al.: ibid. 601 (1979)
96. Gerow, I. H., Davis, M. W. Jr.: Sep. Sci. Technol. *14*, 395 (1979)
97. Reyes, M. J. et al.: J. Inorg. Nucl. Chem. *41*, 1365 (1979)
98. Sevdić, D. et al.: Mikrochim. Acta II, 235 (1975)

99. Sumiyoshi, H. et al.: Talanta *24*, 763 (1977)
100. Sanz-Medel, A. et al.: ibid. *28*, 425 (1981)
101. Takagi, M. et al.: Anal. Chim. Acta *126*, 185 (1981)
102. Szczepaniak, W., Juskowiak, B.: ibid. *140*, 261 (1982)
103. Yoshio, M. et al.: Anal. Lett. A, *11*, 281 (1978)
104. Yoshio, M. et al.: Anal. Lett. *13*, 1431 (1980)
105. Nakamura, H. et al.: Talanta *26*, 921 (1979)
106. Nakamura, H. et al.: Anal. Chem. *52*, 1668 (1980)
107. Pacey, G. E:, Bubnis, B. P.: Anal. Lett. *13* (A12), 1085 (1980)
108. Nakamura, H. et al.: Chem. Lett. 1305 (1981)
109. Nakamura, H. et al.: Bunseki Kagaku *31*, E131 (1982)
110. Nakamura, H. et al.: Anal. Chim. Acta *139*, 219 (1982)
111. Nishida, H. et al.: Mikrochim. Acta *1*, 281 (1981)
112. Nishida, H. et al.: Chem. Lett. 1853 (1982)
113. Danesi, P. R. et al.: J. Inorg. Nucl. Chem. *40*, 1119 (1978)

Crown Compounds as Alkali and Alkaline Earth Metal Ion Selective Chromogenic Reagents

Makoto Takagi and Keihei Ueno

Department of Organic Synthesis, Faculty of Engineering, Kyushu University, Fukuoka 812, Japan

Table of Contents

1 Introduction 40

2 Monoprotonic Crown Ether Dyes 40
2.1 Derivatives of Benzocrown Ethers 40
2.2 N-Side-armed Monoazacrown Ethers 45
2.3 Side-armed Crown Ethers 46
2.4 "Crowned" Phenols 50
2.5 Benzocrown Ethers with Azo-linked Side Arm 51

3 Diprotonic Crown Ether Dyes 53

4 Neutral Crown Ether and Related Dyes 58

5 Conclusion 64

6 References 64

This article deals with the recent development of crown ethers which have added chromophoric functional groups within the molecule (crown ether dyes). These compounds are designed to bring about specific color changes on the interaction with such metal cations as alkali and alkaline earth metals, thus being able to serve as probes or photometric reagents selective for these metal ions. The chromophoric groups can bear a dissociable proton (or protons) or can be nonionic. In the former, the ion exchange between the proton and appropriate metal cations causes the color change, while in the latter, the coordination of the metal ion to the chromophoric donor or acceptor of the dye molecule induces a change of the charge transfer band of the dye. The proton-dissociable crown ether dyes are especially suited for extraction photometric determination of alkali and alkaline earth metal ions. Similarly, the neutral crown ether dyes are potentially useful for such determinations in lipophilic homogeneous media.

1 Introduction

Since the dawn of inorganic analysis and metal complex chemistry, a wide variety of ion selective organic chromogenic reagents have been developed and come to practical use of solve problems in metal analysis [1,2]. Selective chromogenic reagents, alone or in combination with masking agents, have satisfactorily been used for colorimetric determination of most of the commonly encountered metal ions. Some of them are indispensable as metal indicators for standard chelatometric titrations.

While ion selective chromogenic reagents have become popular as early as 1950 for most of the metal ions, attempts to develop those selective to alkali metal ions remained unsuccessful. As to alkaline earth metal ions, the reagents with satisfactory selectivity are still rather rare, since they often suffer competition from transition metals. The discovery of crown ethers in 1967 [3] definitely gave a clue to cultivate this last field of organic reagent-aided metal analysis, but it still took ten years for such alkali and alkaline earth metal-selective chromogenic reagents to receive practical significance. In this review, recent research efforts are summarized for the development of crown ether-based chromogenic reagents which selectively respond to alkali and alkaline earth metal ions. Some critical comments are also added on the use of these reagents for analytical problems of these metal ions.

2 Monoprotonic Crown Ether Dyes

Designing alkali metal selective crown ether dyes is not much different in principle than that of classical chromogenic chelating agents; a monoprotonic chromophore is introduced into a crown ether skeleton in such a proximity to the ethereal function that the dissociation of the chromophoric proton is assisted by the complexation of the positively charged metal ion with the crown ether macrocycle [4]. Conceptually, the function of metal binding and the function of its detection may be separately carried by the crown ether and the protonic chromophore, respectively. However, the anionic chromophore can more or less contribute to the metal binding ability of the crown ether, and thus, in the actual molecule, the two functions can not be separated from each other.

2.1 Derivatives of Benzocrown Ethers

The first crown ether dyes designed according to the above principle are 4′-picrylamino-substituted derivatives of benzo[15]crown-5 (B15C5)[1], *1–3*, reported by Takagi, Nakamura, and Ueno [4,5]. The synthesis is readily attained by first nitrating and then reducing B15C5 leading to 4′-amino-B15C5, which is then reacted with picryl chloride under basic conditions to yield *1*.

The crown ether dye *1* is almost insoluble in water, but is soluble in aqueous alkali

1 Notations frequently used in this report are listed and explained on p. XI

or in common organic solvent such as methanol, chloroform or benzene. The distribution between water and the organic solvent is strongly in favor of the latter, and under ordinary conditions, the amount of *1* dissolved in water which is in equilibrium with an equal volume of chloroform is negligible as compared with that in the organic phase. *1* is appreciably acidic and the colour of the solution turns from orange (λ_{max} = 390 nm; ε, 12,900 M^{-1} cm; 1 M = 1 mol dm^{-3}) to blood red (λ_{max} = 445 nm; ε, 20,000) on dissociation of the amino proton.

When a chloroform solution of *1* (HL) is brought into contact with an alkaline aqueous solution of alkali metal salts, K^+ and Rb^+ are strongly extracted into the chloroform solution and the color of the chloroform solution changes from orange to blood-red, indicating the formation of L^- species in the organic solution. Na^+ is extracted only slightly, while Li^+ not at all. The use of strong alkali such as sodium hydroxide and lithium hydroxide causes the coloration of the aqueous rather than the chloroform phase, indicating that the ML salt of these metal ions is poorly soluble in chloroform; the enforced dissociation of HL by the strong alkali results in the transfer of the dissociated crown ether species (L^-) from the organic to aqueous phase. Thus, the extent of coloration of the organic phase can be used as a measure for alkali metal (potassium and rubidium) present in the original aqueous solution.

In the typical extraction of alkali metal ions, equal volumes of an aqueous solution of alkali metal chloride (10^{-3}–1 M) and a chloroform solution at 2×10^{-3} M in *1* and 1 M in triethylamine are shaken at 25 °C, and the change in optical absorption of the organic layer is measured at 560 nm. The increase in molar absorptivity of HL upon formation of metal complex ML (or ML · HL, see below) is 4580 M^{-1} cm at this wavelength. The extraction is most satisfactorily explained by the reaction in Scheme 1. The extraction constants concerned are defined by Eqs. (1)–(4). The

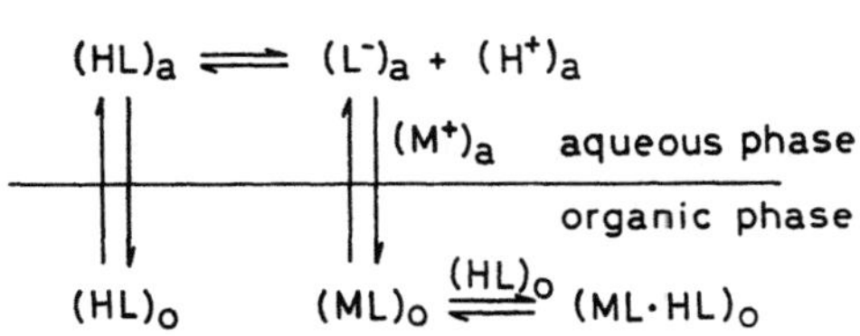

parentheses with suffixes indicate the chemical species (suffix "a" for aqueous phase and "o" for organic phase), and the brackets stand for concentrations (M) in an ordinary sense.

$$(HL)_o + (M^+)_a \rightleftharpoons (ML)_o + (H^+)_a \quad (1)$$

$$K_{ex(ML)} = \frac{[(ML)_o]\,[(H^+)_a]}{[(HL)_o]\,[(M^+)_a]} \quad (2)$$

$$2\,(HL)_o + (M^+)_a \rightleftharpoons (ML \cdot HL)_o + (H^+)_a \quad (3)$$

$$K_{ex(ML \cdot HL)} = \frac{[(ML \cdot HL)_o]\,[(H^+)_a]}{[(HL)_o]^2\,[(M^+)_a]} \quad (4)$$

Table 1. Extraction constants (K_{ex}) of picrylamino-substituted benzocrown ether dyes (HL) for alkali metal ions

Crown ether	pK_{HL} [a]		Li^+	Na^+	K^+	Rb^+	Cs^+
1 [b]	10.55	$pK_{ex(ML)}$	— [c]	~13	—	—	—
		$pK_{ex(ML \cdot HL)}$	—	~10	7.55	8.5	10.4
2 [d]	8.63	$pK_{ex(ML \cdot HL)}$	—	~ 8.8	~ 7.1	~ 7.9	~ 8.3
4 [b]	10.58	$pK_{ex(ML \cdot HL)}$	—	12.49	10.39	10.70	—
		$pK_{ex(ML \cdot HL)}$	—	—	—	—	7.55

[a] $K_{HL} = [H^+][L^-]/[HL]$, water dioxane (9:1, v/v), 0.10 M LiCl, 25 °C; [b] Water chloroform (0.1 M triethylamine), pH 11.46, 25 °C; [c] Dashes indicate no measurable extraction according to this reaction; [d] Water chloroform (0.5 M triethanolamine), pH 10.5, 25 °C

Table 1 summarizes the extraction constants of *1*, *2*, and *4* [6] for alkali metal ions. As for 15C5-type crown ethers, *1* and *2* differ from each other in their acidity, the latter being higher by 2 logarithmic units due to the electron-withdrawing effect of the added nitro group in *2*. Since the extraction constants defined in Eqs. (2) and (4) are a kind of acid dissociation constants and thus implicitly involve the acid dissociation constants in aqueous solution [5], the extraction constants obtained for *2* in Table 1 are appreciably higher than those for *1*. This means that the extraction of alkali metals takes place at lower pH for *2* than for *1*. In agreement with this, the extraction study is made at lower pH for *2* by using triethanolamine rather than triethylamine. However, for practical use of these crown ether reagents for extraction colorimetric determination of alkali metal ions, it is best to use EDTA · Li (lithium salt of ethylenediaminetetraacetic acid) as a pH-buffering agent, since EDTA also serves as a potent masking agent for interfering polyvalent metal ions (see below). Li^+ has no measurable tendency to interact with the crown ethers.

It can be seen from Table 1 that picrylamino-substituted B15C5 ethers extract alkali metal ions predominantly in the form of 1:2 (metal:ligand) complexes (ML · HL). The extraction of the 1:1 complex (ML) is inferred only for Na^+. The 18C6-type dye *4*, on the other hand, extracts alkali metal ions predominantly through the formation of 1:1 complexes, except for Cs^+ which is extracted as a 1:2 complex (CsL · HL). A fitness-relationship between the cavity size of crown ethers and the ionic radius of metal ions has been well documented and discussed in reviews [7,8]. It is widely known that 15C5-type crown ethers readily form 1:2 complexes with metal

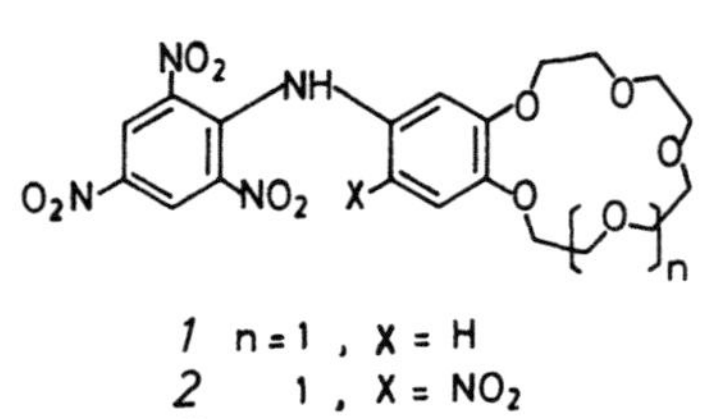

1 n = 1, X = H
2 1, X = NO_2
3 1, X = Br
4 2, X = H
5 2, X = NO_2

ions of larger ionic radius than Na^+ [9,10]. An extensive study has been made on the complex formation and extraction equilibria in water-chloroform system for *1*, *2*, *4*, and *5* [5,6].

It is to be noted that a general comparison of extraction efficiency or extraction selectivity of alkali metals between the 15C5- and 18C6-type crown ethers can not be simply made from the extraction data under particular conditions. This is because the mode of metal extraction differs. Thus, the fraction of the metal extracted, or of the crown ether HL converted to L^- species in organic solution, is quite dependent on the concentration of crown ethers used. However, by using the constants in Table 1 it is easy to calculate these quantities as a function of particular extraction conditions.

An extraction photometric determination of 40–400 ppm and 4–40 ppm K^+ by *1* and *5*, respectively, has been described [5,6]. Potassium in sea water was determined by using *1* in water (0.2 M Li · EDTA, pH 11.46) — chloroform (1 M triethylamine, 1×10^{-3} M *1*). The presence of Na^+ upto 200 fold molar excess of K^+ did not interfere the determination. Li · EDTA masked magnesium and calcium which otherwise interfered with the determination by forming hydroxide precipitates. K^+ in portland cement was similarly determined by using *5* in water (0.1 M Li · EDTA, pH 12.35) — chloroform (1×10^{-4} M *5*).

Under the above determination conditions, usually only serveral percent of K^+ in aqueous solution is transferred into organic solution, and correspondingly, the calibration lines are not entirely linear but often show a curvature. This is in contrast with the traditional concept of colorimetry of metals, where the complete conversion into colored metal complexes is taken as a prerequisite. The extraction constants of the crown ether dyes are much too small compared to the standard of such traditional extraction photometric reagents like dithizone and 8-hydroxyquinoline (for the determination of heavy metals). The low extraction constants necessitate the use of relatively high pH conditions for the effective extraction of alkali metals; a pH control (pH buffer) is also important in order to obtain reproducible results.

Pacey and coworkers described some new variations of picrylamino type benzocrown ethers in which one of the nitro groups of the picryl moiety was substituted by either a cyano or trifluoromethyl group [11,12]. The crown ether dyes *6*–*8* have their general properties in common with dyes *1*–*5*, but improvement is attained in their spectral properties. Thus, while a considerable spectral overlap between HL and ML (or ML · HL) species is observed for *1*–*5*, such overlap is diminished for *6*–*8* (Fig. 1).

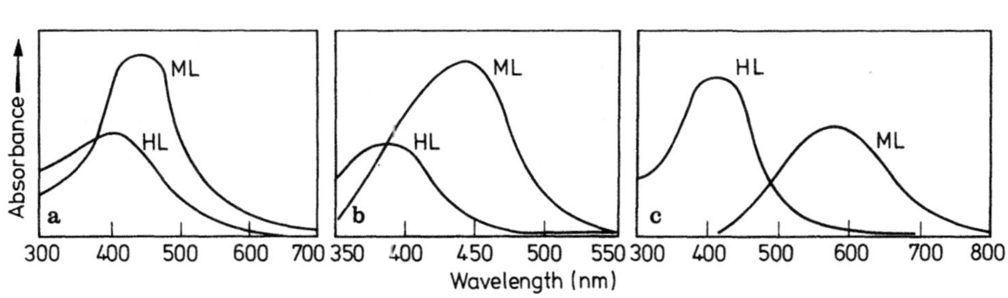

Fig. 1. Absorption spectra of HL and L^- for crown ether dyes *1*, *7*, and *8*. 2×10^{-4} M crown ether dye. **a** *1* in water dioxane (1:1); **b** *8* in water acetonitrile (3:2); **c** *7* in water acetonitrile (3:2) (Reproduced with permission from: G. E. Pacey, Y. P. Wu and B. P. Bubnis, Analyst *106*, 636 (1981))

4′-(2″,4″-Dinitro-6″-trifluoromethylphenyl)amino-B15C5 (*8*) showed the best extraction efficiency among these modified crown ether dyes, and it is claimed that its calibration graph gives a linear range of 5–700 ppm for K^+ in the presence of 3000 ppm of Na^+ [12].

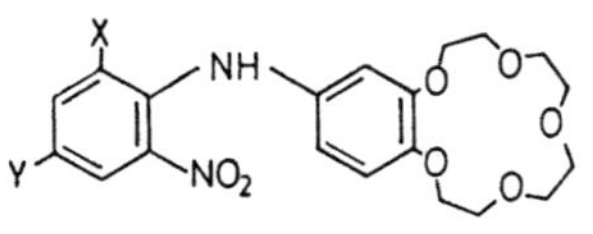

	X	Y
6	X = NO_2	Y = CN
7	NO_2	CF_3
8	CF_3	NO_2

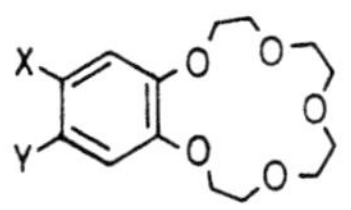

9 X = 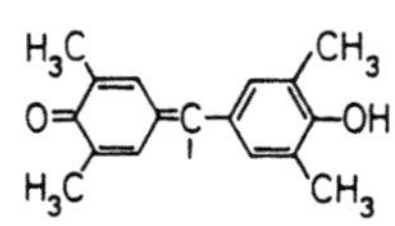 , Y = H

10 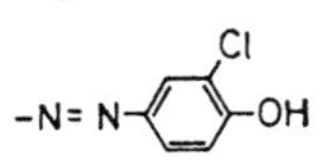 H

11 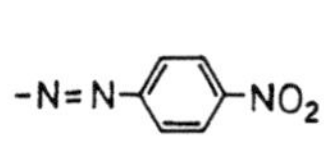 OH

Various other monoprotonic chromogenic crown ethers are readily derived from benzocrown ethers. Dyes *9–11* are such examples. However, few proved to be useful as a reagent for extraction photometry of alkali metals; the metal extraction efficiency was too low [13]. The low extractability of these metal complexes are related to their "intramolecular" ion pair structure in the organic solution, where the alkali metal cation is located in the cavity of the crown ether and the anionic charge is located in the remote chromophoric portion of the molecule. The charge-separated species are not stabilized in non-polar organic solvents. A strong hydration toward the separated ionic charges also precludes the transfer of such species into poorly solvating media. A successful extraction of alkali metal ions by picrylamino and related types of crown ethers *1–8* seems to be due to the effective delocalization of the anionic charge over the entire chromophore (Structure *12*); a steric congestion hinders the hydration around the anionic amino-nitrogen, which is not the case for phenolate anions [13].

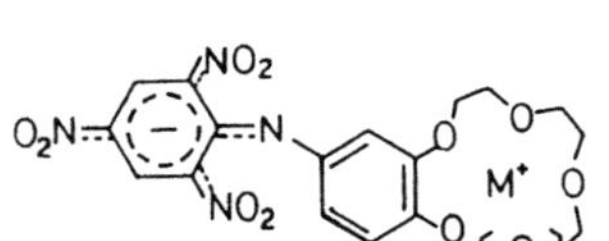

12

2.2 N-Side-armed Monoazacrown Ethers

To enhance the metal extraction efficiency or to increase the extraction constant [Eqs. (2) and (4)] one can introduce a chromophoric side arm into a crown ether skeleton in such a manner that the deprotonated anionic group can interact directly with or coordinate to the metal ion which is bound in the crown ether cavity. Synthetically, this is most readily attained by using azacrown ethers. Thus, p-nitrophenyl type and umbelliferone type azacrown ethers *13–16* (HL) are described by Nakamura, Takagi, Ueno, and coworkers [14, 15]. These are synthesized by reaction of 2-hydroxy-4-nitrobenzyl bromide (*13*, *14*) or by the Mannich reaction of 4-methylumbelliferone (*15*, *16*) with the appropriate monoazacrown ethers.

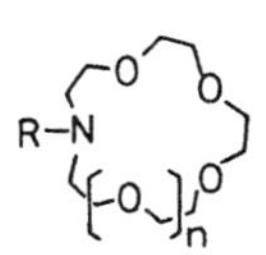

13 n = 1 R = 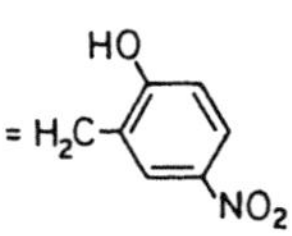

14 2 〃

15 1 R = 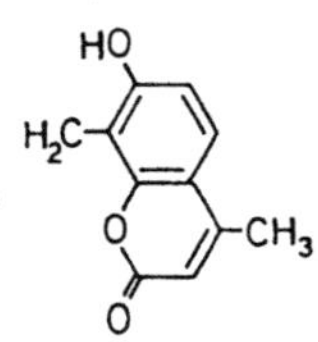

16 2 〃

Like picrylamino-substituted benzocrown ethers, crown ethers *13–16* distribute predominantly into organic solution (water 1,2-dichloroethane system), and selectively extract certain alkali metal ions, causing the color (*13*, *14*) or fluorescence (*15*, *16*) change in the organic phase. In the fully protonated forms (H_2L^+), the dyes have two dissociable protons on the phenolic oxygen and the ammonium nitrogen. The two proton dissociation constants are fairly apart, however, and only the ammonium proton (weaker acid) takes part in the alkali metal extraction reactions which can be formulated in a similar manner to Scheme 1 and Eqs. (1) and (2). In contrast to picrylamino-substituted benzocrown ethers, there was no tendency of 1:2 complex forma-

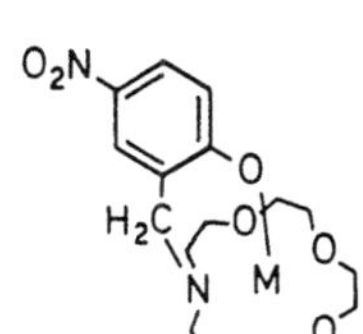

17

tion, suggesting that the residual coordinating site of the crown ether-bound metal ions is occupied from the axial direction by the anionic side arm of the crown ether (Structure *17*).

Table 2 summarizes some equilibrium data for the extraction of alkali metal ions by p-nitrophenyl type reagents *13* and *14*. Similar extraction behavior, especially with

Table 2. Extraction constants (K_{ex}) of N-side-armed monoazacrown ether dyes (HL) for alkali metal ions

Crown ether	pK_{H_2L}[a]	pK_{HL}[a]	$pK_{ex(ML)}$[b]				
			Li^+	Na^+	K^+	Rb^+	Cs^+
13	5.79	9.69	9.15	9.76	9.86	—[c]	—
14	5.77	9.59	10.29	9.46	8.93	9.63	10.62

[a] K_{H_2L} = $[H^+][HL]/[H_2L^+]$, K_{HL} = $[H^+][L^-]/[HL]$, 0.10 M $(CH_3)_4NCl$, 25 °C; [b] Water 1,2-dichloroethane, 25 °C; [c] Dashes indicate no extraction

respect to metal selectivity, is observed with the fluorescent reagents *15* and *16*. In accord with the lower proton dissociation constants (by 0.5–0.7 log units) of *15* and *16* as compared with *13* and *14*, metal extraction constants $K_{ex\ ML)}$ are also lower (by 0.1–1.1 log unit) for *15* and *16* than those for *13* and *14* [15]. It is interesting that while the neutral species of the crown ethers (HL) assume a zwitter-ionic (ammonium phenolate) structure in water, they exist predominantly in an uncharged, aminophenol form in 1,2-dichloroethane or other non-polar solvents. In accordance with this, the maximum absorption of dyes *13* and *14* in 1,2-dichloroethane displaces from 325 and 327 nm to 411 and 417 nm, respectively, on the extraction of alkali metals. Thus the spectral overlap between HL and ML species is very small.

Comparing Table 1 and Table 2, one observes a small increase of extraction constants for azacrown ethers, but the selectivity among the metals is diminished. However, a selectivity of *13* and *15* toward Li^+ is noteworthy. This is probably due to the contribution of the covalent interaction between the phenolate anion and Li^+. Extraction photometric and fluorimetric determinations of ppm level Li^+ are feasible by using *13* and *15*, respectively. As to the latter, the maximum of excitation spectra for the intense fluorescence at 440 nm shifts from 326 (HL) to 380 nm (ML) upon complex formation. Thus, for fluorimetry, the excitation is made at 380 nm, while fluorescence is measured at 440 nm. The 18-membered azacrown ethers *14* and *16* are selective for K^+. None of the crown ethers *13*–*16* is selective for Na^+.

2.3 Side-armed Crown Ethers

Four of this type of crown ether dyes (*18*–*21*, HL) have been described [16,17]. These are prepared via three synthetic reactions starting from commercially available guaiacol glyceryl ether. Crown ethers *18* and *19* are similar in solubility — distribution behavior to those crown ethers discussed above, but *20* and *21* are somewhat different because of their enhanced acidity and water solubility.

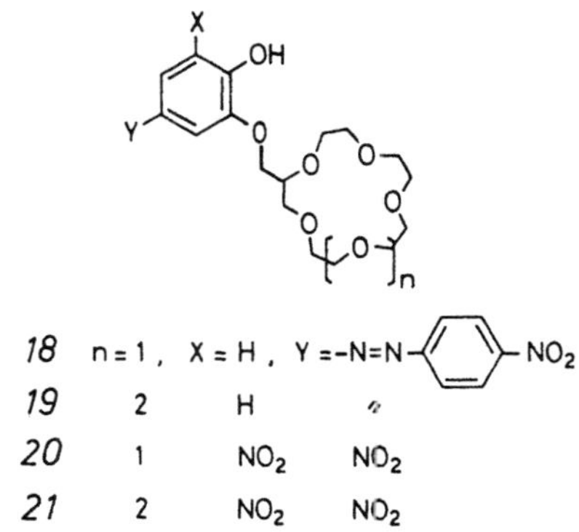

When a 1,2-dichloroethane solution of *18–21* is brought into contact with an aqueous solution at various pH values, the distribution of the reagent species between the two phases is strongly dependent on the pH and the cation present in the aqueous solution. As to *18* and *19*, the distribution of the reagent species in the aqueous phase is negligible under ordinary pH conditions (pH $\leqq 11$). When the aqueous solution contains a salt of the metal whose ionic diameter fits that of the crown ether, the metal becomes extracted as the pH is raised, giving a violet coloration to the organic phase. When the metal does not fit the crown ether, the organic phase remains yellowish. On the other hand, the distribution of *20* and *21* gradually shifts to the aqueous phase as the pH is raised, until the distribution becomes constant at above pH 10. At this pH region, all the reagent species are dissociated to L^- both in the aqueous and organic phases. The distribution of L^- is determined solely by the distribution behavior of the metal salt of the reagent or of the complex (ML), between the aqueous and organic phases. Thus, when the aqueous solution contains the metal ion favorable for complex formation with *20* or *21*, the complex is extracted and the organic phase turns orange. Conversely, if the metal ion does not

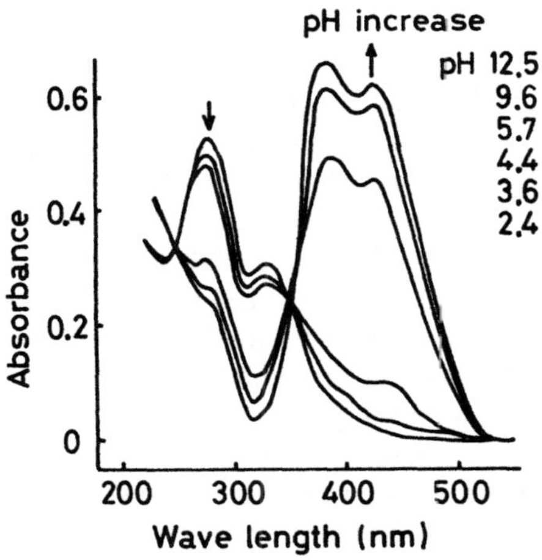

Fig. 2. Absorption spectra of *20* in 1,2-dichloroethane for the extraction of Na^+. $[(20)_0] = 1 \times 10^{-5}$ M and $[(Na^+)_a] = 2 \times 10^{-2}$ M before equilibration

fit to the reagent, the organic phase becomes colorless because the reagent stays only in the aqueous phase. Figure 2 shows the spectral behavior of *20* in the extraction of Na^+.

Tables 3 and 4 summarize some equilibrium constants and spectral properties. The extraction constants of *18–21*, especially of *20* and *21*, are considerably higher than those of reagents *1–8* and *13–16*; large proton dissociation constants of these reagents are obviously contributing to this outcome.

Table 3. Extraction constants (K_{ex}) of side-armed crown ether dyes (HL) for alkali metal ions

Crown	pK_{HL} [a]	$pK_{ex(ML)}$ [b]				
		Li^+	Na^+	K^+	Rb^+	Cs^+
18	7.51	9.8	8.40	9.20	9.55	>10
19	7.54	8.8	9.32	7.11	7.75	>10
20	3.16	>5.6	3.60	4.15	4.76	5.6
21	3.27	4.8	4.01	2.22	2.68	3.76

[a] $K_{HL} = [H^+][L^-]/[HL]$, water dioxane (9:1, v/v), 0.10 M $(CH_3)_4NBr$, 25 °C; [b] Water 1,2-dichloroethane, 25 °C

Table 4. Spectral properties of side-armed crown ether dyes (HL) and their alkali metal complexes

Crown ether	$\lambda_{max}(\varepsilon \cdot 10^{-3})$ [a]			
	HL[b]	L^- [b]	NaL[c]	KL[c]
18	390(18)	518(30)	560(37)	567(37)
19	390(18)	518(30)	571(35)	575(35)
20	273(12)	385(13)	423(14)	426(13)
21	273(10)	385(11)	430(13)	433(12)

[a] λ (nm), ε (molar absorptivity, $M^{-1}\,cm^{-1}$); [b] Water dioxane (9:1, v/v), 0.10 M $(CH_3)_4NBr$, 25 °C; [c] 1,2-Dichloroethane

With regard to the selectivity between Na^+ and K^+, 15C5-type reagents prefer Na^+, whereas 18C6-type reagents prefer K^+. This is simply ascribed to the "size selectivity" of the crown ether macrocycles. However, a closer inspection of Table 3 reveals some interesting facts: Selectivity for K^+, as measured by the extraction constant (Table 3), is higher for *21* than for *19* ($\log K_{KL} - \log K_{NaL} = 1.79$ for *21* and 2.21 for *19*). Similarly, a comparison of selectivity between *20* and *18* reveals that *18* has a higher selectivity for Na^+ ($\log K_{NaL} - \log K_{KL} = 0.80$ for *18* and 0.55 for *20*). This means that for the reagents of the same crown ether skeleton, the more basic pendent anion (*18* and *19*) favors the extraction of Na^+, while the less basic pendent anion (*20* and *21*) favors K^+. This in turn indicates that coordination interaction between the phenolate and the metal is more important for Na^+ in the extraction process. In other words, the structure of the extracted K^+ complexes tends to be more of an intramolecular ion-pair rather than a chelate. In the same notion, sodium complexes have more of a chelate structure. In the light of this concept, Li^+ belongs to a sodium family, while Rb^+ and Cs^+ belong to a potassium family. The difference in the mode of metal — phenolate interaction is schematically illustrated in formulae *22* and *23*.

In order to achieve a high selectivity among alkali metals in solvent extraction with monoprotonic crown ether reagents, two critical factors have to be considered. The first is the well known "size selectivity" by the crown ether macrocycle. The second is the coordination interaction (*22*) and the ion-pair (*23*) interaction of the anionic site with the crown ether-bound metal ion. A localized negative charge (high basicity) strengthens the coordination interaction and favors the extraction of lighter alkali metal ions. A delocalized negative charge (low basicity) favors the formation of ion-pair type complexes and leads to more efficient extraction of the heavier alkali metals.

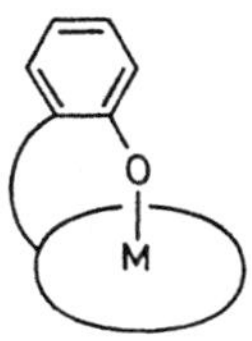

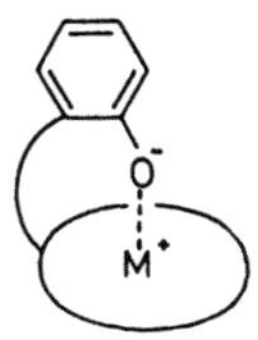

22 (M = Li, Na) *23* (M = K, Rb, Cs)

Reagent *20* is useful for the extraction photometric determination of Na^+ in human blood serum. A acid-deproteinated serum sample (0.1 ml) is mixed with an alkali metal-free tetramethylammonium borate solution (0.1 M, 4 ml) and diluted to 10 ml with distilled water (pH 10.3). The mixture is shaken with a 1,2-dichloroethane solution of *20* (2×10^{-4} M, 10 ml) for 10 min. After phase separation by centrifugation, the absorbance of the organic phase is measured at 423 nm against a reagent blank. A calibration line, which is prepared with aqueous solutions of sodium chloride is linear of Na^+ concentrations between 0 and 6 ppm. K^+ in the blood serum does not interfere with the Na^+ determination since the former concentration is less than one tenth of the latter. The determination of K^+ is similarly attained at ppm level by using *21*. In this case, a presence of ten-fold excess Na^+ does not cause any serious interference.

One of the remarkable features of the dinitrophenol-type reagents *20* and *21* is, as mentioned before, that they do not distribute into the organic phase under photometric conditions unless metal ions of right size are present (pH $\geqq$ 10). This stands in contrast to the reagents *18* and *19* as well as phenolic monoazacrown ethers and picrylamino-substituted benzocrown ethers. These latter reagents stay essentially in the organic phase; if one tries to achieve a higher metal determination-sensitivity by increasing the metal extraction efficiency through the use of higher reagent concentration, the reagent blanks (absorption due to HL species) inevitably increase. Though the spectral overlap between HL and ML is avoided for the phenolic reagents, the blank absorption is not negligible and its increase causes uncertainties when the accurate photometric measurements are required. The absence of blank absorption for *20* and *21* insures an accurate and sensitive determination as well as good reproducibility. Moreover, the determination is not sensitive to pH variation in the extraction procedure. These are the advantages of dinitrophenol-type reagents over the others from the practical point of view.

Extraction photometry is often conveniently replaced by photometry in homogeneous solution by using surfactant micelles as a pseudo-phase for extraction. Reagents such as *18* and *19* which do not dissociate and distribute into an aqueous phase should, in principle, be useful for such purposes. Unfortunately, the alkali metal complexes of the crown ether dyes are not quite lipophilic enough to be concentrated into the core of the micelle, but both HL and ML species seem to be concentrated on the surface. Consequently, the formation of ML and the ordinary proton dissociation of HL (to give M′L, where M′ stands for cationic component of the pH-buffering agent) take place simultaneously at the micelle surface and can not be fully distinguished from each other spectrometrically [18]. The problems involved in homogeneous photometry are yet to be solved.

It is to be added that, again in the extraction with *20* and *21*, the fraction of metal transferred to the organic phase can amount only to 10–20% of the total metal originally present in the aqueous solution under ordinary conditions. Accordingly, the "apparent" molar absorptivity of the extracted metal complex is only 2200 as compared with the "true" value of 14,000 M^{-1} cm^{-1} shown in Table 4. This in turn means that the sensitivity of the determination by *20* or *21* may be increased even more by increasing the reagent concentration. However, due to contamination from glassware, the handling of Na^+ and K^+ below ppm levels requires special precautions [19], and such ultrasensitivity is practically not always worthwhile. With regard to contamination problems, it may also be pointed out on the side that proton-dissociable chromogenic crown ethers discussed above as well as those to be discussed below (extraction of alkaline earth metal ions) can be contaminated with alkali metals through storage in organic solution (1,2-dichloroethane) when kept in glass bottles more than a month. The contaminations, however, are readily removed by washing with aqueous boric acid before usage.

As to the (molecular) absorption photometry of alkali metals in general, another type of extraction photometry is also available. The method is based on ion-pair extraction of crown ether complexes of alkali metals by anionic dyes [19,20]. Ordinary crown ethers as well as bicyclic crown ethers (cryptands) are used, while bromocresol green, resazurin and picrate are the typical of the pairing anions. These crown ethers and cryptands are not colored, but the experimental procedures are essentially similar to those described above for chromogenic crown ethers. Li^+, Na^+, and K^+ can be selectively determined at ppm levels.

2.4 "Crowned" Phenols

Kaneda and coworkers synthesized interesting crown ether dyes in which a phenolic hydroxyl group constitutes an integrated part of the crown ether skeleton (*24*, *25*) [21]. They described that upon addition of crystalline lithium chloride or perchlorate to a yellow solution of *24* in chloroform, a drastic color change to purple red took place when pyridine (800 times molar excess of *24*) was simultaneously added. Other alkali metal salts produced no such change. The coloration is very selective to Li^+, though

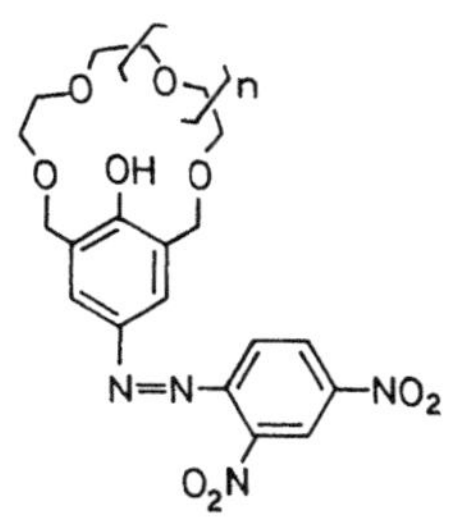

24 n = 1
25 n = 2

it is said that the coloration is sensitive to experimental conditions. Later, Nakashima and coworkers [22] further studied this system and described conditions under which 25 to 250 ppb Li^+ was determined photometrically in chloroform-dimethylsulfoxide triethylamine mixture (94.5 : 5 : 0.5; v/v). Sodium ions did not interfere, but K^+, Rb^+, Ca^{2+}, Sr^{2+}, Ba^{2+}, and Mg^{2+} did and were to be removed before the determination was performed. The coloration of *24* in organic media is no doubt related to the deprotonation of phenolic proton under combined influence of the proton-removing amine and the metal cation stabilizing the resultant phenolate anion, but the complexity of the system does not seem to allow a quantitative approach to the equilibria involved. The high sensitivity claimed for this method is remarkable, but presently a rather unusual medium utilized and strict precautions in sample preparation (exclusion of water) seem to limit its use for general purposes. It is said that reagents *24* and *25* work rather poorly in the liquid-liquid solvent extraction of alkali metals [23].

2.5 Benzocrown Ethers with Azo-linked Side Arm

Shinkai, Manabe and coworkers [24] described an interesting family of crown ether dyes, which responded to light and changed their complexation and extraction selectivity (for this topic see S. Shinkai, O. Manabe, this series 67). The dyes shown by a general structural formula *26* can be photochemically isomerized to *27* according to the *E*/*Z*-isomerization of the azo linkage. The transformation is reversible, the thermodynamically stable *E*-form (*26*) being recovered by a spontaneous,

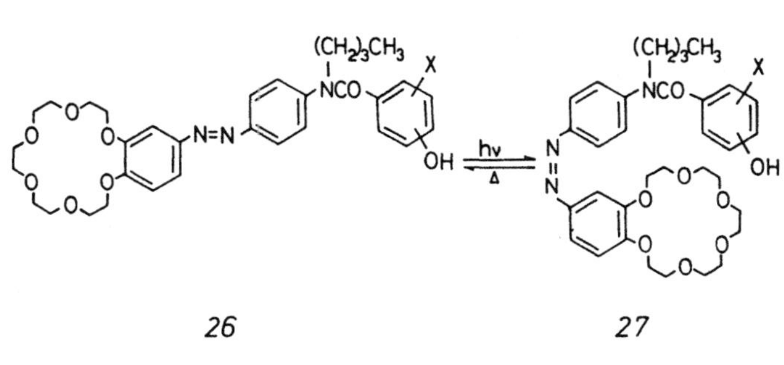

26 *27*

thermal process. Dyes *26* include three isomeric phenol dyes, *o*-, *m*-, and *p*-isomers according to the site of the phenolic hydroxyl with respect to the N-butylcarbamoyl substituent.

As is obvious from structural considerations, *E*-isomers bear no special characteristics other than the benzocrown ethers and phenols, since the phenol function is quite remote from the crown ether function and there can be no cooperation within the molecule for enhanced interaction with metal ions. On the other hand, isomerization to the *Z*-form brings the two functional portions close to each other, making it possible for the phenolate anion to interact directly with the crown ether-bound metal cation from axial directions. The mode of interaction is schematically shown by *28* for an *o*-phenol type dye.

The extraction of alkali metal ions into *o*-dichlorobenzene (with or without added 10% 1-butanol) was studied under strongly alkaline conditions (MOH $\leqq$ 0.50 M). *E*-Isomers show a selectivity which simply follows the order of the size of alkali metal ions and differs from the unique stability of the metal complexes of 18C6. This suggests that alkali metal ions are extracted as a counterion of the phenolate anion and that the crown ether plays no significant role. On the other hand, the extraction with photoisomerized *Z*-compounds shows varying selectivity among alkali metals according to the *o*-, *m*-, and *p*-isomeric phenols. This stands in agreement with the idea of the participation of such complex species like *28*, but in the lack of detailed equilibrium data, the extent of such structural contribution can not be rigorously assessed.

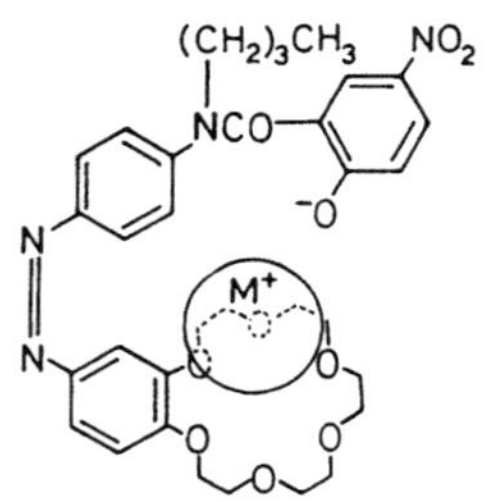

28

The major interest of the work of Shinkai, Manabe and coworkers was the photochemical control of metal transport through organic liquid membranes. The alkali metal transport is attained by consecutive processes of extraction and back-extraction (or stripping) on the high and low pH sides, respectively, of the membrane/aqueous solution interface by using the monoprotonic crown ethers *26* as metal carriers (ionophores). By changing the extraction behavior of the crown ether reagents by photo-irradiation, an acceleration of some of the metal transport rates is realized. In fact, the study on *26* is one of the authors' extensive works on the photoresponsive crown ethers, and is dealt with in more detail in other chapters of this series of monographs.

Reagent *26* can also extract alkaline earth metal ions from their metal hydroxide solutions. A part of the phenolate anion in the organic phase can be replaced by other lipophilic monoanionic dyes such as 2,4-dichlorophenolindophenol (dissociated form, A^-) added externally to the aqueous phase. Taking advantage of this reaction, a counter flow of proton and calcium 2,4-dichlorophenolindophenolate ($Ca^{2+} \cdot A^-$) through the liquid membrane is attained. It is often observed that cationic complexes (ML^+) produced from monoanionic crown ethers (L^-) and alkaline earth metal ions (M^{2+}) distribute to organic phase with the aid of added lipophilic anions by forming ion-pair type association complexes [18].

3 Diprotonic Crown Ether Dyes

The introduction of two of the proton-dissociable chromophores into the crown ether skeleton, according to the similar strategy outlined before, should lead to those dyes which are selective for divalent metal ions — above all to alkaline earth metals.

Dipicrylamino-substituted DB18C6 proved not to extract alkaline earth metal ions [13]. If the picture of intramolecular ion-pair type extraction by picrylamino-benzocrown ethers is correct, the dipicrylamino-substituted crown ether results in the extraction of alkaline earth metal complexes, which involves a charge separation

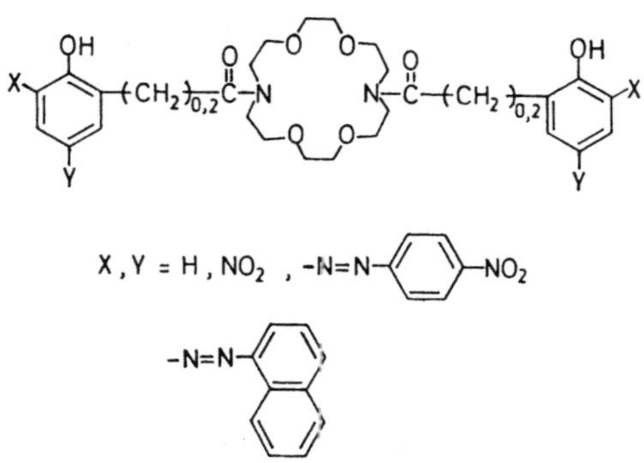

29

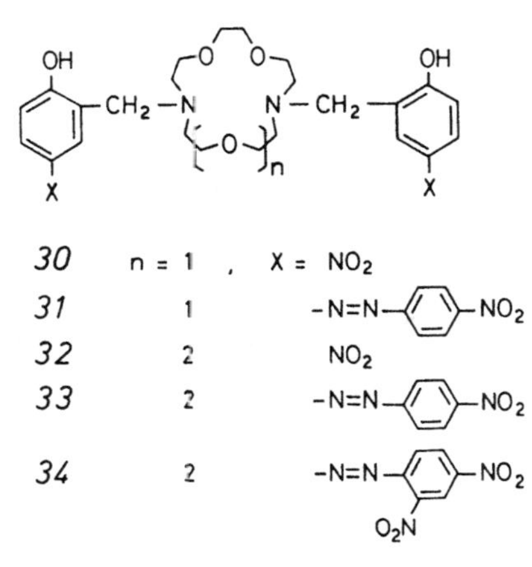

between two anionic single charges and a cationic double charge. Such structures should be highly unfavorable for extraction. Therefore, in order to achieve an efficient extraction of divalent metals, structures of crown ether dyes should be considered in which the two anionic chromophoric groups are capable of interacting directly with the metal ion from axial directions. This type of crown ethers are synthetically most easily obtained by using DA18C6 and other related diazacrown ethers. Some such synthetic products are shown in formulae *29–34*.

Crown ether dyes of structure *29* are poor extraction agents for alkaline earth metal ions, although some do extract these metals as well as Li^+ into chloroform at high metal concentrations (1 M) [13]. A low coordinating ability of the amide nitrogens in the crown ether skeleton is no doubt responsible for this behavior. On the other hand, crown ethers *30–34* unanimously show a strong affinity to alkaline earth metal ions and extract them efficiently into organic solvents such as 1,2-dichloroethane [25,26].

The extraction of alkaline earth metal ions by reagent *30–34* can be formulated by Eqs. (5) and (6). The reagents do not distribute into aqueous phase under ordinary conditions; extraction constants $K_{ex(ML)}$ can be determined photometrically by standard procedures. For reagent *32*, the absorption maximum of the free reagent (H_2L) at 326 nm shifts to 406 nm upon extraction of Ca^{2+} through the formation of CaL species into 1,2-dichloroethane. Similarly, *33* shows a wave length shift from 392 to 518 nm, as illustrated in Fig. 3. The extraction according to Eqs. (5) and (6) is rigorously followed, and there is no indication of other types of complex formation.

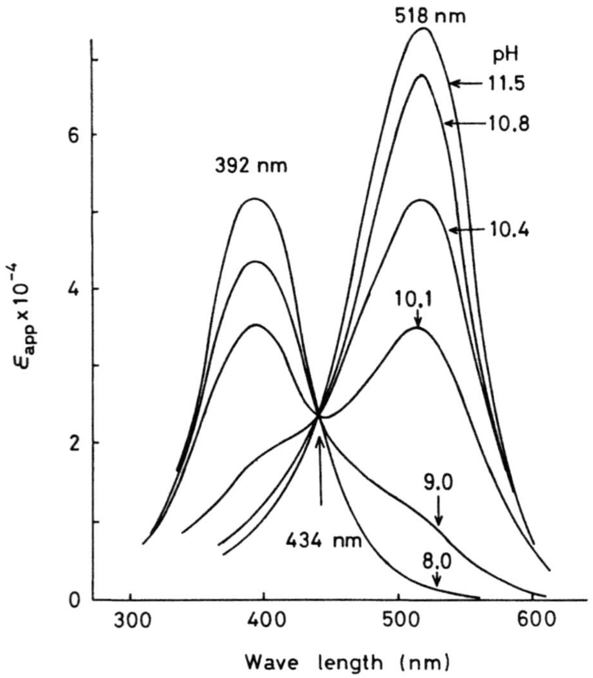

Fig. 3. Absorption spectra of *33* in 1,2-dichloroethane for the extraction of Ca^{2+}. $[(33)_0] = 6.1 \times 10^{-6}$ M and $[(Ca^{2+})_a] = 1 \times 10^{-3}$ M before equilibration

The proton dissociation and the extraction constants as well as some spectral properties are summarized in Tables 5 and 6.

$$(H_2L)_o + (M^{2+})_a \rightleftharpoons (ML)_o + 2\,(H^+)_a \tag{5}$$

$$K_{ML}^{ex} = \frac{[(ML)_o]\,[(H^+)_a]^2}{[(H_2L)_o]\,[(M^{2+})_a]} \tag{6}$$

Table 5. Proton dissociation constants of crown ether dyes (H_2L)[a]

Crown	pK_{H_4L}	pK_{H_3L}	pK_{H_2L}
32	4.03	6.56	9.8
33	5.52	9.00	10.9
34	5.01	8.72	10.3

[a] Water dioxane (9:1, v/v), 0.1 M $(CH_3)_4NBr$, 25 °C; $K_{H_4L} = [H^+]\,[H_3L^+]/[H_4L^{2+}]$; $K_{H_3L} = [H^+]\,[H_2L]/[H_3L^+]$; $K_{H_2L} = [H^+]\,[HL^-]/[H_2L]$

Table 6. Extraction constants (K_{ex}) and spectral properties of crown ether dyes (H_2L)[a]

Crown ether	H_2L		CaL		$pK_{ex(ML)}$		
	λ_{max}	$\varepsilon \times 10^{-3}$	λ_{max}	$\varepsilon \times 10^{-3}$	Ca	Sr	Ba
30	326	24	406	45	15.0	16.8	18.4
31	392	52	513	74	17.6	18.9	20.4
32	326	23	406	45[b]	12.54	13.51	15.1
33	392	52	518	80	16.6	17.9	19.1
34	410	63	574	81	15.0	16.3	18.5

[a] Water 1,2-dichloroethane; spectral data are obtained in 1,2-dichloroethane saturated with water; λ (nm), ε (molar absorptivity, $M^{-1}\,cm^{-1}$), 25 °C; [b] Previously miss-typed as 55 [25)]

Metal extraction selectivity is: $Ca^{2+} > Sr^{2+} > Ba^{2+} \gg Mg^{2+}$ for both types of diazacrown ethers: 18C6- and 15C5-types. Mg^{2+} and alkali metals are extracted only to a minor extent. An especially high extraction constant is exhibited by the 18C6-type diazacrown ether with p-nitrophenol-chromophore (*32*). Unlike the extraction of alkali metal ions with monoprotonic crown ethers, a spectrophotometric concentration of *32* can extract Ca^{2+} quantitatively into 1,2-dichloroethane if the pH of the aqueous phase is adjusted high enough (pH ≧ 10). As seen in Table 5, *32* has much higher proton dissociation constants than *33* or *34*. The constants K_{H_4L} and K_{H_3L} correspond to the dissociation of two phenolic protons, and K_{H_2L} to an ammonium proton. Thus, in aqueous solutions the crown ether dyes *32*–*34* as well as *30* and *31* exist in a zwitter ionic, ammonium phenolate form, while in 1,2-dichloroethane they are present in a neutral, amino-phenol structure.

An extraction photometric determination of Ca^{2+} in blood serum was successfully performed by using *32*. A serum sample (0.1 ml) is deproteinated by 1 ml of 0.33 M trichloroacetic acid, and a 0.5-ml aliquot of the supernatant liquid is diluted to 10 ml with tetramethylammonium borate buffer (pH 10.25). The solution is shaken with 10 ml of 2.5×10^{-4} M *32* in 1,2-dichloroethane. The absorbance of the organic solution is measured at 406 nm. The calibration line is linear at 0–0.8 ppm Ca^{2+} concentrations. As little as 0.1 ppm or 100 ppb Ca^{2+} is easily determined. By using *34*, the determination of Ca^{2+} below 300 ppb is feasible. The molar absorptivity 81,000 for the calcium complex is among the highest known for Ca^{2+} determinations. As to the generality and the convenience of Ca^{2+} determination, however, this type of crown ether reagent (extraction photometry) does not seem to offer particular advantages over many traditional metallochromic chelating agents such as Calmagite [1-(1-hydroxy-4-methyl-2-phenylazo)-2-naphthol-4-sulfonic acid], Chlorophosphonazo-III [2,7-bis(4-chloro-2-phosphonophenylazo)chromotropic acid] and Eriochrome Black T [1-(1-hydroxy-2-naphthylazo)-6-nitro-2-naphthol-4-sulfonic acid], which are able to determine ppm to ppb level Ca^{2+} in aqueous media.

The extraction of other divalent metal ions have not been studied in detail, but Cu^{2+}, Ni^{2+}, Zn^{2+}, Cd^{2+}, and Pb^{2+} are extracted by *32* leading to errors in Ca^{2+} determination. These interferences can be removed by using NTA (nitrilotriacetic acid) as masking agent [26]. Complex formation constants of *32* are not yet determined in homogeneous solutions, but presumably the complexes of alkaline earth metal ions are less stable than those of other divalent metal ions. With ordinary chelating agents, transition metal and related heavy metal ions usually form much more stable complexes than alkaline earth metals do. However, the presence of the crown ether structure in *32* and related crown ether reagents (*33*, *34*) may well be expected to reduce the difference in the stabilities between the two families of metal ions. In fact, the complexation behavior of "crown complexanes" (*35*, H_2L) described by Takagi, Tazaki, Ueno and coworkers [27,28] have partially supported this expectation.

Some of the complex formation constants of *35* with alkaline earth metal ions are quite close to those with divalent transition metal ions. Particularly, compound *35* with X = COOH, m = 1 and n = 2 forms complexes of almost same stability with Co^{2+}, Ni^{2+}, Ba^{2+}, Ca^{2+}, Sr^{2+}, and Zn^{2+} [28]. It may be pointed out on the side that the crystal structure of the Cu^{2+} complex of this ligand indicates that the two acetate groups coordinate in a *trans* configuration to the Cu atom bound in the

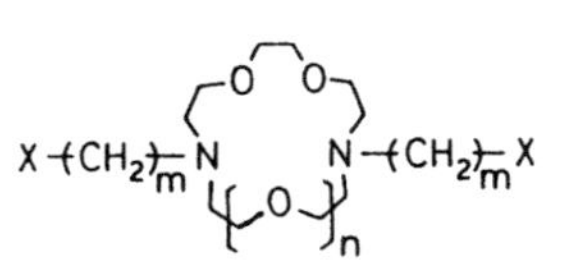

35

distorted diazacrown ether ring [29]. Considering this fact as well as the size-fit of the Ca^{2+} ion and the crown ether cavity, it is very plausible that the calcium complexes of *32*–*34* involve the coordination of two phenolate anions from axial directions onto the metal ion held in the crown ether cavity, well completing an electrically neutral, lipophilic skin around the otherwise highly hydrophilic Ca^{2+}. Few synthetic reagents are known to extract Ca^{2+} with appreciable selectivity among common divalent metal ions (including transition and post-transition metals). In this sense, it is hoped that the crown ether dye of type *30–34* may find a use in the study of such biochemical processes as the transport of alkaline earth metal ions through a lipophilic barrier (cell membranes). The monoprotonic crown ether reagent discussed in the previous sections may be used for similar purposes for alkali metals. It can be readily shown that these reagents mediate the counter-flow of alkali metal ions (or alkali metals ions) and protons across lipophilic membranes (liquid membranes).

Fluorescent versions of the reagent of type *30–34* is easily obtained via a Mannich reaction of diazacrown ethers with 4-methylumbelliferone. Reagent *36* shows a similar metal extraction behavior as *32*, but the extraction constants are lower by 2–2.6 log units than *32*, reflecting reduced proton dissociation constants of *36*. Fluorimetry can, in principle, be more sensitive than absorption photometry. However, the lower extraction constant requires more basic conditions for efficient metal extraction, thus compensating the higher sensitivity of fluorimetry in practical applications for alkaline earth metal determination [15,18].

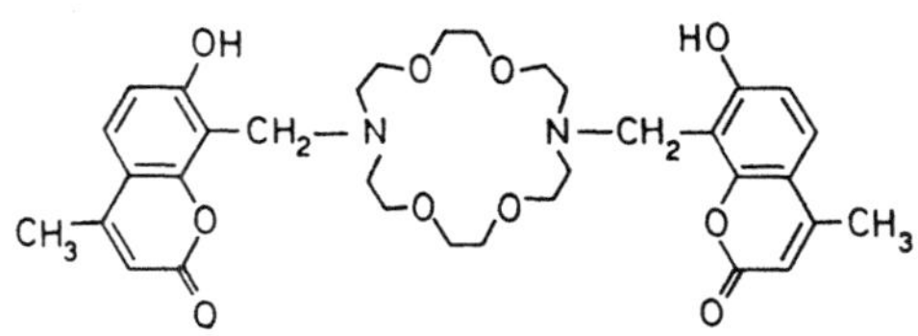

36

Diprotonic crown ethers can also be derived from DA24C8. Dye *37* offers a much larger cavity to accommodate metal ions, and it selectively extracts Ba^{2+} into 1,2-dichloroethane (Table 7) [18]. To our knowledge, *37* is the first photometric reagent

Table 7. Extraction constants of *37* (H_2L) for alkaline earth metal ions[a]

Metal ion	Mg^{2+}	Ca^{2+}	Sr^{2+}	Ba^{2+}
$pK_{ex(ML)}$	$\gg 18$	17.7	17.0	14.3

[a] Water 1,2-dichloroethane, 25 °C

selective for Ba^{2+} capable of determining the ion at ppm levels. Considering the size of the macrocycle, it would also be an interesting reagent for separation and concentration of Ra^{2+}.

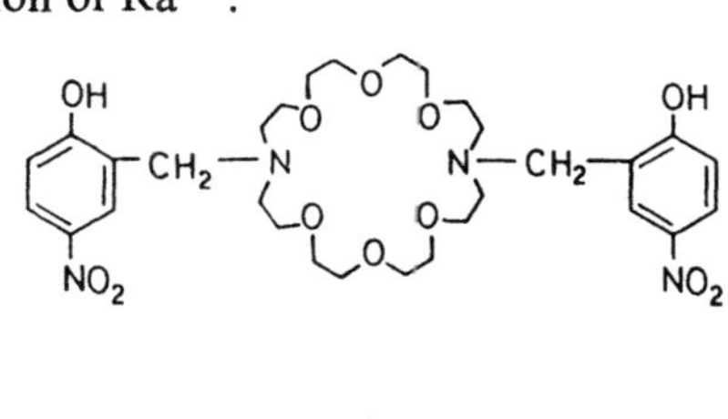

37

4 Uncharged, Neutral Crown Ether and Related Dyes

Shinkai, Manabe, and coworkers synthesized and studied a variety of bis(crown ether)s made by connecting two benzocrown ethers through azo linkage (*38*) [30]. This yields derivatives of azobenzene which, on photo-irradiation, isomerize reversibly from the (thermodynamically stable) *E*- to *Z*-configuration. The two isomers have different metal selectivities depending on the intramolecular cooperation of the two crown ether rings. However, spectral change caused by metal binding is, small as expected, and therefore this type of crown ethers and related bicyclic crown ethers [31,32] will not be discussed in further detail.

Takagi, Ueno, and coworkers described another type of azobenzene crown ethers in which the azo-linkage constitutes a part of the crown ether macrocycle (*39*, *40*) [18,33]. A metallochromism is appreciable when the E-isomers are allowed to interact with excess alkali and alkaline earth metal ions in acetonitrile. However, this is not quite sufficient to be useful as a photometric aid for metal analyses. Quite recently, azobenzenophanes with a oligooxyethylene bridge at 4- and 4′-positions (instead of 2- and 2′-positions in *39* and *40*) have been successfully synthesized and studied [34].

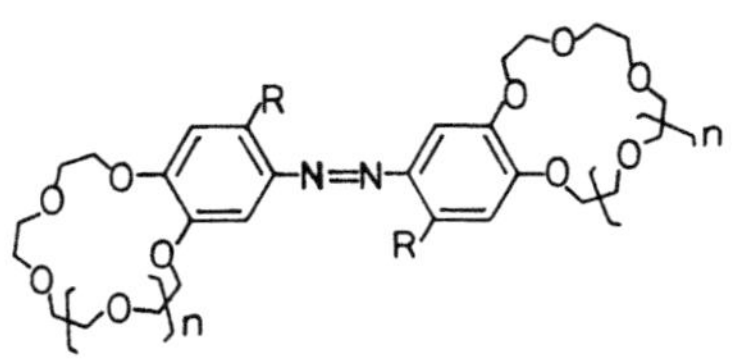

38

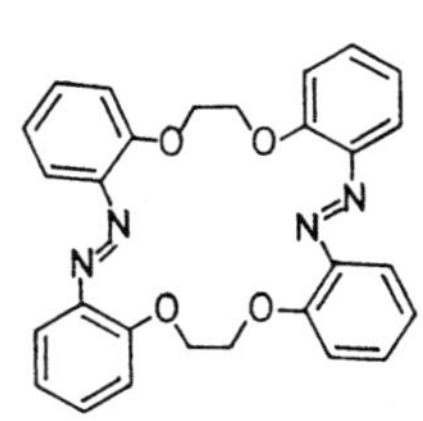

39

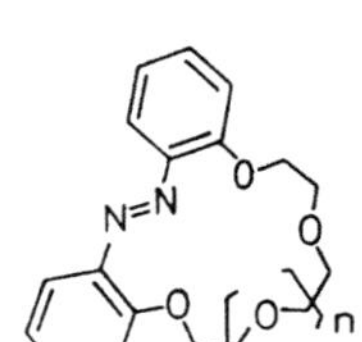

40

The most interesting chromogenic crown ethers have been described by Dix and Vögtle [35-37]. Some 40–50 crown ether dyes of various structures have been synthesized, the typical of which are shown in the structural formulae *41–50*. The leading principle of synthesis is the attachment of crown ethers of various ion selectivity to dyes in such a manner that the interaction with the alkali and alkaline earth metal ions causes color changes. Thus, the basic idea of molecular design is the same as that independently developed by Takagi and coworkers [4]. However, the main interest of Dix and Vögtle was for neutral, uncharged crown ether dyes, and therefore, the chromophores without dissociable protons were of specific concern in their study.

Dyes *41–50* have electron donor and acceptor sites within a molecule so that the charge transfer from the donor to acceptor according to electronic excitation gives rise to their strong visible light absorption. The interaction of metal ions with the dye molecule in a manner to inhibit the charge transfer (stabilization of the localized charge on the donor site) leads to the hypsochromic shift of the charge transfer band. On the contrary, the interaction to assist the charge transfer (stabilization of the charge on the acceptor site) results in a bathochromic shift.

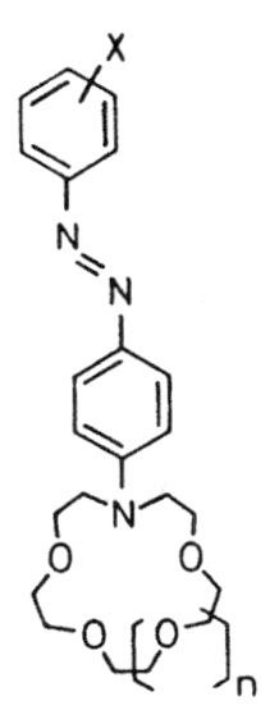

41

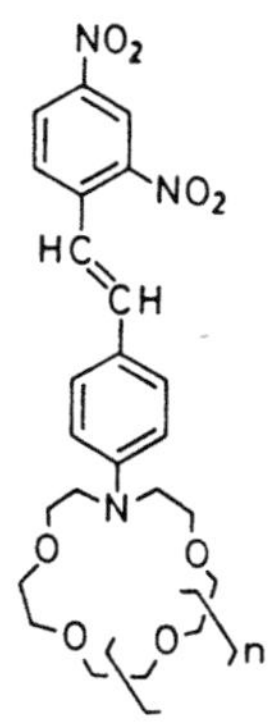

42 n = 1

43 n = 2

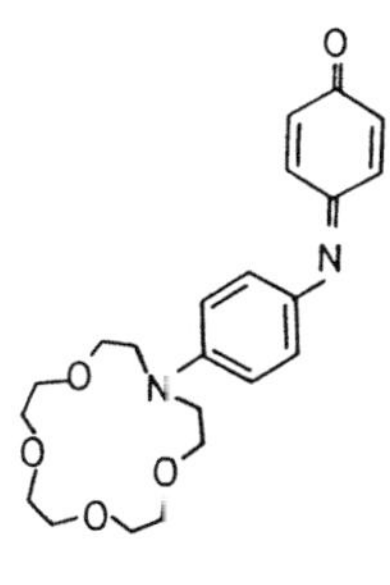

44

In dyes *41–44*, the amino nitrogen which is an integrated part of the crown ether ring is simultaneously involved in the dye chromophore as a vital electron donor site. The two aromatic ethereal oxygens are in the same position in dyes *45* and *46*. Thus, in these crown ether dyes, the interaction with metal cations tends to localize the electronic charge on the crown ether portion causing a hypsochromic shift of the charge transfer band which appears as the longest wave length absorption of these dyes. On the other hand, for dyes *48–50*, the metal cations interact with the acceptor site (carbonyl oxygen) of the dyes giving rise to a bathochromic shift of the charge transfer band (structures *51* and *52*). Figures 4 and 5 exemplify these relations.

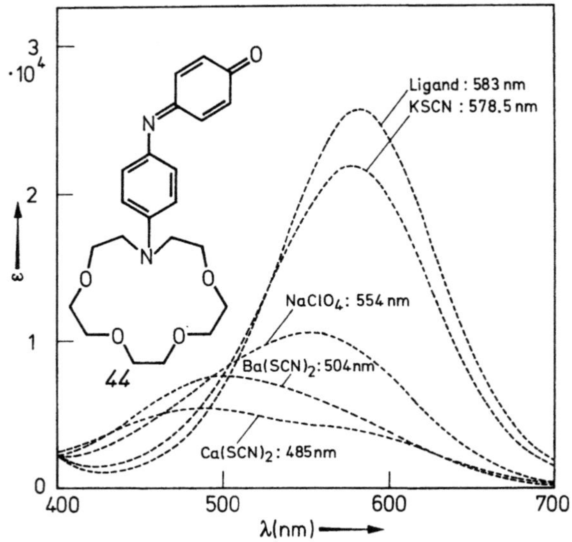

Fig. 4. Absorption spectra of *44* in the presence and the absence of metal salt: *44* > 100 in acetonitrile (Reproduced with permission from: J. P. Dix and F. Vögtle, Chem. Ber. *114*, 638 (1981))

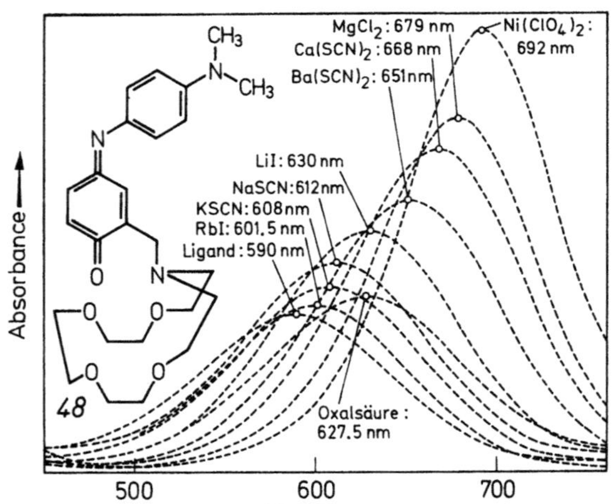

Fig. 5. Absorption spectra of *48* in the presence and the absence of metal salt. Metal salt: *48* > 100 in acetonitrile (Reproduced with permission from: J. P. Dix and F. Vögtle, Chem. Ber. *114*, 638 (1981))

Some of the typical spectral data are summarized in Table 8 [36, 37]. Dye *47* is somewhat peculiar in that both of the donor and acceptor sites of the dye chromophore can simultaneously interact with a metal cation. The dye shows two absorption maxima or shoulders, both of which are subjected to a regular shift on the addition of metal salts in acetonitrile. The long wavelength absorption (λ_1) corresponds to the charge tranfer band which involves the amino nitrogen. The short wavelength absorption (λ_2) corresponds to the benzenoid band [37].

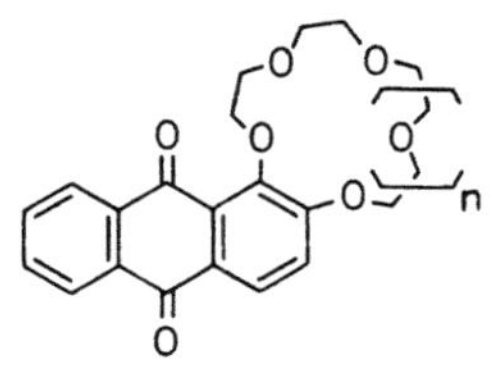

45 n = 1
46 n = 2

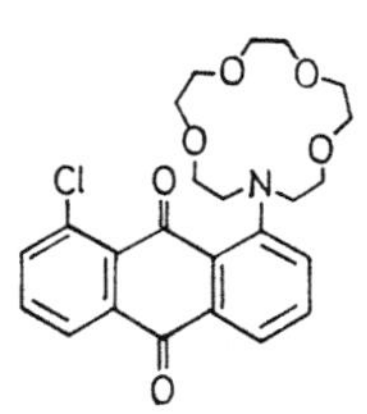

47

The extent of the absorption shift depends on

i the fit between the size of crown ether and metal ion,
ii the surface charge density of metal cations, and
iii the solvent.

The point (i) is most clearly exemplified in Table 8 by *43*. K^+ among alkali metals and Ba^{2+} among alkaline earth metals produce an almost exclusive absorption shift which is as larger as 100 nm in both acetonitrile and methanol. Point (ii) is easily seen in the behavior of *47* and *49* in acetonitrile. The spectral shift is most prominent for Li^+ among alkali metals. Moreover, if one compares alkali and alkaline earth metals, the latter obviously show larger spectral shifts. In this connection, it may be pointed out that a similar observation, though to a much smaller extent, can be made in the extraction of alkali and alkaline earth metal ions with proton-dissociable crown ethers. This is partially presented in Table 4 for the absorption maxima of NaL and KL species. Generally, for phenolic crown ether dyes, absorption maxima shift toward the longer wave lengths in the order, $Rb^+ = K^+ > Na^+ > Li^+$ ($\gg H^+$) and $Ba^{2+} \geqq Sr^{2+} \geqq Ca^{2+}$. As to point (iii), one notices the

Table 8. Absorption maxima (nm) of crown ether dyes and their variation in the presence of metal salts[a]

Metal salt	*43*	*43*[b]	*45*	*47*		*49*	*49*[c]
				λ_1[d]	λ_2[d]		
none	477	—[e]	373	493	~317; s	598	628
LiI	472(—5)[f]	—(0)	—	~430(—63); s	342(+25); s	633.5(+35.5)	628(0)
LiSCN	—	—	361(—12)	—	—	—	—
$LiClO_4$	—	—	—	—	—	—	—
$NaClO_4$	463.5(—13.5)	—(—5)	361(—12)	—	—	608.5(+10.5)	632(+4)
NaSCN	464(—13)	—(—7)	361(—12)	438(—55)	~330(+13); s	608(+10)	—
KSCN	382(—95)	—(—99.5)	368(—5)	487(—6)	~320(+3); s	598.5(+0.5)	634(+6)
RbI	468(—9)	—(—6)	371(—2)	—	—	598.5(+0.5)	633(+5)
$MgCl_2$	—	—	—	~480(—13)	344(+27)	681(+83)	629(+1)
$Ca(SCN)_2 \cdot 4\,H_2O$	381(—96)	—(—1.5)	~345(—28); s	~412(—81); s	349(+32)	676(+78)	667.5(+39.5)[g]
$Ba(SCN)_2 \cdot 2\,H_2O$	371(—106)	—(—99)	350(—23)	~415(—78); s	344(+27)	645(+47)	664.5(+36.5)

[a] Concentration of dyes: 10^{-4} — 10^{-5} M in acetonitrile, molar ratio metal: dye > 100; [b] Measured in methanol, metal: dye > 10; [c] Measured in methanol/water (1/1) with 1.5×10^{-2} M triethylamine; [d] An absorption maximum or a shoulder according to the conditions; [e] Dashes indicate that the value are not reported; [f] The figures in the parentheses indicate the spectral shift (nm) from the free dye; the negative values denote hypsochromic shifts, while the positive values bathochromic shifts; [g] $CaCl_2$ instead of $Ca(SCN)_2 \cdot 4\,H_2O$

enhanced metal selectivity of *43* on going from acetonitrile to methanol [reduction of spectral shift in methanol, particularly for $Ca(SCN)_2$]. However, it should also be pointed out that the shift widths are considerably diminished on going from acetonitrile to aqueous methanol for *49*; maximum shift is 83 nm in acetonitrile, while it amounts to only 39.5 nm in aqueous methanol.

Large spectral shifts which take place ion selectively suggest possible applications to the photometry of some alkali and alkaline earth metal ions in organic solution in a similar manner to that performed for crown ether dye *24*. Unfortunately, the spectral behaviors of dyes *41–50* have been studied only in the presence of excess metal salts, and their response to low concentrations of metal salts has not been reported. For practical application a right choice of solvents has to be made, since most of these dyes are expected to show considerable solvatochromism. It is pointed out that the absorption maximum of dye *49* moves from 598 nm in acetonitrile to 628 nm in aqueous methanol (Table 8). The effect of the anionic component of the salt as well as the interferences from acids and bases need to be carefully assessed.

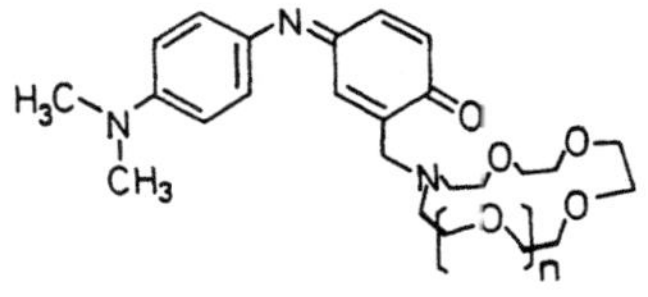

48 n = 1

49 n = 2

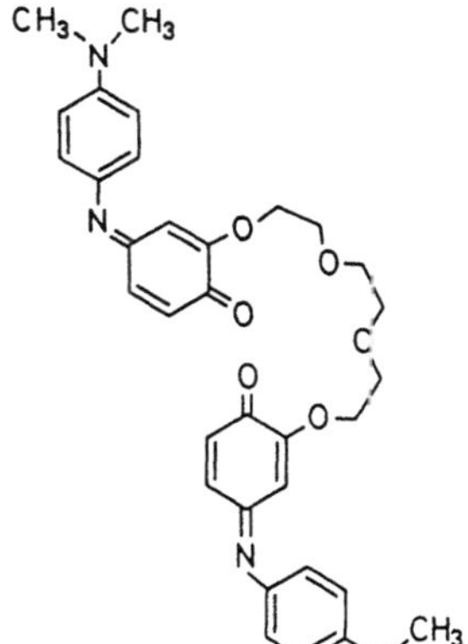

50

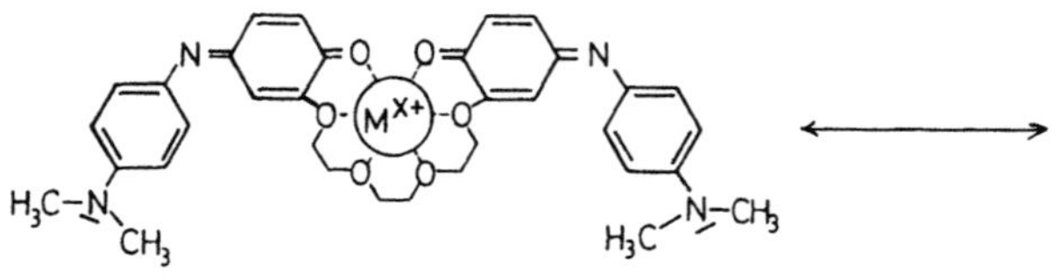

51

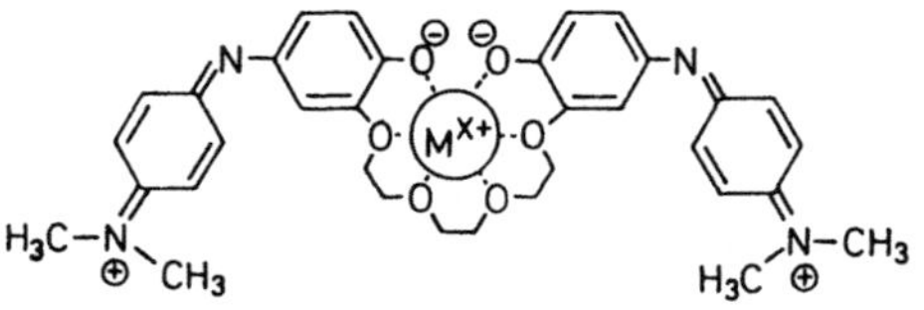

52

5 Conclusion

Crown ether dyes as chromogenic reagents for alkali and alkaline earth metal ions are still in the early stage of development. However, the basic idea of molecular design for such reagents has been fully materialized by a substantial number and variety of proton-dissociable as well as neutral crown ether dyes hitherto synthesized. Some of them are practically usable for extraction photometry of Na^+ and K^+. Molecular absorption photometry is especially suited for automatic flow analysis because of the operational simplicity and stability. The use of chromogenic crown ether reagents, however, may not only be limited to such elemental analysis. As pH indicators are used to check and monitor the acidity of aqueous solutions or even some organic solutions, the crown ether-based metallochromic indicators may eventually be used for the similar purpose on alkali and alkaline earth metal ions. Such indicators may also find a use as photometric probes in biochemical studies such as alkali and alkaline earth metal transport in cell membranes. The present state of arts is yet far from this goal, and a further study in this field would certainly be fruitful and rewarding.

6 References

1. Sandell, E. B.: Colorimetric Determination of Traces of Metals, New York—London, Interscience 1959
2. Cheng, K. L., Ueno, K., Imamura, T.: Handbook of Organic Analytical Reagents, Boca Raton, CRC Press 1982
3. Pedersen, C. J.: J. Am. Chem. Soc. *89*, 7017 (1967)
4. Takagi, M. et al.: Anal. Lett. *10*, 1115 (1977)
5. Nakamura, H. et al.: Talanta *26*, 921 (1979)
6. Nakamura, H. et al.: Anal. Chem. *52*, 1668 (1980)
7. Lamb, J. D. et al.: in: Coordination Chemistry of Macrocyclic Compounds; Melson, G. A. (Ed.), p. 145, New York, Plenum Press 1979
8. Dalley, N. K.: in: Synthetic Multidentate Macrocyclic Compounds, Izatt, R. M., Christensen, J. J. (Eds.), p. 207, New York, Academic Press 1978
9. Mallinson, P. R., Truter, M. R.: J. Chem. Soc., Perkin Trans. II 1818 (1972)

10. Ishizu, K. et al.: Chem. Lett. 227 (1978)
11. Pacey, G. E., Bubnis, B. P.: Anal. Lett. *13*, 1085 (1980)
12. Pacey, G. E. et al.: Analyst *106*, 636 (1981)
13. Yamashita, T. et al.: Bull. Chem. Soc. Jpn. *53*, 1550 (1980)
14. Nakamura, H. et al.: Chem. Lett. 1305 (1981)
15. Nishida, H. et al.: ibid. 1853 (1982)
16. Nakamura, H. et al.: Bunseki Kagaku *31*, E131 (1982)
17. Nakamura, H. et al.: Anal. Chim. Acta *139*, 219 (1982)
18. Takagi, M. et al.: unpublished work
19. Takagi, M. et al.: Anal. Chim. Acta *126*, 185 (1981)
20. Sumiyoshi, H. et al.: Talanta *24*, 763 (1977)
21. Kaneda, T. et al.: Tetrahedron Lett. *22*, 4407 (1981)
22. Nakashima, K. et al.: Chem. Lett. 1781 (1982)
23. Nakashima, K., Akiyama, S.: private communication
24. Shinkai, S. et al.: J. Am. Chem. Soc. *104*, 1967 (1982)
25. Nishida, H. et al.: Mikrochim. Acta I, 281 (1981)
26. Shiga, M. et al.: Bunseki Kagaku, *32*, E293 (1983)
27. Takagi, M. et al.: Chem. Lett. 1179 (1978)
28. Tazaki, M. et al.: ibid. 571 (1982)
29. Uechi, T. et al.: Acta Crystallogr. *B38*, 433 (1982)
30. Shinkai, S. et al.: J. Am. Chem. Soc. *104*, 1960 (1982) and references cited therein
31. Shinkai, S. et al.: Chem. Lett. 499 (1982)
32. Shinkai, S. et al.: Tetrahedron Lett. *20*, 4569 (1979)
33. Shiga, M. et al.: Chem. Lett. 1021 (1980)
34. Shinkai, S., Manabe, O.: J. Am. Chem. Soc. *105*, 1851 (1983)
35. Dix, J. P. Vögtle, F.: Angew. Chem. *90*, 893 (1978); Angew. Chem. Int. Ed. Engl. *17*, 857 (1978)
36. Dix, J. P., Vögtle, F.: Chem. Ber. *113*, 457 (1980)
37. Dix, J. P., Vögtle, F.: ibid. *114*, 638 (1981)

Photocontrol of Ion Extraction and Ion Transport by Photofunctional Crown Ethers

Seiji Shinkai and Osamu Manabe

Department of Industrial Chemistry, Faculty of Engineering, Nagasaki University, Nagasaki 852, Japan

Table of Contents

1 Introduction 68

2 Classification of Photoresponsive Crown Ethers 68
2.1 Capped Crown Ethers 68
2.2 Polymeric Photoresponsive Crown Ethers 74
2.3 Cylindrical Ionophores 77
2.4 Phane-Type Crown Ethers 79
2.5 Azobis (Crown Ethers) 83
2.6 Anion-Capped Crown Ethers 86
2.7 Crown Ethers with Other Responsive Functions 88

3 Light-Driven Ion Transport 93
3.1 Ion Transport Mediated by Photoresponsive Crown Ethers 93
3.2 Photocontrolled Permeability of Liposomal Membranes 98
3.3 Other Ion Transport Systems with Responsive Functions 99

4 Conclusions 101

5 References 102

1 Introduction

Photoresponsive systems are ubiquitously seen in nature: photosynthesis, vision, phototropism, and phototaxis are typical examples. In these systems, light is used as a trigger to cause subsequent life processes. We have been interested in the application of the phenomena to biomimetic systems. To imitate the fundamental functions of such photoresponsive systems, one has to combine within a molecule a photoantenna to capture a photon with a functional group to mediate a subsequent event. Host molecules have drawn attention as simplified enzyme model systems. We assumed that the photocontrol of host molecules, as well as that of enzymes [1-3], might provide a somewhat new field of chemistry.

We selected a crown ether moiety as a functional group because it has the ability to associate with charged and uncharged species and its conformation is easily changeable owing to the flexible nature of the macrocyclic polyether. In other words, one may expect that, provided that the conformation of the crown ether ring is easily changed, both the ion-binding ability and the ion-selectivity would be largely affected.

In the artificial photoresponsive systems, chemical substances with photoinduced structural changes are candidates for the photoantenna. Chromophores offering possibilities as a photoantenna are listed in Fig. 1. Prerequisites for a good photoantenna behaviour are high quantum yield, high reversibility, and large structural change. Among these candidates, (*E*)-(*Z*) isomerism of azobenzene (a), spiropyran-merocyanineequilibrium (d), and dimerization of anthracene (f) have frequently been employed.

The last problem is how to transmit the photoinduced structural changes occurring in the chromophore to the crown ether. The most expeditious method is to connect the crown ether with the chromophore through a covalent bond. In fact, most of the artificial photoresponsive systems used so far follow this concept. However, the utilization of secondary valence forces, ion flux, conformational change of polypeptides, etc., as the natural transducers, may be also interesting.

2 Classification of Photoresponsive Crown Ethers

2.1 Capped Crown Ethers

The most direct and expeditious method to transmit the photo-induced configurational change, occurring at the chromophore moiety, to the crown ether subunit is to construct a "chromophore bridge" at the crown ring. Considerations on CPK models showed that the distance between N(7) and N(16) of the diazacrown ether *1* is almost equal to that between the two carbonyl groups of 3,3′-bis(chlorocarbonyl)-azobenzene *2*. Thus, the first photoresponsive crown ether *3* was synthesized from *1* and *2* in 23% yield using high-dilution conditions [4]. When 4,4′-bis(chlorocarbonyl)-azobenzene was used instead of the 3,3′-isomer, the reaction system resulted in either polymers (e.g. *11–13*) or cylindrical ionophores (section 1.3).

The conformational change of the crown ether moiety of *3* which expectedly occurs in conjugation with (*Z*)-(*E*) isomerism of the azobenzene cap was inferred from

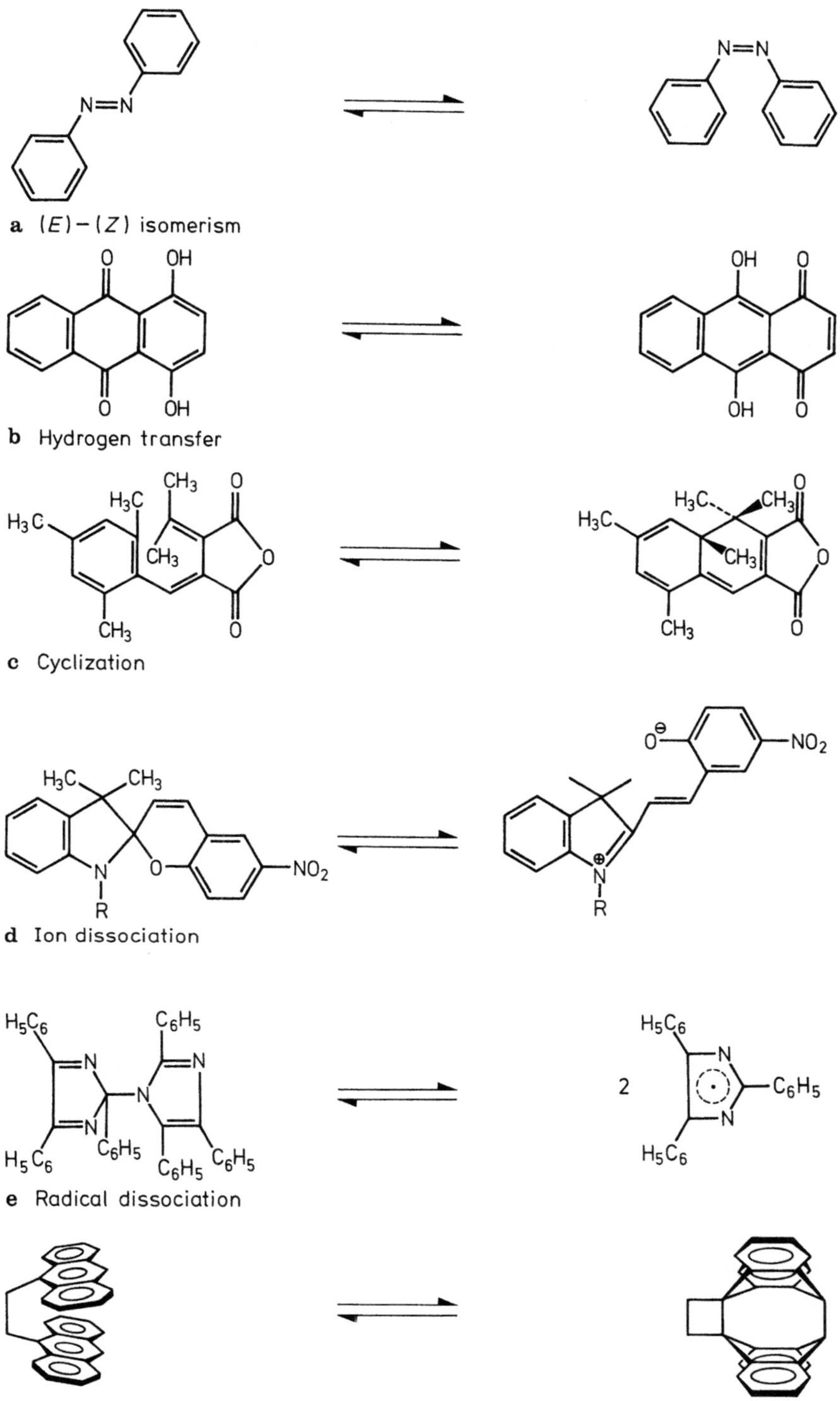

Fig. 1. Candidate chromophores for photoantennas

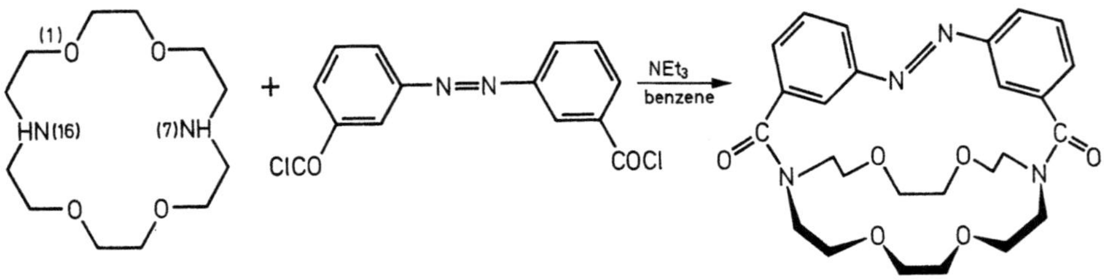

1 2 (*E*)-*3*

the CPK model studies. They show that the azobenzene moiety of (*E*)-*3* projects vertically over the crown ether plane, whereas that of (*Z*)-*3* is almost parallel to the crown ether plane. The structure was later confirmed by an X-ray crystallographic study [5]. The CPK models also suggest that in (*E*)-*3* the *ortho* hydrogen of the azo linkage penetrates partially into the plane of the crown ether. This would probably interfere with deep complexation of alkali metal cations.

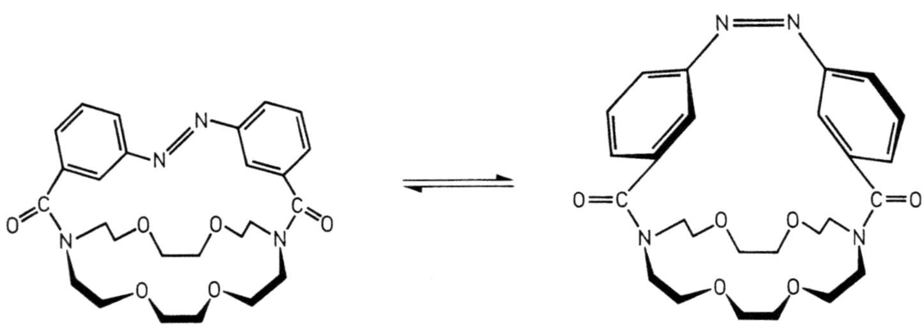

(*E*)-*3* (*Z*)-*3*

Under UV-light irradiation, the absorption band of (*E*)-*3* decreased rapidly and reached a photostationary equilibrium where the ratio of (*E*)-*3*/(*Z*)-*3* was 40/60. The solvent extraction of alkali metal salts of Methyl Orange from water to benzene indicated that (i) the extractability of (*E*)-*3* changes in the order $K^+ > Na^+ > Li^+ > Rb^+, Cs^+$, whereas that of (*Z*)-*3* is in the order $K^+ > Na^+ > Rb^+ > Li^+, Cs^+$, (ii) large alkali metal cations such as Rb^+ and Cs^+ are hardly extracted by (*E*)-*3*, (iii) (*E*)-*3* extracts Li^+ which cannot be extracted by (*Z*)-*3*, whereas (*E*)-*3* cannot extract Rb^+ which is extracted by (*Z*)-*3*, and (iv) Na^+ is better extractable by (*E*)-*3* than by (*Z*)-*3* but the reverse is true for K^+. These findings consitantly suggest that the crown ether ring of (*Z*)-*3* is somewhat expanded by the (*E*)-to-(*Z*) isomerization of the azobenzene-cap.

The azo-linkage allows the coordination of heavy-metal ions, however such a coordination is generally weak [6,7]. For 2,2′-azopyridine, one may expect the complexation with the azo-linkage to be stabilized by neighboring pyridine nitrogens. In fact, Shinkai et al. [8] have found that (*E*)-2,2′-azopyridine acts as a much better ligand than photo-isomerized (*Z*)-2,2′-azopyridine. Based on CPK models (and the X-ray crystallographic study) [5,8], one may predict that pyridine nitrogens of (*E*)-*4* are directed toward the crown ether plane and coordinate metal ions in the crown ether ring, whereas those of (*Z*)-*4* have not such a coordination ability due to the distorted configuration. Thus, this may be an example of a photo-responsive crown-cryptand transformation.

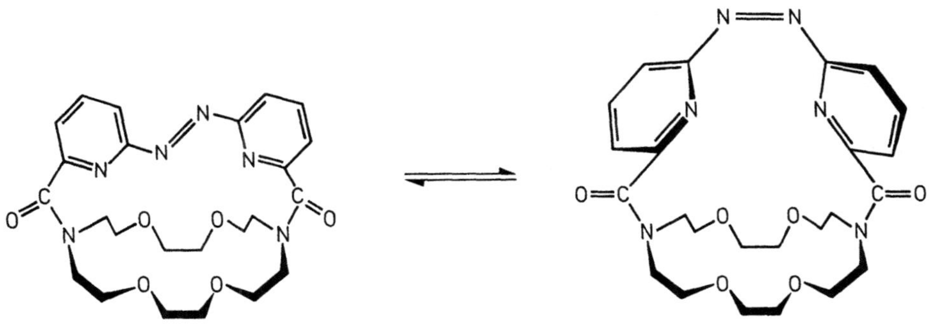

(*E*)-4 (*Z*)-4

The result of solvent extraction with *4* indicated that the photoisomerization of the 2,2′-azopyridine bridge affects the extractability of alkali metal cations only to a smaller extent. In contrast, (*E*)-*4* extracts considerable amounts of heavy metal ions (Cu^{2+}, Ni^{2+}, Co^{2+}, and Hg^{2+}), whereas photoisomerized (*Z*)-*4* scarcely extracts these metal ions [8]. Such a difference in the extractability was not observed between (*E*)-*3* and photoisomerized (*Z*)-*3*. Thus, the affinity of *4* toward heavy metal ions is primarily governed by the configuration of the 2,2′-azopyridine bridge. Shinkai et al. [9] further synthesized the following capped thiacrown ethers, *5* (X=CH) and *6* (X=N).

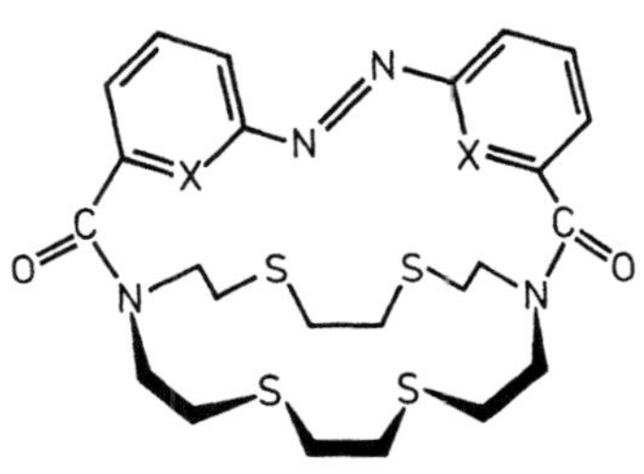

(*E*)-*5* (X=CH) and (*E*)-*6* (X=N)

The thermal (*Z*)-to-(*E*) isomerization of azobenzene has been controversial: the reaction may proceed either via a rotation about the N=N bond or flip-flop inversion of one of the nitrogen atoms (Fig. 2) [10]. The difference between the two mechanisms is that the rotational mechanism proceeds via a dipolar transition state accompanying a large volume change.

Nowadays, it is believed that the thermal isomerization of "usual" azobenzene derivatives proceeds through an inversion mechanism: for example, the lack of major solvent or pressure effect on the rate of isomerization may be cited as an evidence against rotation [10–13]. Whitten et al. [10,14] showed, however, that sizable kinetic solvent effects are observed for azobenzene with push-pull substituents (e.g., dimethylamino and nitro) stabilizing the dipolar transition state of the rotational mechanism. Since (*Z*)-*3* has an azobenzene covalently bridged to an azacrown ether and the rotation of the benzene rings is sterically restricted, this molecule provides the "standard" thermodynamic parameters for the inversion isomerism [15]. Thus, it was demonstrated that the thermal isomerization of most (*Z*)-azobenzenes occurs via an inversion mechanism, the activation parameters ($\Delta H^{\ddagger}$ and $\Delta S^{\ddagger}$) including those of (*Z*)-*3* being subject to a good $\Delta H^{\ddagger}$—$\Delta S^{\ddagger}$ compensation relationship. The pressure effect also supported the inversion mechanism because the $\Delta V^{\ddagger}$ was too small (ca. —0.4 to —0.7 ml mol^{-1}) to consider the rotational mechanism [16]. Interestingly, simple azobenzenes have small negative $\Delta V^{\ddagger}$ values, whereas only (*Z*)-*3* has a small positive $\Delta V^{\ddagger}$ value (2.0 ml mol^{-1}).

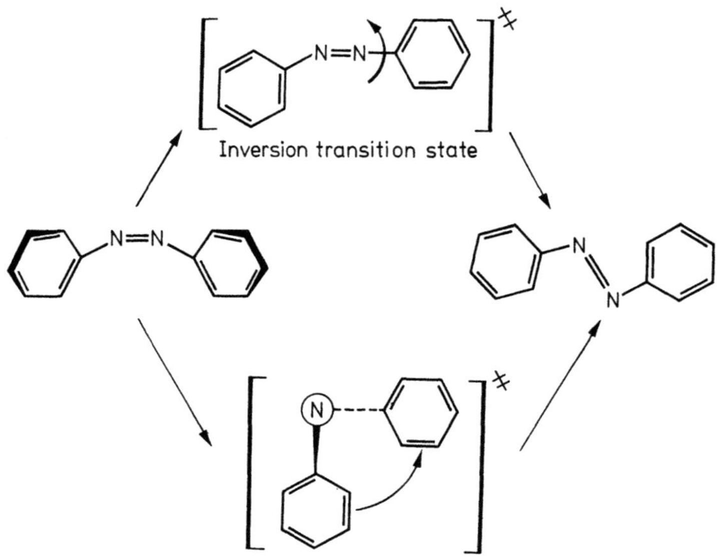

Fig. 2. Inversion vs. rotational mechanism for the thermal (*Z*)-to-(*E*) isomerization of (*Z*)-azobenzene

The finding implies that the crown ether ring is expanded in the inversion transition state, the increase in the void volume into which solvent molecules cannot penetrate being comparable with that of cyclohexane. Thus, the photo-induced configurational change occurring in the azobenzene chromophore is transmitted as a dynamic energy to the crown ring through the covalent bond. A similar idea to discriminate the thermal isomerization mechanisms on the basis of the azobenzenophane molecules such as *7* and *8* has been reported by Rau and Lüddecke [17].

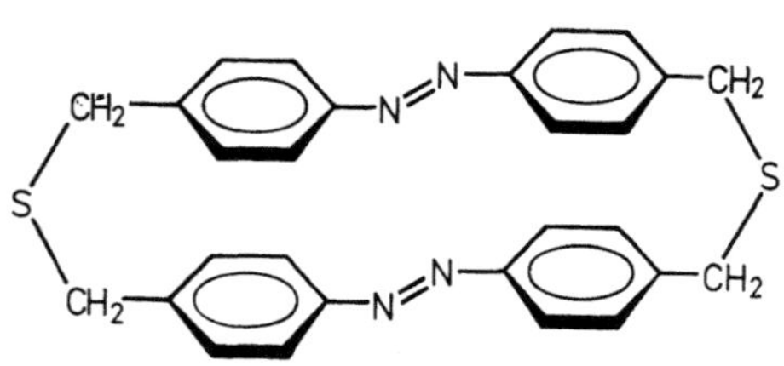

(*E*)-*7*

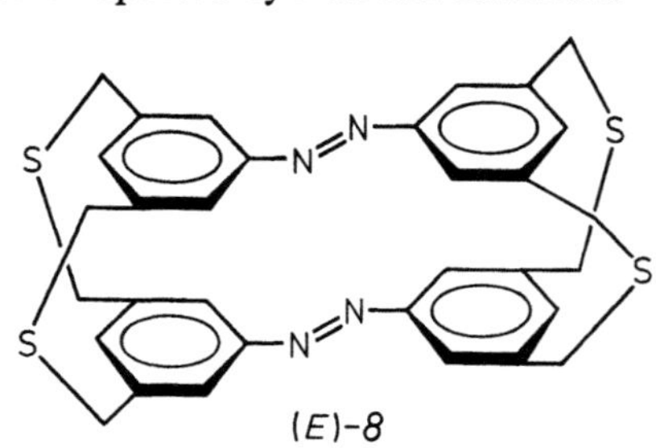

(*E*)-*8*

Shinkai et al.[4] have found that the rate of the thermal isomerization of (*Z*)-*3* is suppressed by K^+ and ammonium ions. The crown ether ring of (*Z*)-*3* must expand once in the transition state[16], so that the interaction between the crown ether and the complexed cation must be disrupted (or weakened) temporarily. This requires an additional free-energy of activation for the isomerization process. Probably, the transition state species of *3* exhibits the ion-affinity and the ion selectivity quite different from those in the ground state.

Cyclodextrin is another interesting host-molecule. Ueno et al.[18] synthesized azobenzene-capped β-cyclodextrin *9* to regulate the binding ability of β-cyclodextrin (β-CD). They found that *9* presents an induced circular dichroism band in the azobenzene π—π^* region (355 nm) before irradiation, whereas it shows another band in the azobenzene n—π^* region (445 nm) after irradiation. Since both circular dichroism bands nearly disappeared on the addition of guest molecules in large excess, the association constants could be determined from the change in the band intensity.

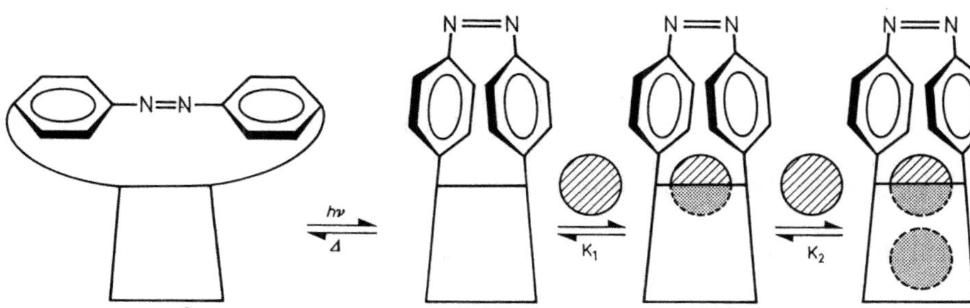

(*E*)-*9* (*Z*)-*9*

The examination of the association constants revealed that (i) (*E*)-*9* has the association constants generally smaller than β-CD, whereas (*Z*)-*9* has the greater binding ability than β-CD and (ii) (*Z*)-*9* tends to include two guest molecules. These results allow to consider the poor binding ability of (*E*)-*9* to be associated with the "shallow floor" concept suggested by Emert and Breslow[19]. On the other hand, the improved binding ability of (*Z*)-*9* stems from the expansion of the cavity due to the capping with (*Z*)-azobenzene. The circular dichroism spectrum of *9* shows in some cases an abrupt change on the addition of guest molecules with ionic nature[20]. The non stoichiometric behavior was regarded as evidence for a conformational transition of the cyclodextrin moiety caused by outside binding of the guest molecules.

The rate of hydrolysis of *p*-nitrophenyl acetate (PNPA) catalyzed by *9* according to Eq. (1) is accelerated by photoirradiation (eq. 1)[21].

The reaction proceeded according to the Michaelis-Menten-type kinetics: $9 + \text{PNPA} \underset{k_{-1}}{\overset{k_1}{\rightleftharpoons}} 9 \cdot \text{PNPA} \xrightarrow{k_2}$ products. The maximum rate constant k_2 for (*Z*)-*9* was smaller than that for (*E*)-*9*, indicating the unfavorable geometry of the (*Z*)-*9* · PNPA complex to hold the substrate molecule in the correct catalytic position. Probably, PNPA is included too deeply in the expanded cavity of (*Z*)-*9* to attain a favorable attack of the alkoxide ion of the cyclodextrin moiety at the carbonyl group of PNPA. On the other hand, (*Z*)-*9* has a considerably smaller K_m value ($= k_{-1}/k_1$) than (*E*)-*9*, indicating that the substrate is included deeply in the expanded hydrophobic cavity of (*Z*)-*9*. These situation leads to an increased overall hydrolysis rate (k_2/K_m) for (*Z*)-*9*; it was about five times greater than that for (*E*)-*9*.

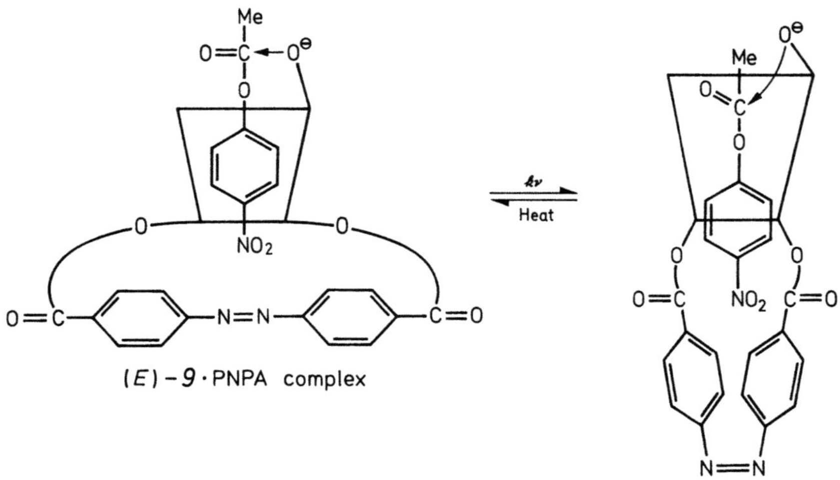

(*E*)-*9*·PNPA complex (*Z*)-*9*·PNPA complex (1)

2.2 Polymeric Photoresponsive Crown Ethers

Photoregulation of the conformation of synthetic polymers has been attempted in several systems to elucidate the function of photobiological transducers and also to construct photoreceptor analogs. The basic idea for regulating the conformation of synthetic polymer chains is to prepare a polymer system containing a chromophore capable of transforming the light energy into a conformational change of the polymer chain. Thus, a number of polymers containing either azobenzene or spiropyran were synthesized (e.g., *10–12*) [22–26]. With these photoresponsive polymers, the solution viscosity, pH change, polarity change of film surface, etc. was photocontrolled. These phenomena are all due to the photoinduced changes of the chromophoric groups.

The combination of these photoresponsive behaviors of synthetic polymers with a crown ether group may lead to novel photofunctional systems. Shinkai et al. [4] synthesized *13* by the polycondensation of *1* and 4,4′-bis(chlorocarbonyl)azobenzene. Conformational behavior of *13* was conveniently assessed by viscosity measurements. The intrinsic viscosity of (*E*)-*13* (0.057 dl/g) in *o*-dichlorobenzene decreases to 0.047 dl/g under UV-light irradiation [4]. The result implies that the (*Z*)-azobenzene unit produced by photoirradiation changes the polymer conformation to a more compact one.

The solvent treatment with (*E*)-*13* showed the extractability of $Na^+ > K^+ > Cs^+ > Li^+, Rb^+$ [4], indicating that (*E*)-*13* also belongs to a class of polymers with crown ethers in the main chain, because the crown polymers with a crown moiety in the side chain show a higher affinity for alkali metal cations with large ion radii. The crown polymers with the crown group in the main chain also exhibit a cation affinity analogous to that of the monomeric analogs [27–30]. As expected, the extract-

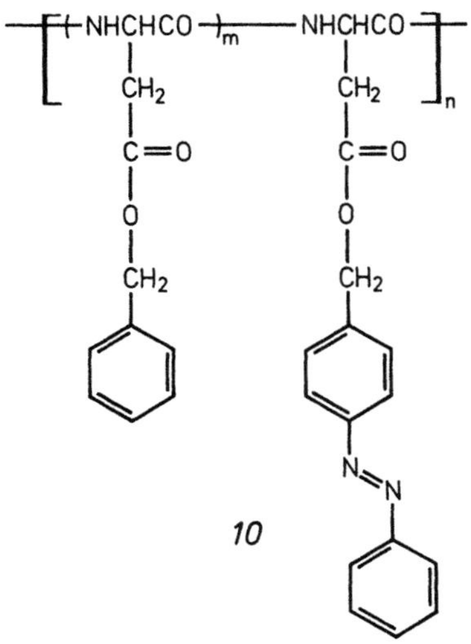

10

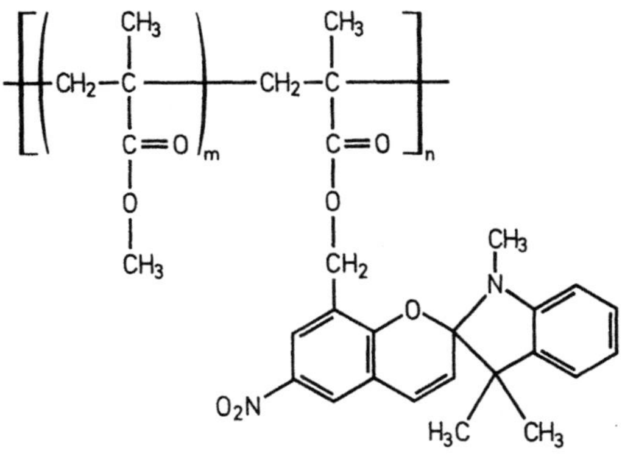

11

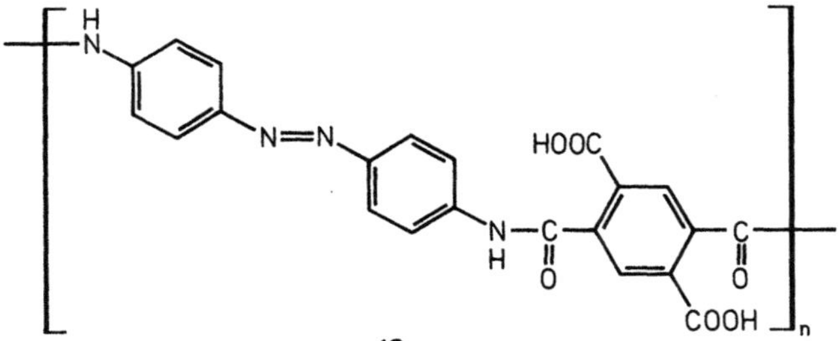

12

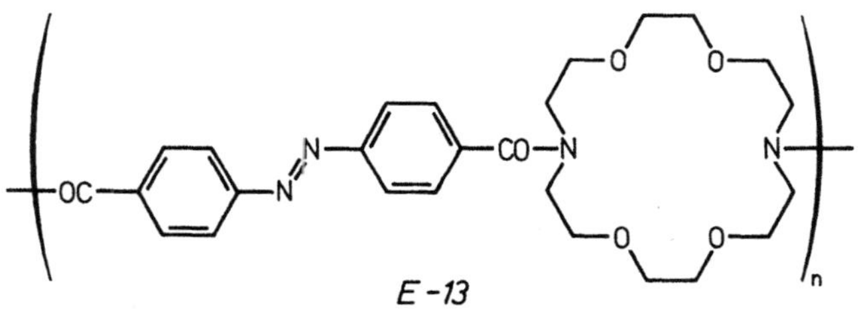

E-13

ability of *13* was affected by UV-light irradiation: under photoirradiation the extractability of K^+ was somewhat enhanced but that of Na^+ was markedly reduced. It was also found that (*E*)-*13* extracts dipotassium salt of phthalic acid in preference to those of isophthalic acid and terephthalic acid and the extractability decreases under photoirradiation [4].

Polymer supports carrying various functional groups have attracted much attention in recent years. In particular, immobilization of crown ethers in polymer supports is one of the most noteworthy achievements demonstrating the usefulness of polymer supports. In the photoresponsive system, polymer supports may be useful as a "fixed point" to induce the conformational change of crown ethers by light. The distance between the 4- and 4′-positions of (*E*)-azobenzene is 9.0 Å, while that of (*Z*)-azobenzene is 5.5 Å. When a molecule containing both a crown ether and an azobenzene is immobilized in a polymer support, the contraction of the azobenzene moiety due to the photo-induced (*E*)-to-(*Z*) isomerization is compensated by the elongation of the "flexible" crown ether moiety. *14* immobilized in the cross-linked polystyrene beads shows this behaviour [31].

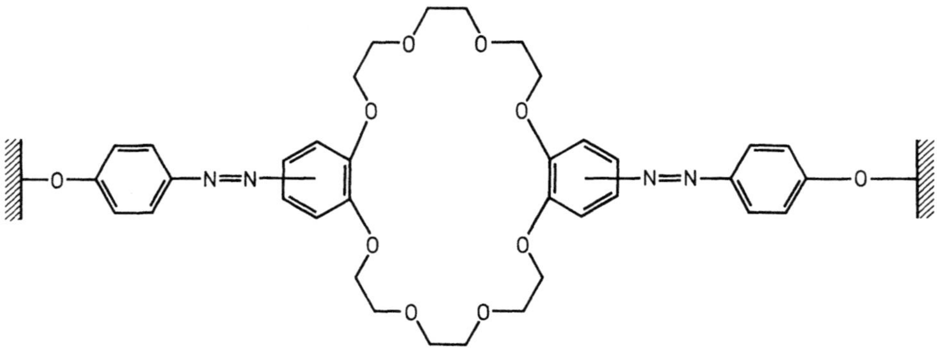

14

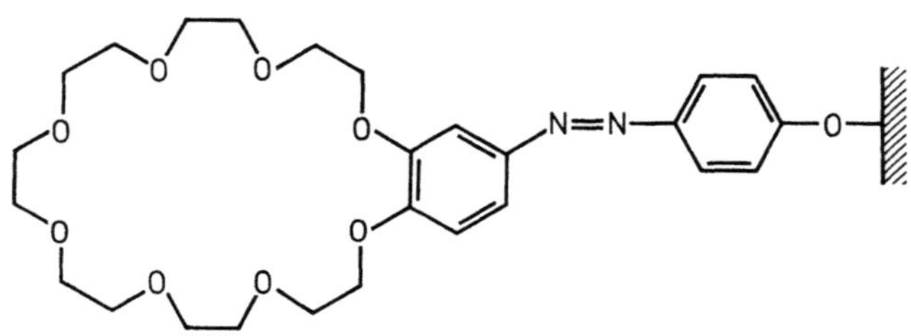

15

When *14* was mixed with the DMF solution of cesium *p*-nitrobenzoate in the dark, the concentration of Cs^+ in the solution decreased rapidly and reached an equilibrium within 20 min. When this solution was subjected to photoirradiation by a UV-lamp inducing the (*E*)-to-(*Z*) isomerization of the azobenzene moiety, the concentration of Cs^+ in the solution gradually increased. On the other hand, such a photoresponsive ion-adsorption was hardly observed for *15* in which only one terminal azophenoxide group was immobilized in the polymer support [32]. These results consistently suggest that, as shown in Fig. 3, the photoinduced (*E*)-to-(*Z*) isomerization of the azobenzene moiety is capable of changing the conformation of the crown ether into a more stetched one which has a poor ion-binding ability relative to the "normal" crown ether.

Several exciting facts were found by measuring the rate of the thermal (*Z*)-to-(*E*) isomerization occurring in the polymer beads [31,32]. In the 3.2% crosslinked polystyrene beads bearing *14*, the rate of the ion-adsorption to the polymer beads was almost comparable with that of the thermal (*Z*)-to-(*E*) isomerization of the azobenzene unit. In the 9.0% cross-linked polystyrene beads, however, the former was much slower than the latter. Probably, diffusion of Cs^+ ion in the polymer matrix is involved in the rate-limiting factor in the highly crosslinked polystyrene beads. It is also interesting that the thermal isomerization in the 9.0% crosslinked polystyrene beads is faster by about one order of magnitude

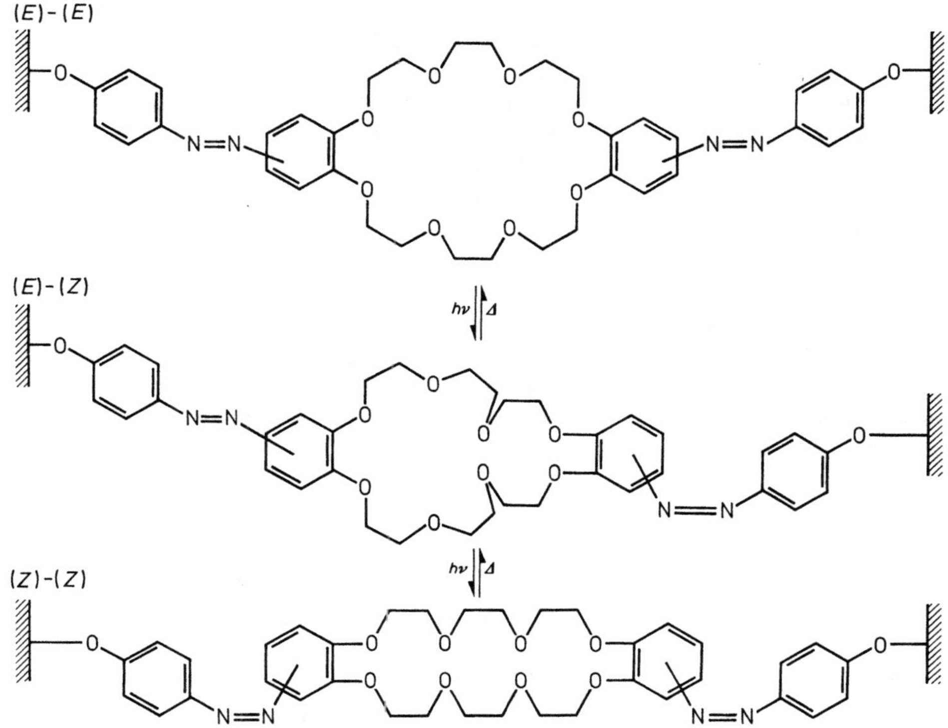

Fig. 3. Photoisomerism of the azobenzene-crown-azobenzene bridge immobilized in the polymer support

than that in the 3.2% crosslinked polystyrene beads. The most fascinating rationale is that the photoinduced dynamic tension being stronger in the highly cross-linked polymer facilitates the thermal isomerization (i.e., the relaxation). If so, this may be an interesting example for the conversion of dynamic into chemical energy.

Similar phenomena were observed for the adsorption of K^+ on the immobilized [18]crown-6 compounds *16* [32].

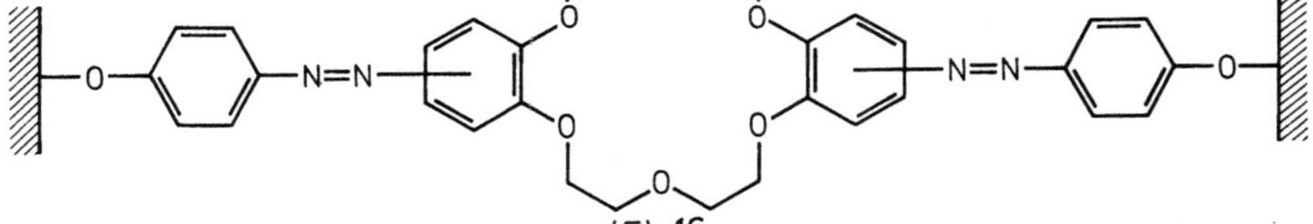

(E)-*16*

2.3 Cylindrical Ionophores

A cylindrical ionophore, in which two macrocyclic ligands are linked by two (or more than two) pillars, represents an interesting type of receptor molecule which allows to bind diammonium ions and more than one metal ion. Since the metal affinity of cylindrical ionophores is crucically governed either by the pillar length or by the ring size of ligands, one may expect that the binding affinity and the binding selectivity is readily controlled by changing the length of the two-pillars. Thus, ionophores with azobenzene pillars act as "elastic" cyclindrical ionophores in response to photo-induced (*E*)-(*Z*) isomerization of the azobenzene pillars.

The first photoresponsive cylindrical ionophore synthesized by Shinkai et al. [33] is shown by formula *17*. As described before, the reaction of *1* with 3,3′-bis(chlorocarbonyl)azobenzene under high dilution conditions yielded *3* (i.e., 1:1 adduct), whereas the reaction of *1* with 4,4′-bis(chlorocarbonyl)azobenzene gave *17* (i.e., 2:2 adduct). The difference can be rationalized in terms of the steric factor that the distance between N(*7*) and N(*16*) of *1* is almost equal to that between two carbonyls of the 3,3′-isomer but significantly shorter than that between two carbonyls of the 4,4′-isomer. Similarly, *18* and *19* were obtained by condensation under high-dilution conditions.

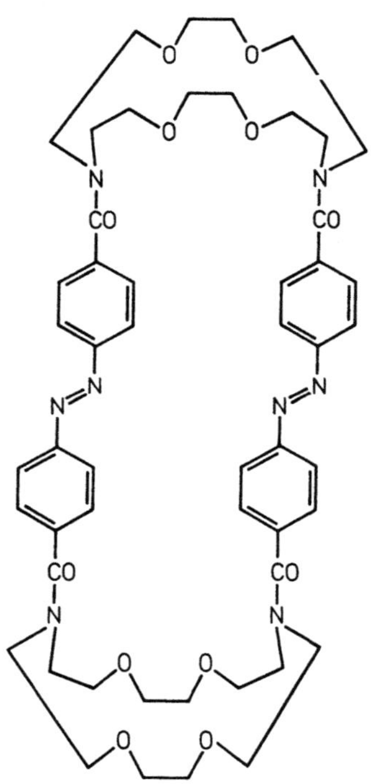

(*E*)-*17*

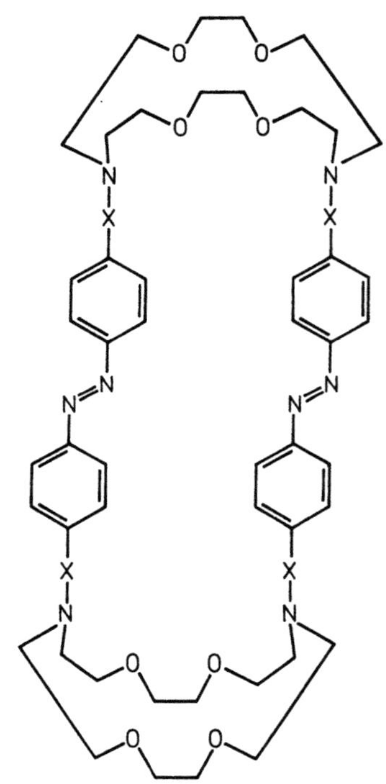

(*E*)-*18* (X = OCH_2CO)
(*E*)-*19* (X = $CON(CH_3)CH_2CH_2$)

From CPK model considerations it was speculated that, assuming that the polymethylene chain adopts an extended conformation, the distances between the two crown rings to (*E*)-*17* and (*Z*)-*17* (two azobenzenes are (*Z*)-forms) are almost equal to those between the two terminal ammonium groups of decamethylenediamine and hexamethylenediamine, respectively. It was shown that (*E*)-*17* extracts decamethylene and dodecamethylene diammonium salts efficiently but hardly extracts tetramethylene and hexamethylene diammonium salts, whereas hexamethylene diammonium salt is most extractable under UV-light irradiation [33]. The results are basically compatible with the prediction from CPK model studies. Similarly, *18* allows to extract diammonium salts with long methylene units but scarcely extracts those with short methylene units [34].

(*E*)-*19* having four tertiary amine functionalities strongly binds alkali metal ions, but UV-light irradiation has almost no effect on the binding ability. This is probably due to the too long distance between the two crown ether rings. Instead, *19* acts as a photoresponsive host molecule in aqueous solution [35]. (*E*)-*19* is very soluble in water at pH 5 by protonating the azacrown nitrogens and strongly associates hydrophobic guest molecules such as Methyl Orange. When the two azobenzene pillars were photo-isomerized to the (*Z*)-form, the complex was dissociated and the increase in the free Methyl Orange was apparently correlated to the concentration of (*Z*)-*19*. The result indicates that the association ability of the host molecule is crucially subject to the molecular shape.

2.4 Phane-Type Crown Ethers

In order to design a photoresponsive crown ether exhibiting a large change in the ion-binding ability, one should choose a large geometrical change without steric hindrance. In fact, however, these two requirements are quite opposing factors. Crown ethers containing a photoresponsive chromophore like a phane unit might satisfy the two opposing requirements.

Desvergne et al. [36] utilized the photo-dimerization of anthracene to design a "switched-on" crown ether *21*. In the presence of a lithium salt photoirradiation of *20* gave the photocyclo-isomer *21*, which was fairly stable with the lithium salt as in Eq. (2). When the Li^+-*21* complex was shaked in a polar solvent such as acetonitrile, *21* realily reverted to the open form *20* by competing solvation of Li^+ with a polar solvent. The remarkably enhanced stability of *21* in the presence of Li^+ ion was rationalized in terms of a "cation-lock".

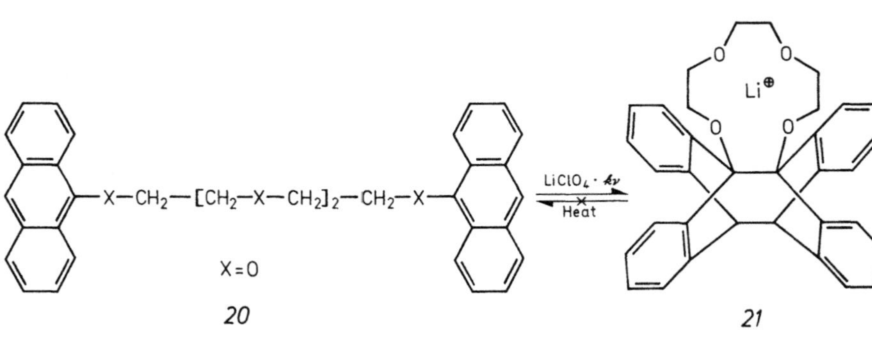

20 *21* (2)

Yamashita et al. [37] synthesized *22* in which the intermolecular photodimerization of anthracene, which may possibly take place in *20*, is completely suppressed. In order to assess the effect of metal ions on photochromic process, half-life time of *22* in the photo-reaction and of *23* in the thermal reaction (3) were measured with and without metal ions. The half-lifetime of the dark reaction *23*→*22* was prolonged by metal ions compared to that of ion-free case, and the effect on the half-life time was in the order $Na^+ > K^+ > Li^+$. On the other hand, the photo-reaction *22*→*23* is not dependent on metal ions. This may be attributed to much larger energy of the photo-transition than the stabilization energy due to the ion complexation.

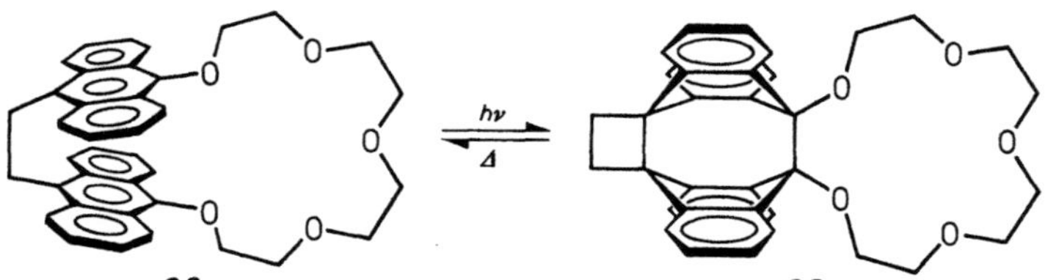

22 *23* (3)

The foregoing successful results suggest that the photochemistry of unsaturated crown ethers appears to as a novel application. Eichner and Merz [38] attempted to synthesize a basket crown *25* or a cage crown *26* by photo-reaction with *24* (4). How-

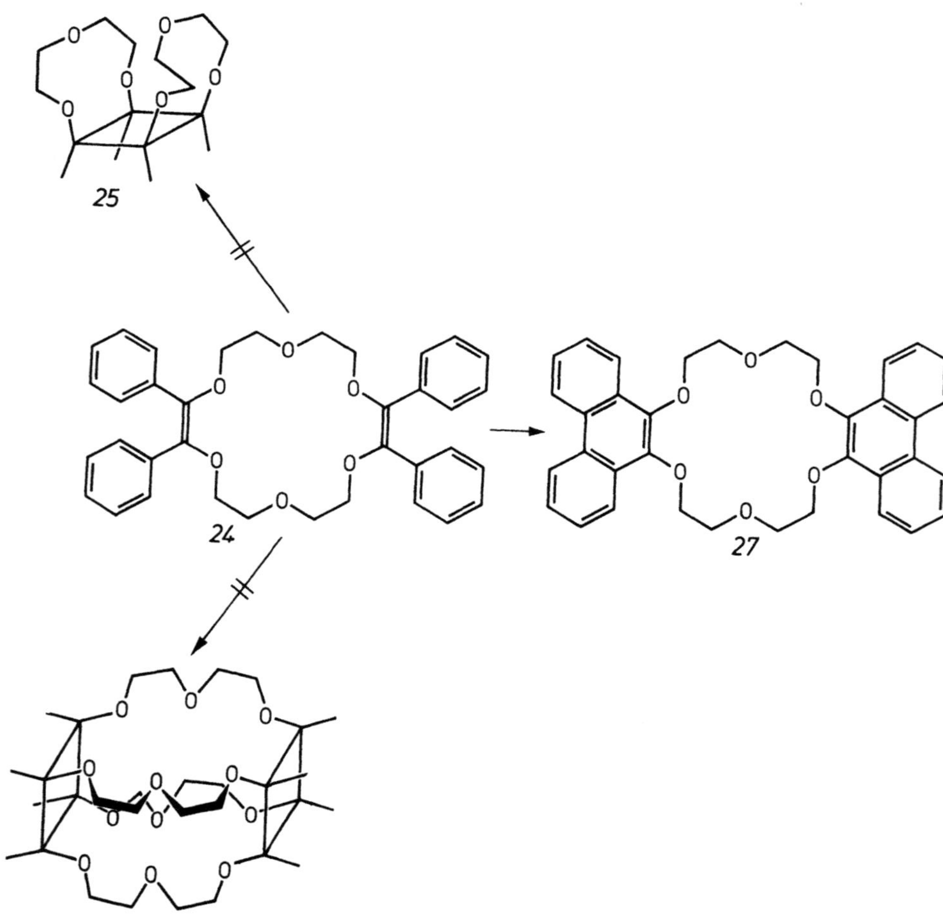

25 *24* *27* *26*

ever, when *24* was irradiated in benzene with catalytic amounts of iodine and oxygen bubbling through the solution, *27* was obtained as the only reaction product in 50–60% yield. Thus, the photo-oxidation to phenanthrene exceeded the photo-dimerization to cyclobutane.

Shinkai et al. [34] synthesized (*E*)-*28* by condensation of the corresponding diamine with 4,4'-bis(chlorocarbonyl)azobenzene under high-dilution conditions. The CPK model of *28* suggested that the polyoxyethylene chain is almost linearly extended when the azobenzene segment adopts the (*E*)-configuration, whereas that of *28* with the (*Z*)-configuration forms a crown-like loop. The result of the solvent extraction with *28* revealed that (*E*)-*28* completely lacks the affinity with alkali metal cations, while photoisomerized (*Z*)-*28* is capable of extracting large alkali metal cations such as Rb^+ and Cs^+. Thus, an ionophoric loop appears in response to photoirradiation; the loop size is roughly comparable with that of [21]crown-7 (21C7)[1].

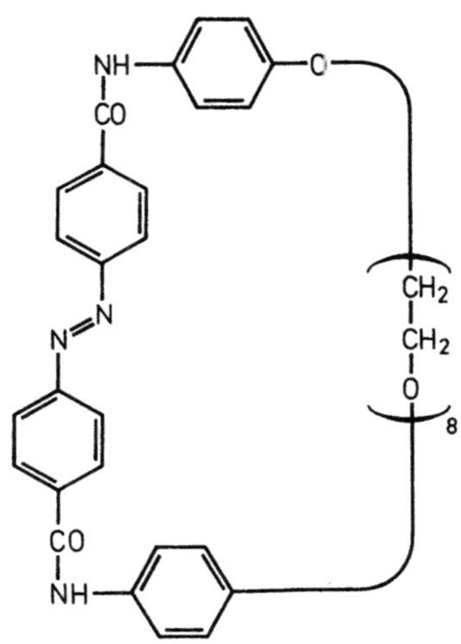

(*E*)-*28*

A large change in the ionophoric nature of the crown-like loop is effected if one can incorporate the azobenzene unit directly into the ring structure (i.e., as azo-benzenenophane). Shiga et al. [39] synthesized the azobenzenophanes (*E*)-*29* and (*E*)-*30*. (*E*)-*29* exhibited the affinity to alkali metal cations such as Na^+ and K^+ similar to "regular" crown ethers. In contrast, photoisomerized (*Z*)-*29* completely lacked the affinity to the metal ions. As speculated from the CPK model, the con-formation of the crown ring of (*Z*)-*29* is highly strained and unfavorable for the interaction with metal ions. However, both the photo- and thermal isomerization of *29* and *30* are extremely slow and the attempt to (*Z*)-isomerize more than 20% leads to an irreversible spectral change. This is probably due to the large steric crowding expected for the transition state.

In the example above, the photoresponsive "all-or-nothing" change was attained but the good reversibility was not. To attain both at the same time, one has to utilize

1 Notations frequently used in this report are listed and explained on p. XI

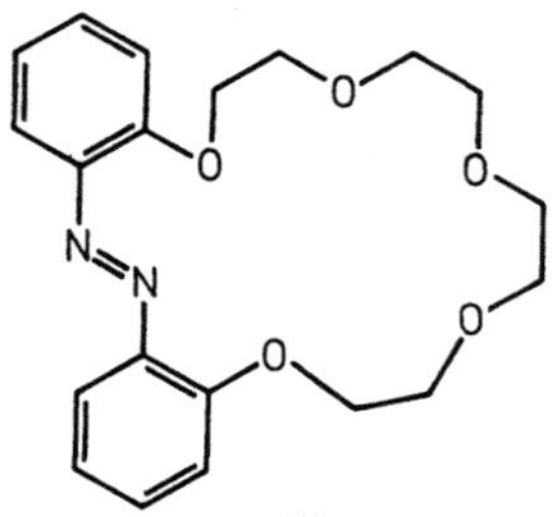

(E)-29

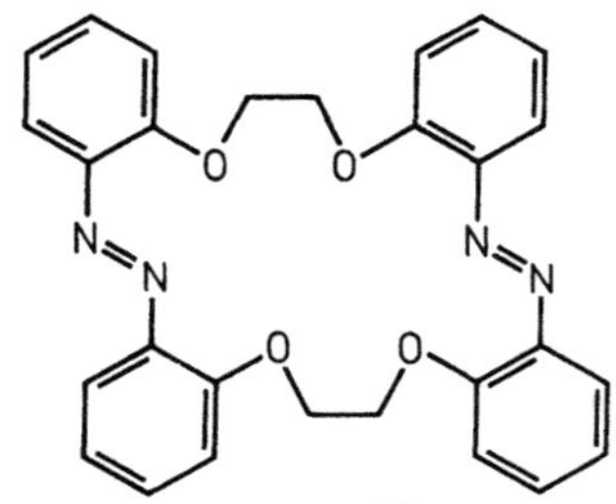

(E)-30

a large geometrical change without inducing any steric strain. Shinkai et al. [40] synthesized azobenzenophane-type crown ethers *31* (n = 1, 2, 3), with 4,4′-positions of azobenzene linked by a polyoxyethylene chain.

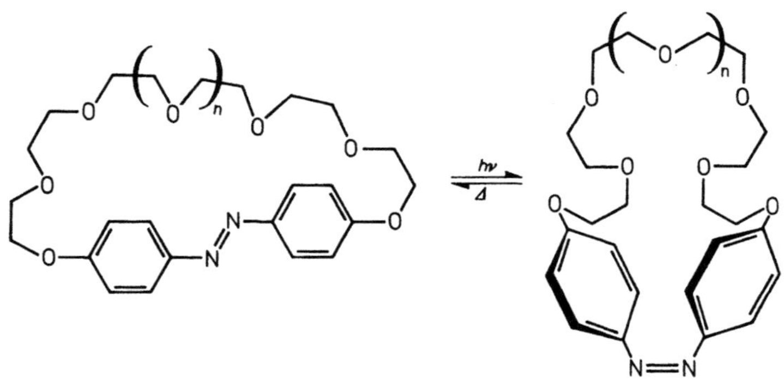

(E)-31 (Z)-31

The CPK model induced that (i) the polyoxyethylene chain of the (*E*)-homologs with 6 to 10 ethylene number is extended almost linearly on the azobenzene plane, (ii) there is no (or little) significant steric restriction in the course of the (*Z*)-(*E*)-isomerization, and (iii) most importantly and interestingly, the number of oxygen atoms which can contribute to the formation of a crown-like loop in the (*Z*)-homologs is the "total oxygen number" minus 2 (for two phenolic oxygen atoms): for example, (*Z*)-*31* (n = 1) forms a polyoxyethylene loop analogous to [15]crown-5 (15C5). The (*Z*)-(*E*) isomerism (5) was reversible.

The solvent extraction of alkali metal cations with *31* indicated that the predictions based on CPK models are reasonable. The (*E*)-isomers totally lacked the affinity for metal ions, whereas, the (*Z*)-isomers extracted considerable amounts of alkali metal cations to the organic phase. The maximum extractability was observed for Na^+ in (*Z*)-*31* (n = 1), K^+ in (*Z*)-*31* (n = 2), and Rb^+ in (*Z*)-*31* (n = 3). The result implies that *31* satisfies the reversible "all-or-nothing" change in the ion-binding ability, and the spheric recognition patterns, typical of crown ethers in solution. The ion selectivity supports that the ring size of (*Z*)-*31* (n = 1), (*Z*)-*31* (n = 2), and (*Z*)-*31* (n = 3) correspond to those of 15C5, 18C6, 21C7, respectively. The rate of the thermal (*Z*)-to-(*E*) isomerization is also suppressed by the complexed metal ions. The association constants of the (*Z*)-homologs were thus estimated by analyzing the plots of the rate vs. the metal concentration.

2.5 Azobis (Crown Ethers)

In the preceding excamples, the photocontrol of the crown ether function is achieved on the basic idea that the conformational change in the crown ether ring is induced by the photoresponsive configurational change in the intramolecular chromophore. Here, the idea is that the spacial distance between two crown ether rings changes in response to photoirradiation. It has been established that alkali metal cations exactly fitting the size of the crown ether form a 1:1 complex, whereas those which have larger ion radii form a 1:2 complex. This view was substantiated clearly by using bis(crown ethers)[41,42], polymeric crown ethers[29,30] and was also shown by the crystal structure of crown-alkali metal cation complexes[43]. For example, 15C5 and its analogs form a 1:1 cation/crown complex with Na^+, whereas a 1:2 cation/crown complex is formed with K^+. Kimura et al.[41] reported that the maleate derivative ((*Z*) form) of benzo[15]crown-5 B15C5 extracts K^+ from the aqueous phase 14 times more efficiently than the fumarate derivative ((*E*) form) owing to the formation of the intramolecular 1:2 complex. The result suggests that the bis(crown ether) in which the C=C double bond is replaced by the azo-linkage exhibits a photoresponsive ion-extraction behavior.

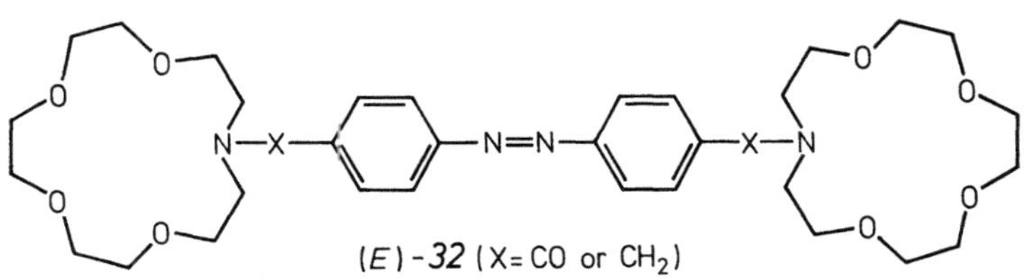

(*E*)-*32* (X = CO or CH_2)

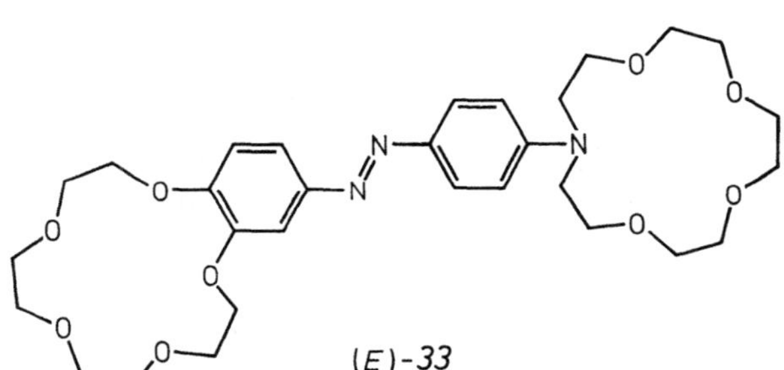

(*E*)-*33*

Shinkai et al.[44–46] synthesized a series of azobis(crown ethers) called "butterfly crown ethers" such as *32* (X=CO or CH_2), *33*, *34* (n = 1, 2, 3), and *35* (n = 2, 3). It was found that the relative concentration of (*Z*)-*34* at the photostationary state is sensitively affected by added alkali metal cations. For example, the content of (*Z*) in *34* (n = 2) (52% in the absence of metal ion) was scarcely affected by added Na^+ and K^+, but remarkably increased with increasing concentrations of Rb^+ and Cs^+[45,46]. In particular, the (*Z*)-isomer was enhanced up to 98% at high Rb^+ concentrations. Similarly, the rate of the thermal (*Z*)-*34* (n = 2) to (*E*)-*34* (n = 2) isomerization was markedly suppressed by Rb^+ and Cs^+, Rb^+ exhibiting the greatest inhibitory effect. The photostationary state and the thermal isomerization rate of *34* (n = 3) were also affected by Rb^+ and Cs^+, but the Cs^+ provided on influences greater than Rb^+ [45,46]. These novel findings were attributed to the "tying effect" of two crown ethers by one alkali metal cation, which is expected only for the sandwich-type 1:2 cation/crown complexes (e.g., *36*).

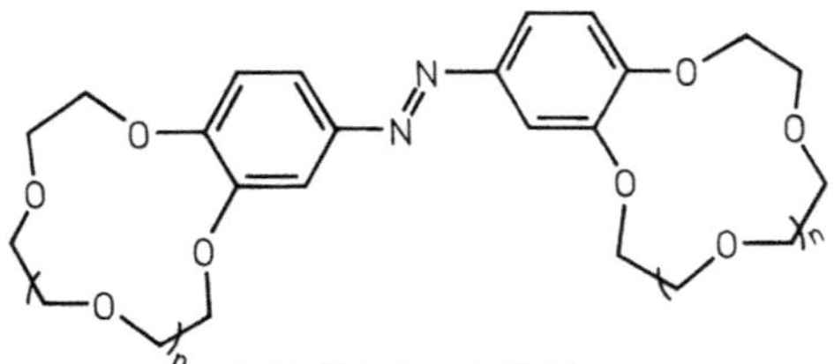

(*E*)-*34* (n = 1, 2, 3)

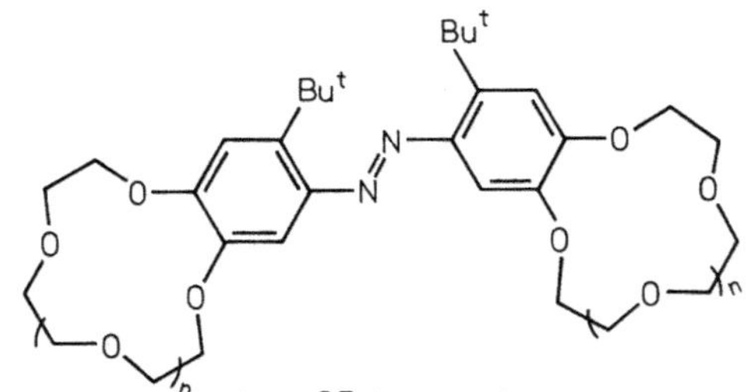

(*E*)-*35* (n = 2, 3)

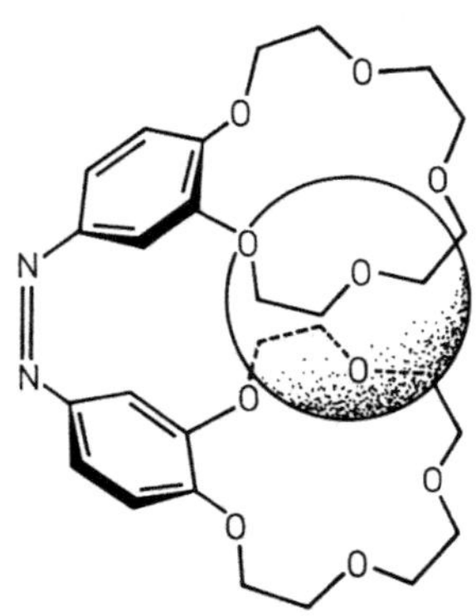

36

The influence of added alkali metal cations on the photostationary state and the isomerization rate were also observed for *35* (n = 2, 3); but when comparing crown series *35* with *34*, the changes of *35* occurred with smaller alkali metal cations: for example, K^+ increases the (Z)% of *35* (n = 2), whereas the (Z)% of *34* (n = 2) was hardly affected [46]. The result suggests that the inner cavity between two crown ether rings of (*Z*)-*35* might be somewhat smaller than that of (*Z*)-*34*. The difference between crown series *34* and *35* is ascribed to the buttressing effect of the *tert*-butyl groups. (*Z*)-Azobenzenes are sterically crowded; the introduction of *tert*-butyl groups into the ortho positions of the azo linkage would further increase the steric crowding of *Z-35*.

The influence of the photoirradiation on the ion-binding ability of *34* and *35* was quite remarkable: for example, (*E*)-*34* (n = 2) extracted Na^+ 5.6 times more efficiently than (*Z*)-*34* (n = 2), whereas (*Z*)-*34* (n = 2) extracted K^+ 42.5 times more efficiently than (*E*)-*34* (n = 2) [45]. The selectivity is expressed by the ratio ((*E*)/(*Z*) of extractability for Na^+ against that for K^+ and was found 238-fold. Conceivably, Na^+ was extracted as a 1:1 cation/crown complex, while K^+, Rb^+ and Cs^+ (the ion radii of which are somewhat greater than the size of B15C5) were extracted as 1:2 cation/crown sandwich-type complexes as *36*. A systematic investigation of the crown series *34* and *35* [45,46] established that the extractibility of the (*Z*)-forms follows the same order as the inhibitory effect on the thermal isomerization rate and the (*Z*) percentage at the photostationary state. Conceivably, both orders are associated with the stability of the sandwich-type complex with the *Z*-isomers.

Shinkai et al. [44] also found that (*Z*)-*32* has a higher extractability for K^+ and Rb^+ than (*E*)-*32*. This was also rationalized by the formation of a sandwich-type complex with (*Z*)-*32*. Being different from the crown series *34* and *35*, however, neither the photostationary state nor the thermal isomerization of *32* was affected by added alkali metal cations. The contrasting behavior may be related to the difference in the "rotational freedom" of the crown ether groups.

Compounds *34* and *35* employ benzocrown ethers as ion-binding groups, so that the geometrical change induced by the isomerization of the N=N double bond is readily trasmitted to the crown ethers. On the other hand, *32* has two free-rotating single bonds between the crown ether and the azobenzene, which may act as an absorber of the geometrical change occuring at the N=N bond. As a result, the isomerization of the N=N bond may start without disruption of the interaction between the cation and crown ethers. In contrast, the extractability of *33* was scarcely changed by photoirradiation, indicating that the two crown ethers act quite independently [44]. Of course, neither the photostationary state nor the thermal isomierzation was affected at all by added alkali metal cations [44]. Probably, these effects are associated not only with the rotational freedom but also with the molecular symmetry. As shown in *37*, (*Z*)-*33* has an unsymmetrical cavity between disordered crown caps and thus becomes disadvantageous to catch "round" metal cations.

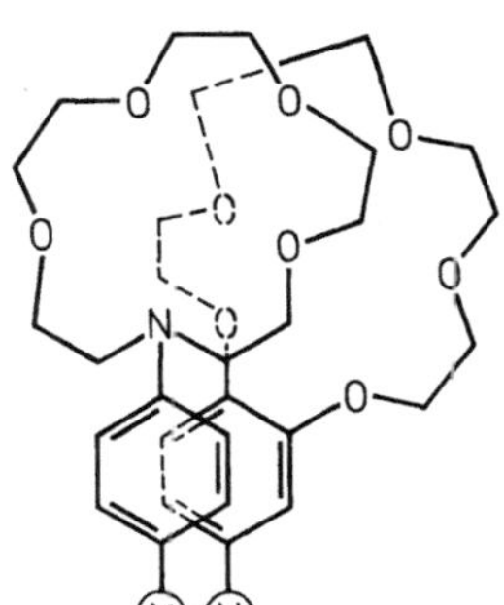

37

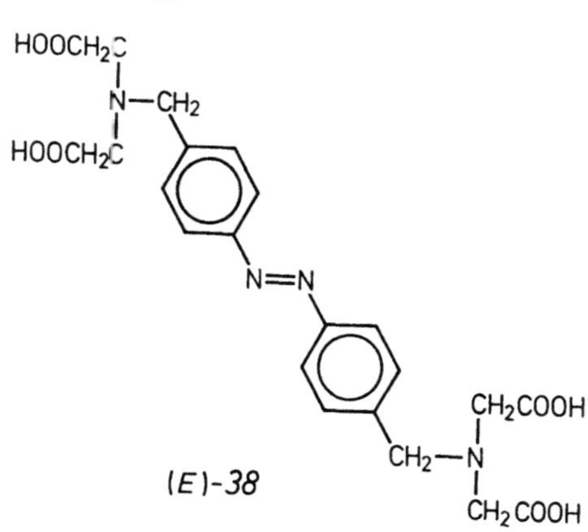

(*E*)-*38*

Blank et al. [47] reported on an EDTA-like photoresponsive ligand *38*. The planar (*E*)-*38* could not bind Zn^{2+}. On exposure to light of 320 nm, (*E*)-*38* was converted to a nonplanar chelating agent (*Z*)-*38*, which, because of two cooperative iminodiacetic acid groups, bound Zn^{2+} with a binding constant estimated to be 1.1×10^5 M^{-1}. The interconversion between (*E*)- and (*Z*)-*38* is reversible, suggesting possible application to photo-driven ion pumps.

2.6 Anion-Capped Crown Ethers

In a biochemical field, the more often employed ionophores for alkaline earth metal cations are the polyether antibiotics such as monensin and nigericin representing monobasic, carrier-type ionophores. They feature coupled-transport of cations and protons [48]. It is known that coupled-transport can be imitated in an artificial system using anion-capped crown ethers such as *39*, *40*, and *41*; their ion-binding ability is associated with the correctly placed anionic group on the crown ether ring [49, 50]. These results imply that an ionophore having a crown ring and an anionic cap in a suitable geometrical position acts as an efficient receptor for alkaline earth metal cations.

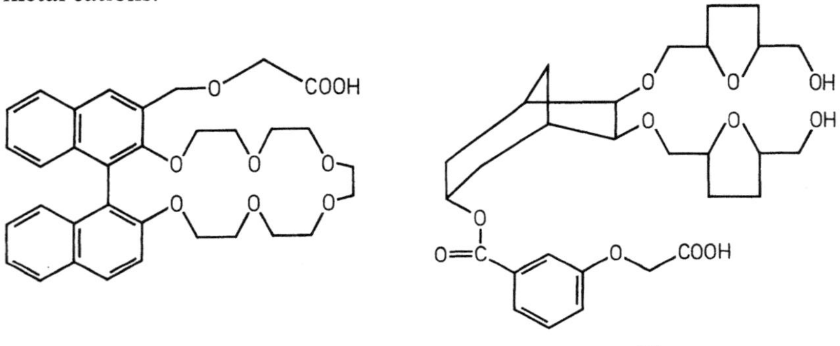

39 *40*

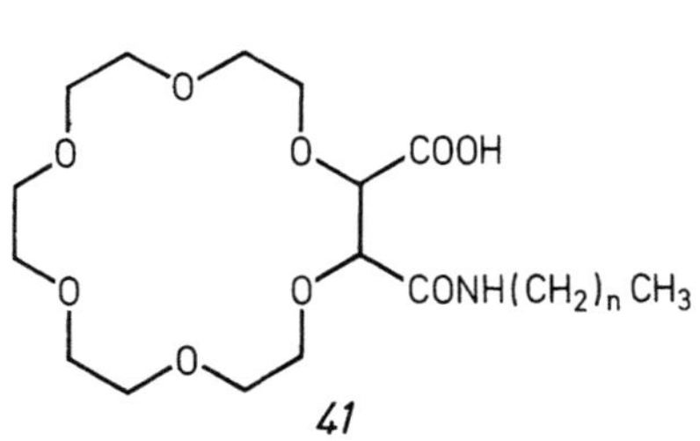

41

The above concept is directly applicable to the design of photoresponsive anion-capped crown ethers. If one synthesizes a crown ether putting on and off an anion-cap in response to photoirradiation, it would lead to the photocontrol of the binding of alkaline earth metal cations. Structure *42* was the first synthetic attempt to design such a molecule [51]. It was found that photo-isomerized (*Z*)-*42* binds K^+ more effectively than (*E*)-*42*. This was attributed to the effect of the carboxylate group capped on the crown ether ring. However, the extractabity was generally low owing to the hydrophilic nature of *42*.

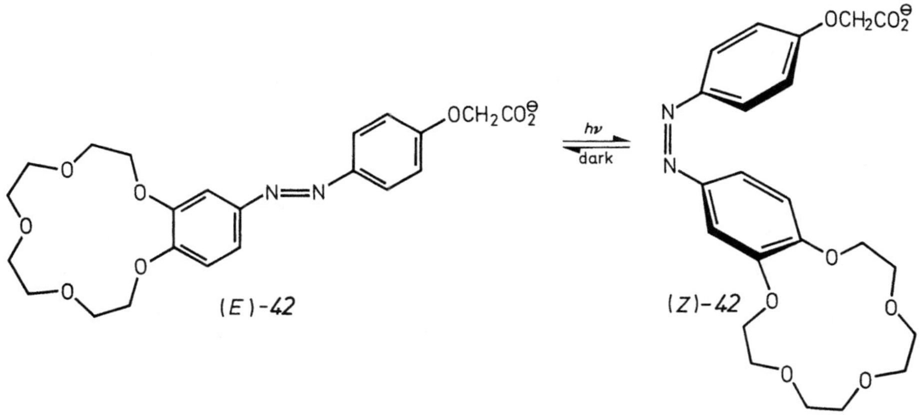

Shinkai et al.[52] subsequently synthesized *43* (*m*-OH and *p*-OH) and *44*. In order to enhance the hydrophobic nature, the carboxylate group was substituted with the phenolate group and the *n*-butyl group was newly introduced. The 5-nitro group in *44* was introduced to lower the pK_a of the phenol group so that (*Z*)-*44* may

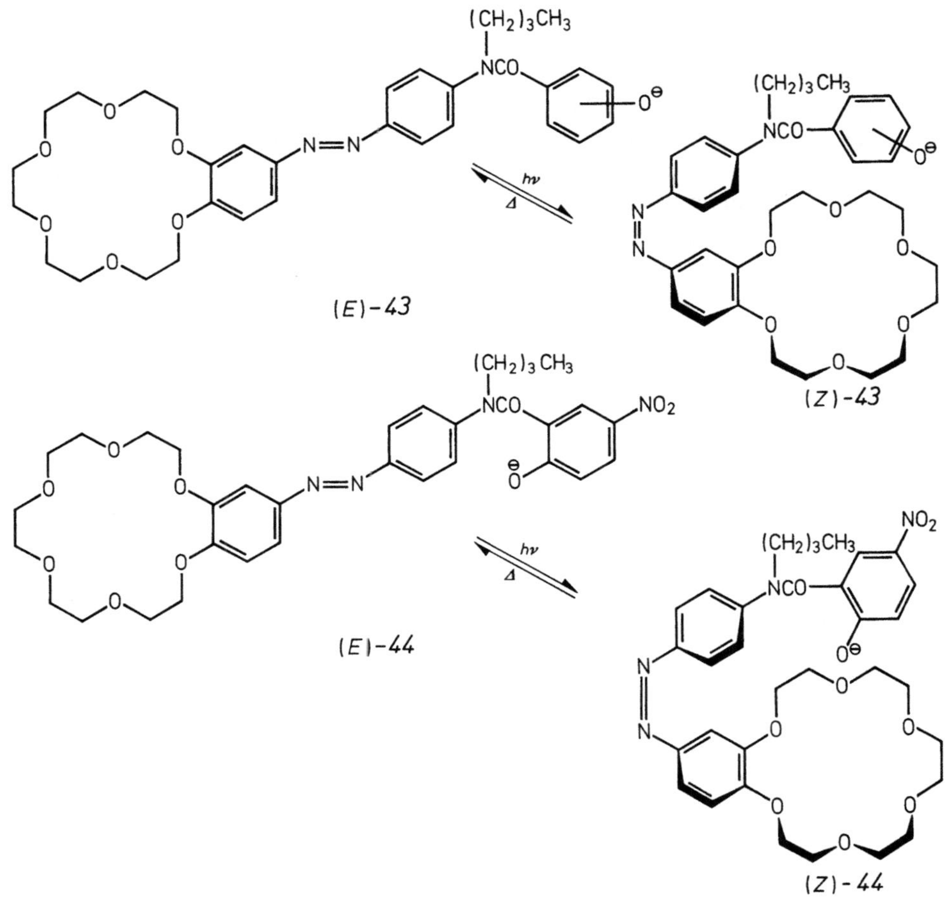

extract Ca^{2+} ion at neutral pH. Examinations of CPK models revealed that the formation of the complex sandwiched between the crown ring and the phenoxide cap is possible in case of (*Z*)-*43* (*m*-OH) and (*Z*)-*44* but not with (*Z*)-*43* (*p*-OH), because in (*Z*)-*43* (*m*-OH) and (*Z*)-*44* the phenoxide exactly sits on the top of the crown ring, while that of (*Z*)-*43* (*p*-OH) is located parallel to the crown ether plane.

The extraction with (*Z*)-*43* gave the maximum selectivity for Rb^+ and (*Z*)-*44* efficiently extracted K^+, Rb^+, Cs^+, and Ca^{2+}. The extractability was improved in most cases by UV-light irradiation, indicating that the photoisomerized anionic group contributes to the ion-binding. Unlike *34* and *35*, the sharp ion-selectivity was not observed in anion-capped systems.

2.7 Crown Ethers with Other Responsive Functions

The preceding chapters surveyed photoresponsive crown ethers "excited" by a light trigger. In that class of the crown ethers, the light energy is transferred in the order, light→chromophore→covalent bond→crown ether. One may consider other trigger energies or other transfer routes which may collaborate in "responsive crown ethers".

The oldest example is a class of *pH-responsive crown ethers* which change their ion-binding ability in response to medium pH. Anion-capped crown ethers such as *39–41* belong to this group [49, 50], and *42–44* are pH-responsive as well as photoresponsive [52]. As an attempt to imitate the functions of the polyether antibiotics such as monensin and nigericin, a number of acyclic polyethers bearing both a hydroxyl and a carboxylic group at the chain ends have been synthesized. Typicals examples are *45–48* [53–57]. Similarly, a polyether *49* with a quinolyl group instead of a hydroxyl group acts as a pH-responsive receptor for Na^+ [58].

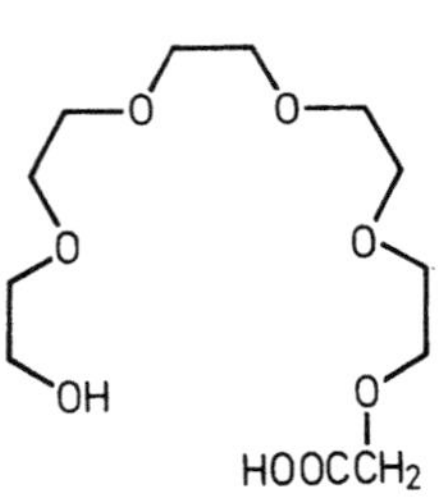

45

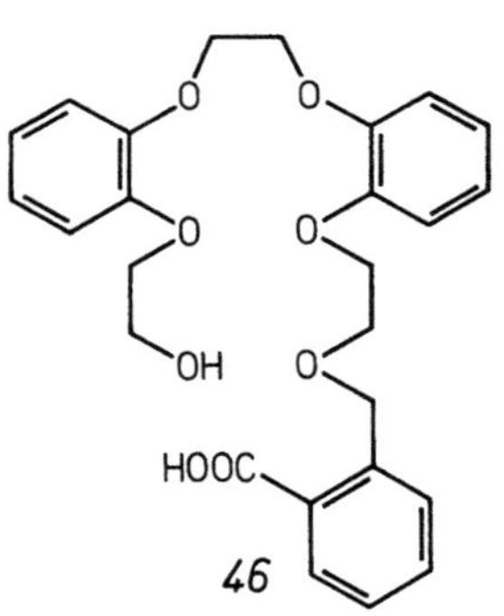

46

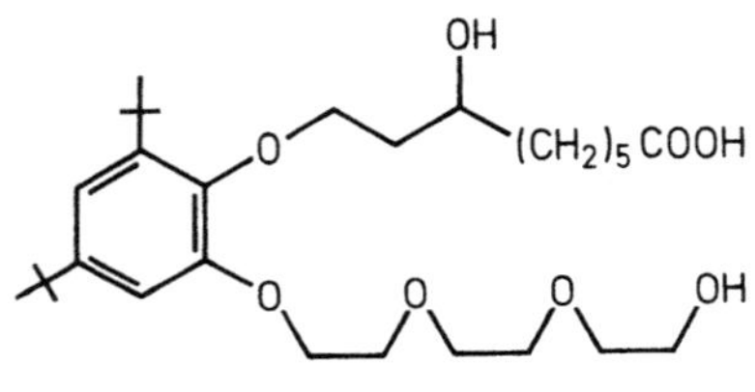

47

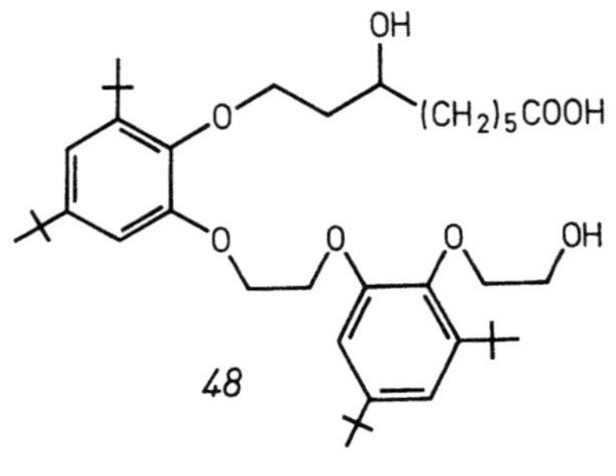

48

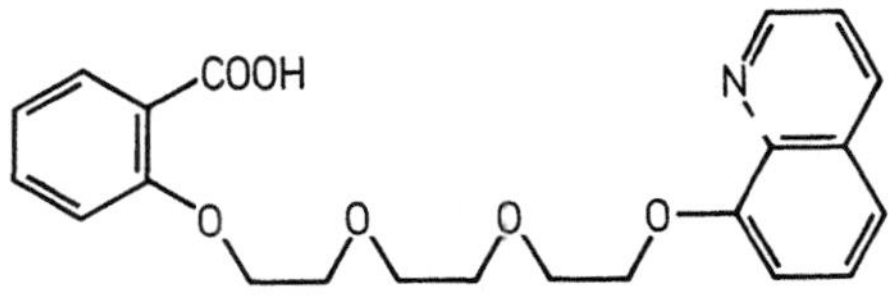

49

In *50* two binding sites are present: the crown ether to bind alkali metal and ammonium ions and the 2,2'-bipyridyl function to bind other metal nuclei [59]. These sites, though separated, are not expected to behave independently because chelation of metals at the bipyridyl function forces the aromatic nuclei toward coplanarity; this leads to a restricted conformational freedom of the crown ring. Thus, the ion-binding ability of the crown site of *50* is "controlled" by the complexation of the 2,2'-bipyridyl site [59]. This relation reminds us of the allosteric effects which frequently appear in enzyme chemistry.

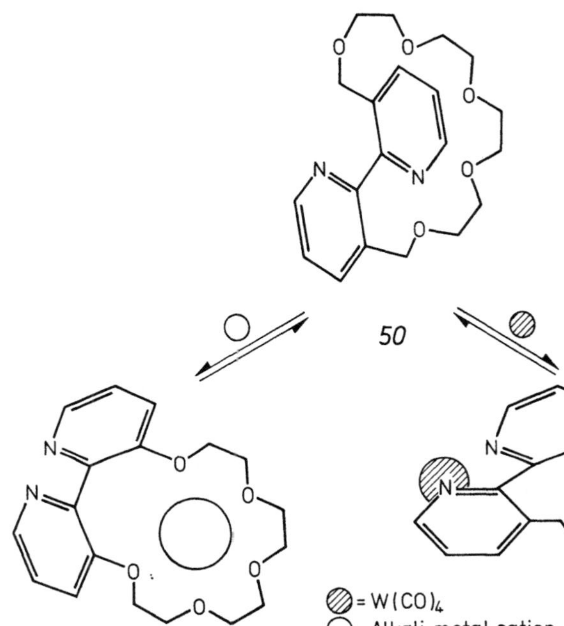

It was found from NMR measurement, association constants, and ion-transport that *50* itself binds K^+ in preference to Na^+, whereas Na^+ is bound more efficiently in the presence of $W(CO)_4$. The rapid racemization of the structure forces the benzylic oxygens into conformations in which only one can participate in the formation of a crown-like cavity.

Other iteresting examples are *redox-responsive crown ethers*. The redox functions in the enzyme-chemical field are frequently mediated by prosthetic groups. Since most prosthetic groups are capable of catalyzing, although weakly, the enzyme-mediated reactions even in the absence of apoenzymes, they have attracted much attention to bioorganic chemists. In contrast to holoenzymes having both the catalytic site and the recognition site, however, they consist only of the catalytic site. One may thus expect that the prosthetic molecules bearing the recognition site within a molecule behave as a more attractive enzyme model.

The ability of crown ethers to associate with charged and uncharged substrates resembles to early reaction steps in enzyme-mediated reactions. Therefore, crown ethers are useful as a potential recognition site for this purpose. Compounds *51–54* represent such a type of molecular assembly containing within the same molecule both the prosthetic group and the crown ether ring. Among them *51* and *52* may act as crown ether NADH mimics [60–62]. E.g. substrates (sulfonium salts or pyridinium salts containing an extra ammonium group, etc.) in the initial step complex with the crown ether moiety and are then reduced by the dihydropyridine function through a pseudo-intramolecular process.

Crown Compound *53* belogs to the flavins (vitamin B_2 family). The redox properties of *53* are sensitively affected by complexed alkali metal cations [63]. In *54* the B15C5 moieties are appended to a porphyrin skeleton at the methine posi-

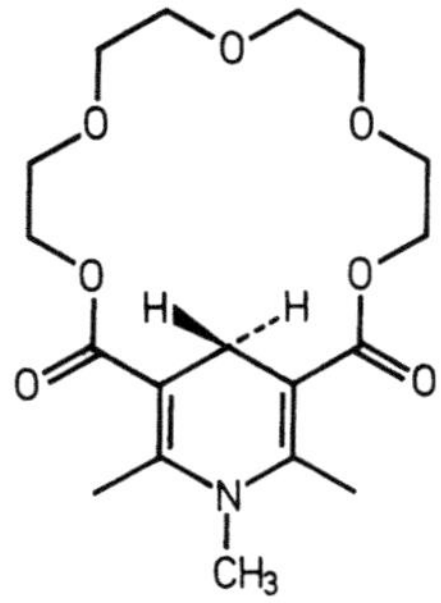

51

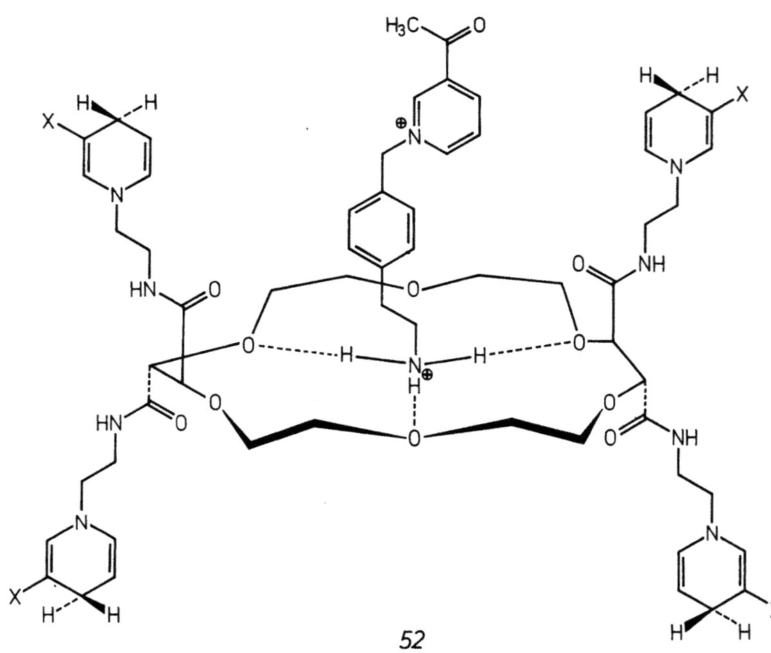

52

tions[64]. It was found that the cations (K^+, Ba^{2+}, and NH_4^+) capable of forming 1:2 cation/crown complexes efficiently quench the fluorescence of the porphyrin. These examples show the crown ether moieries to play a role of a recognition site of these prostetic molecules. It may be said paradoxically, however, that the ion-binding ability of the crown ether moieties is "controlled" by the redox state of the prosthetic molecules. However, there exist few examples for the application of prosthetic molecules to control the crown functions in which the crown ether moieties play a primary role.

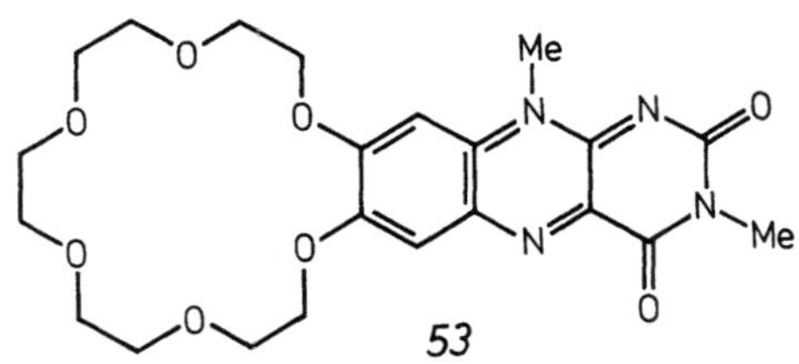

53

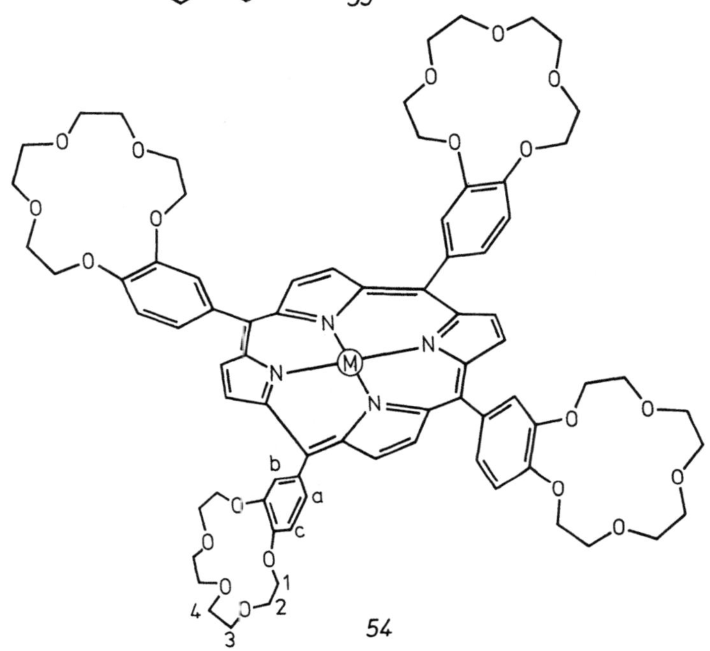

54

Crown ethers bearing thiol functions have been synthesized[65-67]. The object of these investigations is again to imitate the function of thiol-dependent enzymes and to utilize the crown ethers as a recognition site. Recently, Shinkai et al.[68,69] attempted to control the ion-binding ability by the redox function of thiol groups.

In *55*, the interconversion between monocrown and biscrown was thus achieved by redox reaction, and the oxidized form with a methylene spacer ($X=CH_2$) formed a 1:2 cation/crown sandwich-type complex with K^+ and Rb^+. On the other hand, the interconversion between lariat crown and cryptand may be effected in *56* by redox reaction of the thiol caps.

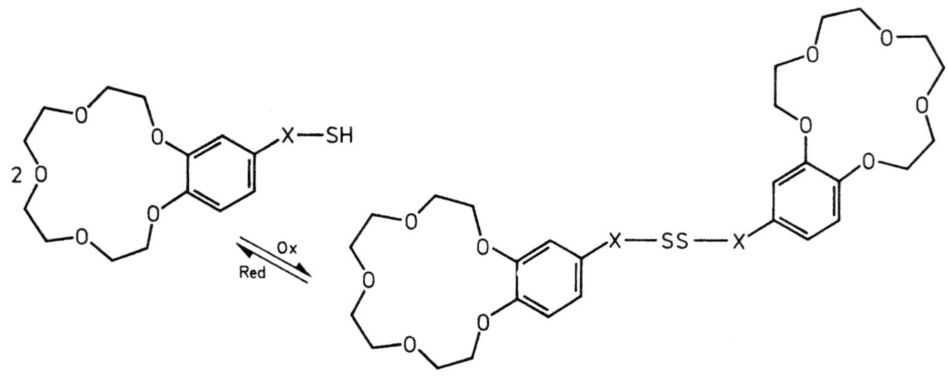

55

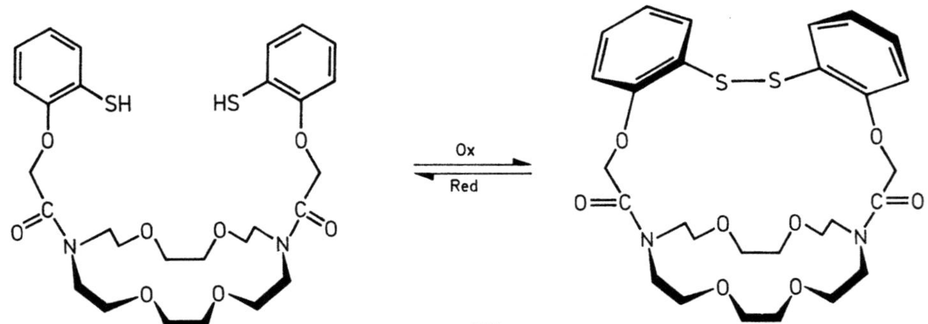

56

Recently, a temperature-responsive crown ether has been exploited by Warshawsky and Kahana [70]. ΔH° and ΔS° values for complexation between cation and crown ether are usually negative and small. Consequently, the sign and value of the free energy, ΔG°, may depend on the absolute temperatures. In homogeneous systems

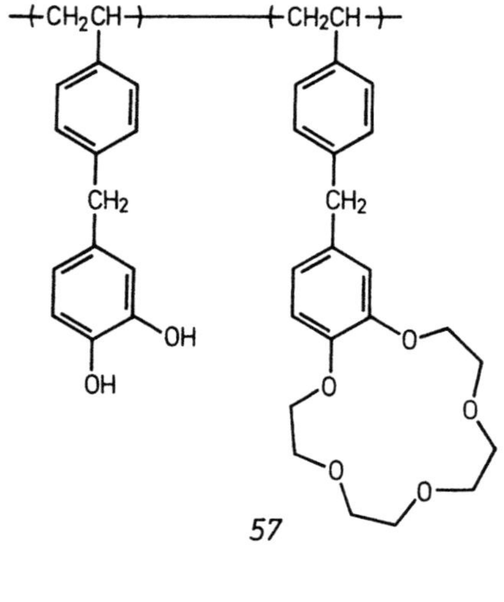

57

this has little significance. In heterogeneous systems such as equilibria between insoluble polymer and solution, however, it is reasonable to assume that the opposing effects of ΔH° and ΔS° could be exploited to induce temperature-regulated release of salts from their insoluble polymeric crown complexes.

A polymeric crown unit *57* in crosslinked polystyrene beads [70] in a column could be saturated with KCl; a sudden heating to 40 °C caused a spontaneous elution by "thermal shock", and a 3-fold increase in the eluant of the original ion concentration. The authors plan the application to control the temperature of phase transfer catalysis and to thermoregulate polymeric delivery systems of Na^+/K^+. The latter, however, is more easily exploited by responsive crown ethers.

Also, the functionalization of a "rope-skipping" crown ether *58* [71] and a "breathing" crown ether *59* [72] may lead to an application of novel responsive crown ethers.

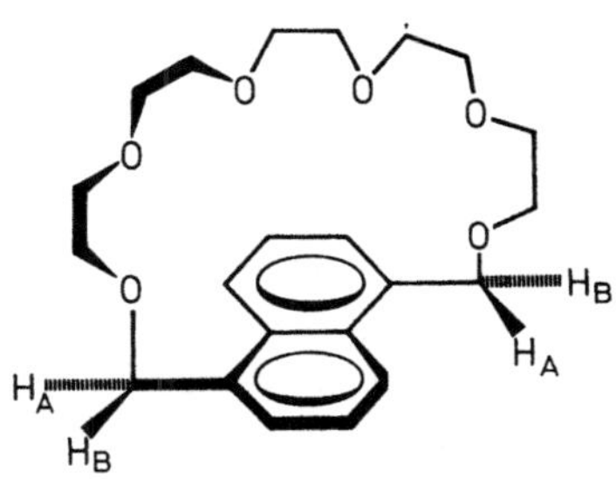

58

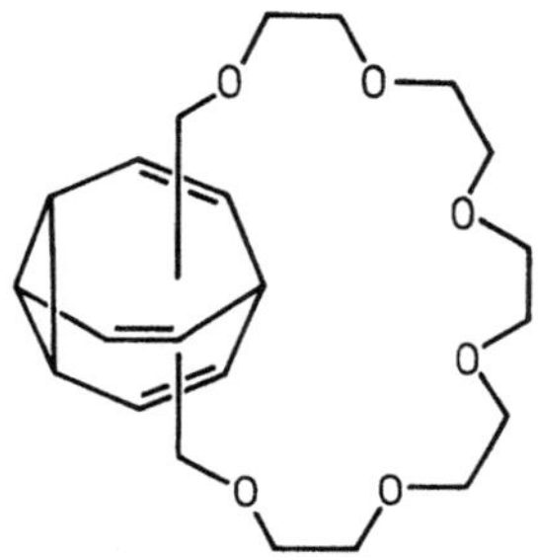

59

3 Light-Driven Ion Transport

3.1 Ion Transport Mediated by Photoresponsive Crown Ethers

Cations are known to be transported through membranes by synthetic macrocyclic polyethers as well as by antibiotics. In particular, some polyether antibiotics feature the interconversion between the cyclic and acyclic form in the membrane phase, a feature which is believed to lead to the high functionality of these antibiotics as ion carriers [48,49,73,74]. For example, the reduction of the ion-affinity through the interconversion to the acyclic form is responsible for the rapid release of ions from the ion-carrier complex in the membrane phase to the second (OUT) aqueous phase. Kobuke et al. [75] and Kirch and Lehn [76] independently established by examination of synthetic macrocyclic polyethers that the best carrier for the ion transport is a ligand giving a moderately stable rather than a very stable complex. This general rule was further corroborated by Lamb et al. [77]. The rule clearly indicates a dilemma occurring in the artificial transport system: the very stable complex, which may rapidly extract ions into the membrane phase, cannot release the ion efficiently from the complex. As a result, the plots of K (association constant) vs. transport rate provide a maximum at a moderate K value. In other words, the rule implies that the ion transport with simple synthetic ion carriers cannot exceed a maximal.

ceiling velocity. Thus, the method of the reversible interconversion of the ionophore structure by which the natural ionophores skillfully break the dilemma becomes very attractive. The idea thus arises that, provided that the binding ability of the ion carrier can be changed by light, it would lead to acceleration or control of the ion-transport rate by an on-off light switch.

Shinkai et al. [44] found in the transport of alkali (K^+, Rb^+, and Cs^+) picrates with *32* (X=CO or CH_2) across a liquid (*o*-dichlorobenzene) membrane in a U-tube that the rate is significantly accelerated by UV-irradiation which isomerizes the (*E*)-forms to the (*Z*)-forms. When UV and visible irradiation were conducted alternately, the interconversion between (*Z*)- and (*E*)-forms occurred rapidly. They independently confirmed that the (*Z*)-forms of *32* (X=CO or CH_2) are rapidly isomerized to the (*E*)-forms by visible light [44]. However, the rates of the ion transport were smaller than those under UV irradiation. These results imply that the ion-release to the second (OUT) aqueous phase is not involved in the rate-limiting step and the light-driven ion transport is solely attributed to the enhanced binding ability of the (*Z*)-forms which extracts cations from the first (IN) aqueous phase to the liquid membrane phase. In other words, the rate enhancement observed for *32* (X=CO or CH_2) is simply correlated to the enhancement of the (*Z*) concentration in the liquid membrane phase, whereas the reversible interconversion between the (*Z*)- and the (*E*)-forms has little significance. As described above, the complexes with (*Z*)-*32* (X=CO or CH_2) are not stable enough to render ion-release to the rate-limiting step [44]. One may say, therefore, that in order to utilize visible light as well as UV light as a source of the ion transport accelerator a bis(crown ether) has to be employed, the (*Z*)-form of which forms a stable complex with metal cations.

The (*Z*)-forms of *34* form relatively stable 1:2 cation/crown sandwich-type complexes with large alkali metal cations [45,46]. At the K^+ transport with *34* (n = 2), Shinkai et al. [45,78] found that the rate with picrate as counter-anion is rather retarded by UV irradiation, whereas that with less hydrophobic 2-nitro-diphenylamine-4-sulfonate as counter-anion is significantly increased. The result suggests, as shown in Fig. 4, that when a less hydrophobic counter-anion is used, the ion extraction from the IN aqueous phase to the membrane phase is rate-limiting and UV irradiation facilitates the ion extraction to enhance the overall transport velocity. On the other hand, when a fully hydrophobic counter-anion (e.g., picrate) is used, the ion extraction is no longer the rate-limiting step and the relatively slow K^+-release to the OUT aqueous phase from the complex with (*Z*)-*34* (n = 2) becomes rate-limiting. In other words, the thermal (*Z*)-to-(*E*) isomerization is not fast enough to eliminate ion-release from the rate-limiting step. As a result, the rate of the ion transport is retarded under UV irradiation.

One may expect that if the conversion of (*Z*)-*34* (n = 2) to (*E*)-*34* (n = 2) is efficiently accelerated by visible light, the ion-release step would no longer be rate-limiting and the overall rate of the K^+ transport would be significantly enhanced even when a hydrophobic counter-anion is used. It was found that the (*Z*)-to-(*E*) isomerization of *34* is effectively accelerated by visible light, but the rates are, like the thermal isomerization [45,46], suppressed by added alkali metal ions; for example, the inhibitory effect on the photoisomerization of (*Z*)-*34* (n = 2) is in the order of $Rb^+ > K^+ > Na^+$ [79]. As expected, alternate irradication of the membrane by UV and visible light significantly accelerated the rate of K^+ and Rb^+ transport with

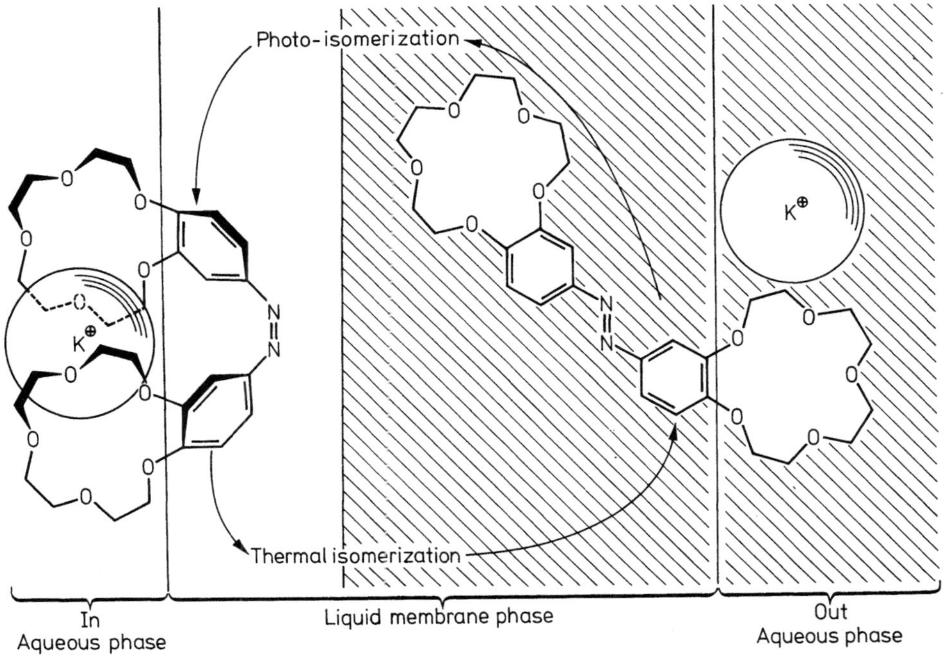

Fig. 4. Schematic representation of light-driven ion transport with *34* (n = 2)

picrate as counteranion, that of K^+ transport being accelerated more efficiently than that of Rb^+ transport [79]. The transport phenomenon may be illustrated schematically as Fig. 5.

In contrast, the rate of Na^+ transport was accelerated by UV irradiation and was retarded by alternate irradiation. The trend is similar to the transport with *32* (X=CO or CH_2) in which the ion extraction with the (*Z*)-forms is exclusively involved in the rate-limiting step.

As described above, the best carrier for ion transport is a ligand that gives a moderately stable rather than a very stable complex. It is of interest that the light-driven ion-transport system provided a similar conclusion. Thus, for less-stable complexes such as Na^+[(*Z*)-*34* (n = 2)], the rate is proportional to the concentration of the (*Z*)-formed carrier, irradiation by visible light which efficiently mediates (*Z*)-to-(*E*) isomerization retarding the transport rate. In contrast, for very stable complexes such as Rb^+[(*Z*)-*34* (n = 2)], the extraction is markedly facilitated by UV light which mediates (*E*)-to-(*Z*) isomerization, but the stable metal-(*Z*)-form complexes cannot be isomerized rapidly by visible light. Hence, the most efficient light-driven ion-transport is realized in moderately stable complexes such as K^+[(*Z*)-*34* (n = 2)].

Niwa et al. [80] prepared a polymer membrane of poly(vinyl chloride) containing *34* (n = 2) and a liquid crystal compound *60* and examined the photoresponsive K^+ transport across this membrane. Since *60* has the strong absorption band at UV region, the (*E*)-to-(*Z*) photoisomerization of *34* (n = 2) takes place at the surface of the polymer membrane. On the other hand, the membrane is transparent in the visible region, so that the visible-light-mediated (*Z*)-to-(*E*) isomerization would be possible

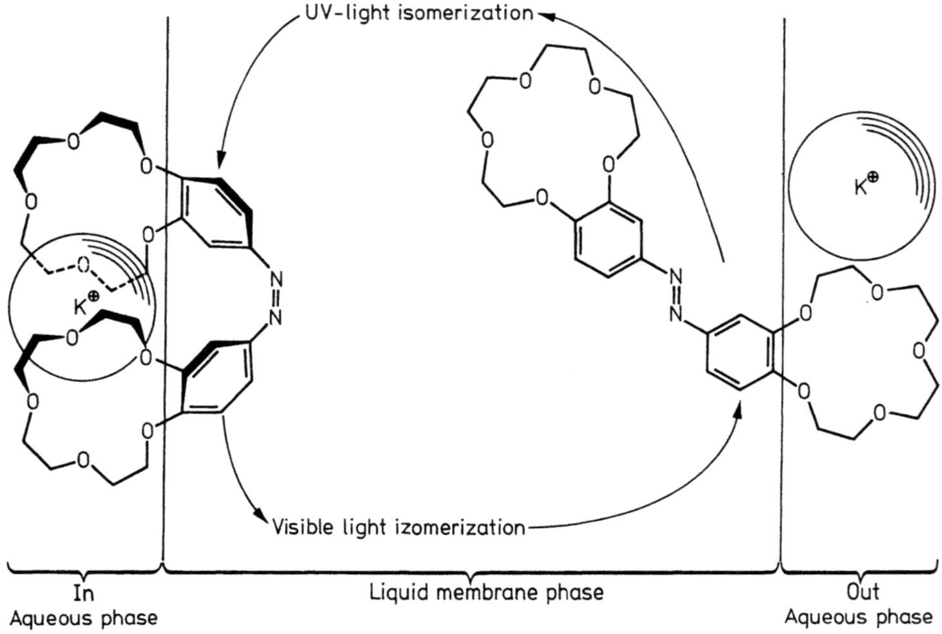

Fig. 5. Schematic reprsentation of ion transport accelerated by alternate irradiation of UV and visible light

anywhere in the membrane phase. In the dark, KCl slowly permeated through the membrane from the KCl-containing IN aqueous phase to the OUT aqueous phase. When UV light and visible light were irradiated alternately from the side of the IN aqueous phase and from the side of the OUT aqueous phase, respectively, the KCl flux rapidly increased (about two fold). When a hydrophobic salt (potassium *p*-toluene sulfonate) was used, the permeation in the dark was almost negligible. The permeation of this salt through membrane was readily induced by alternate irradiation of two different lights. When the lights were switched off, the flux again became negligibly small. In a separate study, they confirmed spectrophotometrically that the photoinduced (*E*)-(*Z*) intercoversion occurred in the polymer membrane. The result shows that the permeation through the polymer membrane is photocontrolled using *34* as a photofunctional valve.

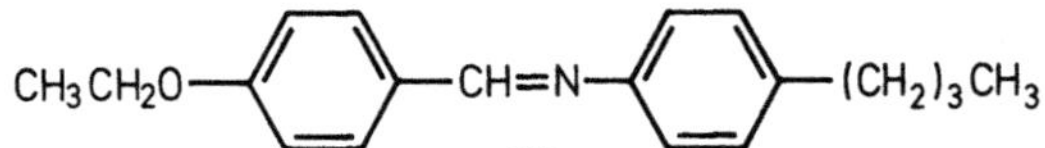

60

The crown ether *44* with its photoresponsive anionic cap carries alkali metal cations without counter-anion and alkaline earth metal cations with a counter-anion. Furthermore, the counter-flow of protons plays an important role. In a U-tube transport apparatus containing a liquid (90 vol% o-dichlorobenzene + 10 vol% *n*-butyl alcohol) membrane, the rates of Na^+ and K^+ transport were acclerated by UV irradiation [52]. Similarly, UV irradiation was effective for the Ca^{2+} transport with

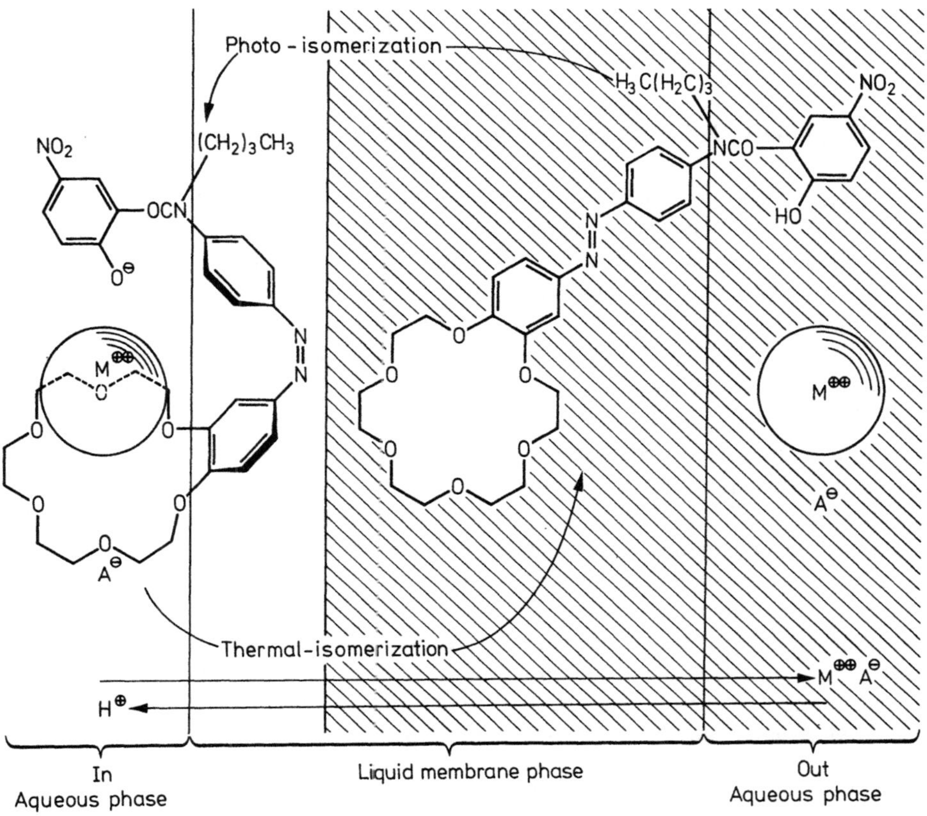

Fig. 6. Schematic representation of membrane transport of alkaline earth metal cations mediated by UV light and pH gradient

picrate as a counter-anion. When the OUT aqueous phase was water, Na^+ was transported by *44* whereas Ca^{2+} was scarcely transported in spite of the increasing Ca^{2+} concentration in the membrane phase. When the OUT aqueous phase was substituted by an 0.01 N HCl aqueous solution, the rates of Na^+ and Ca^{2+} were enhanced by 4.0-fold and 276-fold, respectively. The difference in the magnitude of the rate acceleration suggests that neutralization of the phenoxide group of *44* is important for the release of Ca^{2+} from the Ca^{2+}-(*Z*)-*44* complex. Therefore, the transport of Ca^{2+} is mediated both by UV irradiation and by the counter-flow of protons, the rate increase achieved by these two effects being 760-fold. Summarizing these observations, one may illustrate the Ca^{2+} transport phenomenon as Fig. 6.

(*E*)-*4* is capable of transporting Cu^{2+} across a liquid membrane [8]. When the liquid membrane phase was subjected to UV light irradiation for 1.5 h, the rate of Cu^{2+} transport was suppressed gradually and a new rate of Cu^{2+} transport was obtained. The magnitude of the rate decrease was approximately equal to the concentration of photoisomerized (*Z*)-*4*. When the liquid membrane phase was subjected to visible light irradiation which mediates the (*Z*)-to-(*E*) isomerization, the transport rate was enhanced again. The results clearly demonstrate that the rate of Cu^{2+} transport is reversibly controlled by an on-off light switch and (*Z*)-*4* in the membrane phase scarcely contributes to the transport of Cu^{2+}.

3.2 Photocontrolled Permeability of Liposomal Membranes

In biological systems, there are several pigments which receive light signals to control the physiological phenomena. The typical example is a visual excitation which is initiated upon the light absorption by the 11-(*Z*)-retinal chromophore of rhodopsin (photoresponsive compound) which isomerizes to the all-(*E*) form. This photo-induced configurational change is followed by a conformational change of opsin (protein), Ca^{2+} permeation across membrane, and a conductance change in retinal receptor membrane [81]. Rhodopsin-phospholipid membrane vesicles were studied to localize initial transduction events in a reconstituted system [82–85]. It was found that unexposed rhodopsin-containing phospholipid vesicles are sealed to ion movement while they become permeable after brief exposure to blue-green light. Selected ions (Ca^{2+}, Mn^{2+}, Co^{2+}, Ni^{2+}, and Mg^{2+}) were thus photo-released from the interior of loaded membrane vesicles [84]. The number of ions released/rhodopsin bleached was dependent on the light intensity, and high yields (40–160) fo Ca^{2+}/rhodopsin bleached were observed. The results indicate that rhodopsin spans the phospholipid bilayer membrane, and are consistent with an increase in the permeability of the membrane initiated by light excitation of rhodopsin [84,85].

Light-induced release from phospholipid versicles was achieved in a more simplified system using an azobenzene-containing surfactant *61* [86].

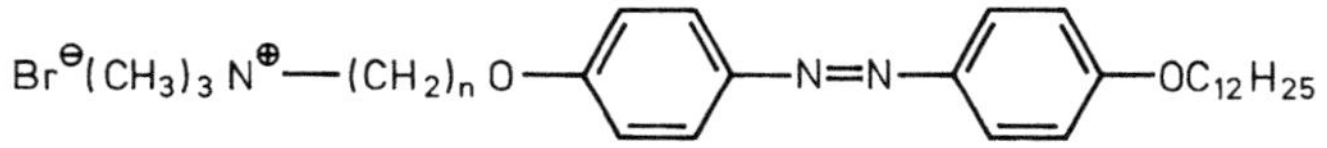

(*E*)-*61* (n = 2,4)

Embedding (*E*)-*61* in phospholipid vesicles provided the enhancement of the liposome shrinkage. This was attributed to the perturbation of the membrane structure to yield small channels for the water permeation. The photoinduced configurational change of (*E*)-*61* to (*Z*)-*61* enhanced the osmotic shrinkage of the embedded liposomes. The result shows that (*Z*)-*61* further enlarges the channel. The photoresponsive change in the channel size was further demonstrated by the data on the release of bromothymol blue (BTB) [86]: the rates of the BTB release from the (*Z*)-*61* loading liposomes were markedly larger than those from the (*E*)-*61* loading liposomes. Conceivable, linear-shaped (*E*)-*61* disturbs the membrane structure to a smaller extent relative to distorted (*Z*)-*61*.

The transport of amino acids across biological membranes is assumed nowadays to be coupled with Na^+ transport [87]. The artificial transport of amino acids has been investigated by serveral groups [88–91], but they mostly employed protected amino acids instead of biologically relevant "free" amino acids.

An efficient transport system of freee amino acids was reported by Behr and Lehn [88]; they demonstrated that a cationic surfactant is useful for the transport from the basic IN aqueous phase to the acidic OUT aqueous phase, while an anionic surfactant is useful for the transport from the acidic IN aqueous phase to the basic OUT aqueous phase. The difficulty of the amino acid transport stems from the poor lipophilicity of their zwitterionic structure.

Sunamoto et al. [92] utilized photoinduced interconversion between spiropyran *62* and merocyanine *63* to mediate the transliposomal amino acid release. Compound *62* showed a normal photochromism in bilayer membranes (i.e., the photocoloration and thermobleachihg cycle). UV irradiation for 20 min followed by visible light irradiation for 5 min on the liposome suspension brought about significant transfer of phenylalanine (Phe) from the interior of the liposomes to the exterior. In the dark under the same conditions no Phe was spontaneously released. Thus, as shown in Fig. 7, *63* is capable of extracting zwitterionic amino acids into the membrane phase by neutralizing the charges each other.

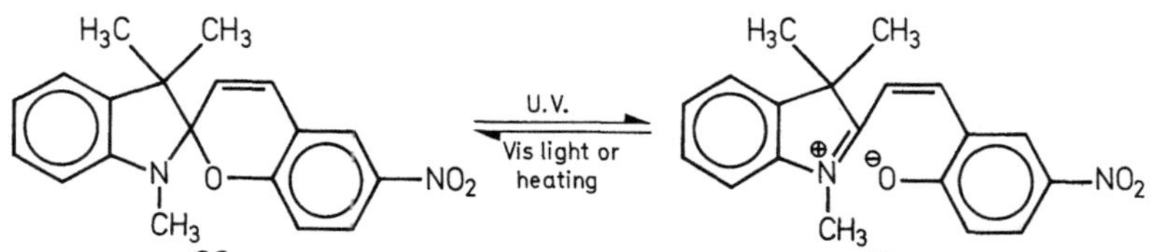

62 *63*

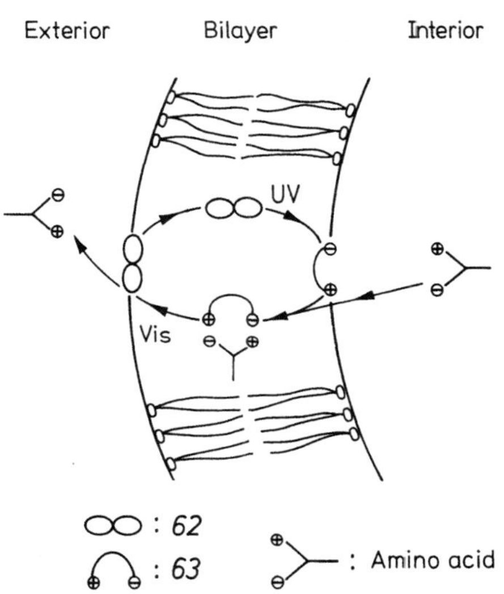

Fig. 7. Transport of amino acids across liposomal membranes with *62*

3.3 Other Ion Transport Systems with Responsive Functions

Here, we wish to survey the ion transport systems switched by energies other than light energy. The most widely investigated energy source is the acid-base neutralization, but this will be omitted from this article as the method has frequently been reviewed. Rebek and Wattley [93] used the remote-controlled crown ether *50* with the bipyridyl unit as a second metal binding site. The ion transport across a liquid (chloroform) membrane indicated that *50* carries K^+ 3.8 times more efficiently than Na^+ while the *50*-$W(CO)_4$ complex carries Na^+ 1.3 times more efficiently than K^+. The remarkable inversion of the K^+/Na^+ transport preference demonstrates that these variables are subject to control within a single carrier by remote binding forces. Therefore, the binding-induced conformational change in the crown ether ring is responsible for the regulation of the ion transport rate.

Another interesting and useful energy source is a redox system. Shinbo et al. [94] designed a liquid membrane system consisting of $K_3Fe(CN)_6$ (aqueous phase)/ N,N,N',N'-tetramethyl-*p*-phenylenediamine (TMPD) (dichloroethane)/ascorbic acid (aqueous phase). As the redox reaction in the interfaces TMPD $\rightleftarrows$ $TMPD^+_{ox}$, picrate anion was carried from the oxidative aqueous phase to the reductive aqueous phase against its concentration gradient. They also demonstrated that in a liquid membrane system consisting of $K_3Fe(CN)_6$ (aqueous phase)/dibutylferrocene (dichloroethane)/ $K_4Fe(CN)_6$ (aqueous system) the redox reactions enforce a coupled counter-flow of protons and anions (perchlorate ions) [94].

The above examples imply that in order to demonstrate coupled cation transport by redox energy, for example, it is necessary to choose a carrier which is negative when reduced and neutral when oxidized. Grimaldi and Lehn [95] satisfied the prerequisite with a nickel bisthiolene carrier *64*. In the presence of dicyclohexano[18] crown-6 (DCH18C6), *64* carried a cation (K^+) and an electron from the reductive ($S_2O_4^{2-}$) to the oxidative ($K_3Fe(CN)_6$) aqeous phase. Similarly, Matsuno et al. [96] succeeded in the coupled transport of Cu^{2+}, an electron, and an anion from the reductive (H_2NOH) to the oxidative (O_2) aqueous phase using bathocuproine *65*.

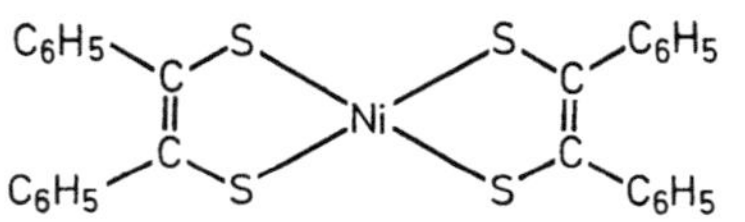

64

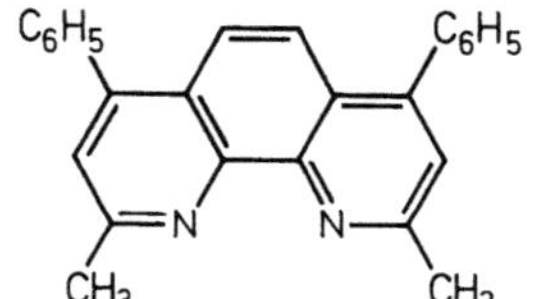

65

Redox-driven ion transport was also achieved by using *55* ($X=CH_2$) [69]. Since the oxidized bis(crown ether) has a large affinity relative to the reduced mono(crown ether), K^+ was rapidly extracted from the oxidative aqueous phase and released to the reductive aqueous phase. It was demontrated that the interconversion between the oxidized and the reduced form of *55* can be mediated by an electrochemical method [69]. Recently, an ion gate membrane in which redox-responsive polymers such as poly-(vinylferrocene), poly(tetrathiafulvalene), and (poly(pyrrole) are coated on a gold minigrid sheet was developed by Burgmayer and Murray [97]. The impedance of the membrane changes in response to the voltage. and the permeability of anion (Cl^-) can be controlled by an electrochemical energy source.

A pH difference may be utilized in an indirect manner for the pH-induced structure change in carrier molecules. Shimidzu et al. [98] synthesized a polymer membrane *66* containing a butyrolactone unit. The membrane exhibited active and selective transport of alkali metal cations [99]. This is due to the pH-dependent cyclic→acyclic interconversion of the lactone ring (Fig. 8). Thus, the active transport was effectively promoted by the pH gradient.

Similar pH-dependent active transports were attained in the polymer membrane sustaining a phenolphthalein analogue [100]. Active transports of anions was also

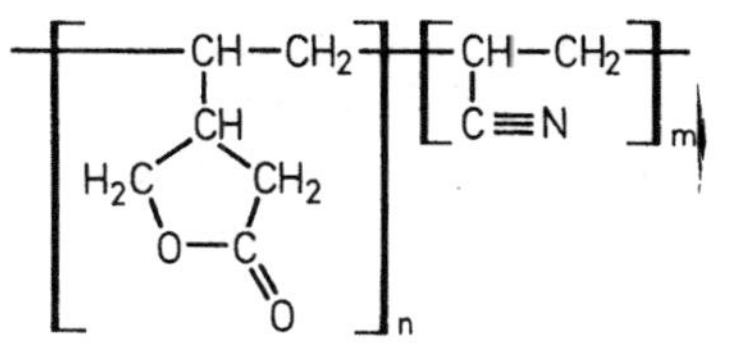

66

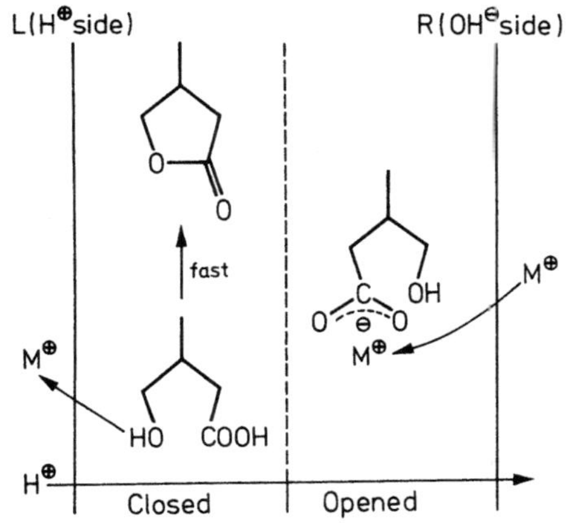

Fig. 8. Active transport achieved in the lactone-containing polymer membrane

demonstrated using the pH-induced structural change. It is known that N-hydroxyethyl amide group *67* exhibits tautomersim and transforms very rapidly into aminoethyl ester *68* in an acid solution. When a polymer having pendant N-hydroxyl groups is synthesized to form a membrane and the membrane is fixed in a cell as a partition film, in which one side solution is acidic and the other side solution is basic, it is expected that an anion may be carried out through the membrane owing to the tautomerism. Thus, the active transport of Cl^- was achieved from the acidic to the basic aqueous solution across the polymer membrane [101].

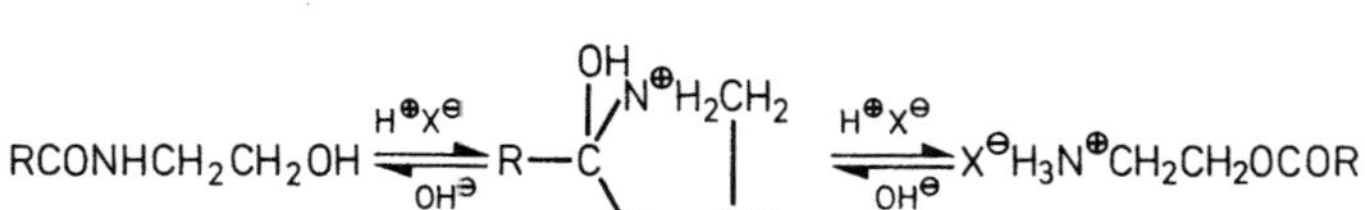

67 *68*

Electron transport systems with switch functions are another fascinating topic in relation to the objects of the present article. It seems impossible, however, to cover this widely extended field here.

4 Conclusions

Since the unexpected discovery of dibenzo[18]crown-6 by Pedersen in 1967, the chemistry of crown ethers has rapidly been established as a new field. In the initial stage of the crown chemistry the syntheses of the crown compounds and the characterizations of their metal complexes were the main objects of the investigation, the functional facet of crown ethers has been left less exploited for a long time. To let the crown ether work — this is the basic idea for the design of functional crown ethers. Natural ionophores actually "work" in biological systems. The basic idea of the photoresponsive crown ethers is originating of the responsive action of polyether antibiotics. The novel phenomena were attained because they combine within the same molecule a photoantenna which acts as a photoresponsive trigger and a functional group (crown ether) which causes a subsequent event. The trigger may be substituted by a component sensitive to pH, redox potential, electron current, spectral pattern, etc. The functional group may be substituted by, for example, cyclodextrin, micelles, membranes, polypeptides, etc. Furthermore, there exist many methods other than the covalent bond to transduce the energy in the antenna to the functional group. We expect that chemistry of responsive systems, including that of photoresponsive crown ethers, will be further advanced as an engineering application to switch-functioned systems.

5 References

1. Bieth, J. et al.: Proc. Natl. Acad. Sci. U.S.A. *66*, 850 (1970)
2. Berezin, I. V. et al.: FEBS Lett. *30*, 329 (1974)
3. Karube, I., Nakamoto, Y., Suzuki, S.: Biochim. Biophys. Acta *445*, 774 (1976)
4. (a) Shinkai, S., Ogawa, T., Nakaji, T., Kusano, Y., Manabe, O.: Tetrahedron Lett. *1979*, 4569, (b) Shinkai, S. et al.: J. Am. Chem. Soc. *102*, 5860 (1980)
5. Ammon, L., Shinkai, S.: to be submitted
6. Fisher, D. P. et al.: J. Am. Chem. Soc. *99*, 2811 (1977)
7. Nakamura, A., Aotake, M., Otsuka, S.: ibid. *96*, 3456 (1974)
8. (a) Shinkai, S., Minami, T., Kouno, T., Kusano, Y., Manabe, O.: Chem. Lett. *1982*, 499, (b) Shinkai, S. et al.: J. Chem. Soc. Perkin 2 *1982*, 2741
9. Shinkai, S., Shigematsu, K., Honda, Y. Manabe, O.: to be submitted
10. Wildes, P. D., Pacifici, J. G., Irick, G., Whitten, D. G.: J. Am. Chem. Soc. *93*, 2004 (1971) and references cited therein
11. Harberfield, P. et al.: J. Am. Chem. Soc. *97*, 5804 (1975)
12. (a) Asano, T.: J. Am. Chem. Soc. *102*, 1205 (1980), (b) Asano, T. et al.: ibid. *104*, 4900 (1982)
13. Nishimura, N. et al.: Bull. Chem. Soc. Jpn. *49*, 1381 (1976)
14. Schanze, K. S., Mattox, T. F., Whitten, D. G.: J. Am. Chem. Soc. *104*, 1733 (1982)
15. Shinkai, S. et al.: Chem. Lett. *1980*, 1303
16. Asano, T. et al.: J. Am. Chem. Soc. *103*, 5161 (1981)
17. Rau, H., Lüddecke, E.: ibid. *104*, 1616 (1982)
18. (a) Ueno, A. et al.: ibid. *101*, 2779 (1979), (b) Ueno, A. et al.: Chem. Lett. *1979*, 841
19. Emert, J., Breslow, R.: J. Am. Chem. Soc. *97*, 670 (1975)
20. Ueno, A. et al.: Chem. Lett. *1979*, 1007
21. Ueno, A. et al.: J. Chem. Soc. Chem. Commun. *1981*, 94
22. Irie, M., Hayashi, K.: J. Macromol. Sci.-Chem. **A13**, 511 (1979)
23. Irie, M. et al.: Macromolecules *14*, 262 (1981)
24. Blair, H. S., et al.: Polymer *21*, 1195 (1980)
25. Irie, M. et al.: J. Polym. Sci. Polym. Lett. Ed. *17*, 29 (1979)
26. Negishi, N. et al.: Kobunshi Ronbun Shu *37*, 293 (1980)
27. Blasius, E., Maurer, P.-G.: Makromol. Chem. *178*, 649 (1977)
28. Bormann, S. et al.: Tetrahedron *31*, 2791 (1975)
29. Smid, J. et al.: J. Am. Chem. Soc. *97*, 5932 (1975)

30. Wong, L., Smid, J.: ibid. *99*, 5637 (1977)
31. Shinkai, S., Kinda, H., Manabe, O.: ibid. *104*, 2933 (1982)
32. Shinkai, S. et al.: unpublished result
33. Shinkai, S., Honda, Y., Kusano, Y., Manabe, O.: J. Chem. Soc. Chem. Commun. *1982*, 848
34. Shinkai, S. et al.: Bull. Chem. Soc. Jpn. *56*, 1700 (1983)
35. Shinkai, S., Ueda, K., Honda, Y., Manabe, O.: to be submitted
36. (a) Desvergne, J.-P., Bouas-Laurent, H.: J. Chem. Soc. Chem. Commun. *1978*, 403,
(b) Bouas-Laurent, H. et al.: Pure & Appl. Chem. *52*, 2633 (1980)
37. Yamashita, I. et al.: Tetrahedron Lett. *1980*, 541
38. Eichner, M., Merz, A.: ibid. *22*, 1315 (1981)
39. Shiga, M., Takagi, M., Ueno, K.: Chem. Lett. *1980*, 1021
40. (a) Shinkai, S., Minami, T., Kusano, Y., Manabe, O.: Tetrahedron Lett. *23*, 2581 (1982),
(b) Shinkai, S. et al.: J. Am. Chem. Soc. *105*, 1851 (1983)
41. Kimura, K. et al.: Chem. Lett. *1979*, 611
42. Maeda, T., Kimura, K., Shono, T.: Bull. Chem. Soc. Japan *55*, 3506 (1982)
43. Mollison, P. R., Turter, M. R.: J. Chem. Soc. Perkin 2 *1982*, 1818
44. Shinkai, S. et al.: J. Chem. Soc. Perkin 1 *1981*, 3279
45. (a) Shinkai, S., Ogawa, T., Kusano, Y., Manabe, O.: Chem. Lett. *1980*, 283,
(b) Shinkai, S. et al.: J. Am. Chem. Soc. *103*, 111 (1981)
46. Shinkai, S. et al.: J. Am. Chem. Soc. *104*, 1960 (1982)
47. Blank, M. et al.: Science *214*, 70 (1981)
48. (a) Choy, E. M. et al.: J. Am. Chem. Soc. *96*, 7085 (1974),
(b) Cussler, E. L.: Am. Inst. Chem. Eng. J. *17*, 1300 (1971)
49. Wierenga, W. et al.: J. Am. Chem. Soc. *101*, 1334 (1979)
50. Frederick, L. A. et al.: J. Chem. Soc. Chem. Commun. *1980*, 1211
51. Shinkai, S. et al.: Tetrahedron Lett. *21*, 4463 (1980)
52. Shinkai, S., Minami, T., Kusano, Y., Manabe, O.: J. Am. Chem. Soc. *104*, 1967 (1982)
53. Yamazaki, N. et al.: Tetrahedron Lett. *1978*, 2429
54. Kubokawa, H. et al.: Chem. Lett. *1982*, 1937
55. Yamazaki, N. et al.: J. Macromol. Sci.-Chem. **A13**, 321 (1079)
56. Gardner, J. O., Beard, C. C.: J. Med. Chem. *21*, 357 (1978)
57. Hiratani, K. et al.: Chem. Lett. *1980*, 477
58. Hiratani, K.: ibid. *1981*, 21
59. Rebek, J., Jr. et al.: J. Am. Chem. Soc. *101*, 4333 (1979)
60. Van Bergen, T. J., Kellogg, R. M.: ibid. *99*, 3882 (1977)
61. De Vries, J. G., Kellogg, R. M.: ibid. *101*, 2759 (1979)
62. Behr, J.-P., Lehn, J.-M.: J. Chem. Commun. *1978*, 143
63. Shinkai, S. et al.: Tetrahedron Lett. *24*, 1539 (1983)
64. Thanabal, V., Krishnan, V.: J. Am. Chem. Soc. *104*, 3643 (1982)
65. Chao, Y., Cram, D. J.: ibid. *98*, 1015 (1976)
66. Matsui, T., Koga, K.: Tetrahedron Lett. *1978*, 1115
67. Lehn, J.-M., Sirlin, C.: J. Chem. Soc. Chem. Commun *1978*, 949
68. Minami, T., Shinkai, S., Manabe, O.: Tetrahedron Lett. *1982*, 5167
69. Shinkai, S. et al.: Chem. Lett. *1983*, 747 and unpublished results
70. Warshawsky, A., Kahana, N.: J. Am. Chem. Soc. *104*, 2663 (1982)
71. Brown, H. S. et al.: J. Org. Chem. *45*, 1682 (1980)
72. Schroder, G., Witt, W.: Angew. Chem. Int. Ed. Engl. *18*, 311 (1979)
73. Pressman, B. C.: Ann. Rev. Biochem. *45*, 501 (1976)
74. Westly, J. W.: Ann. Rep. Med. Chem. *10*, 246 (1975)
75. Kobuke, Y. et al.: J. Am. Chem. Soc. *98*, 7414 (1976)
76. Kirch, M., Lehn, J.-M.: Angew. Chem. Int. Ed. Engl. *14*, 555 (1975)
77. Lamb, J. D. et al.: J. Am. Chem. Soc. *102*, 6820 (1980)
78. Shinkai, S. et al.: J. Chem. Soc. Chem. Commun. *1980*, 375
79. Shinkai, S., Shigematsu, K., Sato, M., Manabe, O.: J. Chem. Soc. Perkin 1 *1982*, 2735
80. Kumano, A. et al.: Chem. Lett. *1983*, 1327
81. Ebrey, T. G., Honig, B.: Quat. Rev. Biophys. *8*, 129 (1975)
82. Montal, M.: Biochim. Biophys. Acta *559*, 231 (1979)

83. Kaupp, U. B. et al.: ibid. *552*, 390 (1979)
84. O'Brien, D. F.: Photochem. Photobiol. *29*, 679 (1979)
85. Tyminski, P. N. et al.: Biochemistry *21*, 1197 (1982)
86. (a) Kano, K. et al.: Chem. Lett. *1980*, 421,
(b) Kano, K. et al.: Photochem. Photobiol. *34*, 322 (1981)
87. Ring, K.: Angew. Chem. Int. Ed. Engl.*9*,345(1970)
88. Behr,J.-P.,Lehn,J.-M.:J.Am.Chem.Soc.*95*,6108(1973)
89. Newcomb,M.etal.:ibid.*101*,4941(1979)
90. Maruyama, K. et al.: Tetrahedron Lett. *22*, 2001 (1981)
91. Tsukube, H.: ibid. *22*, 3981 (1981)
92. Sunamoto, J. et al.: J. Am. Chem. Soc. *104*, 5502 (1982)
93. Rebek, J., Jr., Wattley, R.: ibid. *102*, 4853 (1980)
94. (a) Shinbo, T. et al.: Nature *270*, 277 (1977),
(b) Shinbo, T. et al.: Chem. Lett. *1979*, 1177
95. Grimaldi, J. J., Lehn, J.-M.: J. Am. Chem. Soc. *101*, 1333 (1979)
96. Matsuo, S. et al.: Chem. Lett. *1981*, 1543
97. Burgmayer, P., Murray, R. W.: J. Am. Chem. Soc. *104*, 6139 (1982)
98. Shimidzu, T. et al.: Makromol. Chem. *178*, 1923 (1977)
99. Shimidzu, T. et al.: Macromolecules *14*, 170 (1981)
100. Shimidzu, T. et al.: Polym. J. *12*, 363 (1980)
101. Ogata, N. et al.: J. Polym. Lett. Ed. *25*, 1419 (1980)

Macroheterocyclic Ligands on Polymers

Johannes Smid and Roger Sinta

Polymer Research Institute, Chemistry Department, College of Environmental Science and Forestry, State University of New York, Syracuse, NY 13210, USA

Table of Contents

A Introduction 106

B Linear Polymeric Macroheterocyclic Ligands 108
B.I Macroheterocycles in Polymer Backbones. 108
B.II Macroheterocycles in the Side Chain of Linear Polymers 114
B.II.1 Synthetic Procedures and Polymer Structures 115
B.II.2 Intra- and Intermolecular Polycrown-Cation Complexes . . . 117
B.II.3 Extraction Equilibria and Ion Transport 123
B.II.4 Soluble Poly(crown ether) Catalysts 127
B.II.5 Polysoap-type Properties of Poly(crown ether)s in Aqueous Media 129

C Macroheterocycles Anchored to Insoluble Supports 136
C.I Applications as Exchange Resins in Chromatography. 136
C.II Binding of Ion Pairs to Crown Ether Networks 141
C.III Catalytic Properties 146

D Conclusions 151

E Acknowledgements 152

F References 152

A Introduction

More than fifteen years after the pioneering work of Pedersen [1,2] on the synthesis and cation-binding properties of macrocyclic polyethers the literature still abounds with reports on new synthetic developments and potential applications of these interesting compounds. Aided by the use of CPK models, careful design of the macrocycles including cryptand-type ligands [3] has yielded a host of materials that exhibit complexation specificity not only towards metal ions but also towards ionic and neutral organic molecules. Initially, the effort was chiefly directed at modifying the ring structure of the macrocycle, for example, its size and the nature of the binding sites some of which can be part of a subheterocyclic structure like a pyridine, furan or thiophene moiety. Subsequent work focused more on the design and synthesis of compounds with structural units arranged around the periphery of the macrocyclic cavity. These units may or may not contain additional binding sites. The extensive host-guest complexation studies of Cram [4], Lehn [5], Reinhoudt [6], Stoddart [7] and other investigators have provided numerous examples of host molecules of high specificity. They include the 1,3-xylyl-type crown ethers with functional group binding sites anchored at the 2-position of the xylyl ring [4,6,8], macrocycles with oligo-oxyethylene chains attached to rigid or semi-rigid structures like binaphthyl and paracyclophanyl units [9], ligands with chiral recognition properties [4,5,7] and the "lariat" crown ethers of Gokel [10] which possess additional binding sites on flexible chains attached to the crown ring for added stabilization of the cation-macrocycle complex.

In nearly all the crown ether and cryptand-type ligands the complexing ion or neutral solute converts the thermodynamically stable conformation of the free ligand into a structure that is more conducive to yield a stable complex. Cram and his coworkers have recently reported on a class of ion-chelating compounds of high specificity called spherands and hemispherands in which a favorable orientation of binding sites (usually as part of anisole or urea units) is attained during the actual synthesis of the ligands and *prior* to complex formation [11]. Another interesting development is the synthesis by Weber [12,13] of polytopic cation receptors coined "multiloop crown compounds" in which two to four individual macrocycles differing in chain flexibility, ring size and binding sites are joined by spiro-type linkages. These multisite ligands, like the earlier synthesized bis(crown ethers) [14,15] or bis-tren macrobicyclic ligands [16] may complex more than one of the same cation when identical macrocycles are involved, or they may simultaneously bind different cations depending on the binding characteristics of the individual crown ether components. Mention should also be made of the concurrent progress in the field of cation-binding acyclic ligands. The so called "podands", developed in recent years by Vögtle and Weber [17] have generally a more flexible structure than the macroheterocycles, but proper design and selection of substituents can make them into effective complex formers, and they may sometimes be preferred because of their easier synthesis.

Macroheterocyclic ligands have found application in diverse fields of science. These include phase-transfer catalysis (for this topic see F. Montanari et al., this series *101*, 147), mediated ion transport through solid and liquid membranes (cf. G. R. Painter, B. C. Pressman, this series *101*, 83), chromatographic separation of ionic and neutral solutes (Lit. 130), mimicry of enzyme systems using chiral crown ethers or cryptands with catalytic sites (see R. M. Kellogg, this series *101*, 111), ion-selective electrodes (cf. Lit. 130), chromogenic reagents for photometric analysis of metal ions (see, M. Takagi, this series *121*, 39), systems with macrocycles coupled to photoresponsive chromophores where photoirradiation can alter the conformation of the macrocycle or its distance to a neighboring ligand (see S. Shinkai, O. Manabe, this series *121*, 67), etc. The interested reader is referred to several comprehensive reviews published in recent years which contain a wealth of information on the synthesis, properties and actual or potential uses of crown ethers, cryptands and podands [3–8,19–24].

For several applications it is necessary to anchor the macroheterocyclic ligand to an insoluble support like silica gel or a network polymer. However, the binding characteristics of a macrocycle could change significantly when it becomes incorporated into a polymer backbone or anchored as pendant moiety to a macromolecular chain. For example, close proximity of two macrocycles will facilitate their cooperation in the formation of stable 2:1 macrocycle-cation complexes. This is particularly encountered with polymers having pendant ligands and where the cation diameter is larger than that of the macrocycle cavity. Also, copolymerization with different co-monomers will yield different micro-environments around the macrocyclic binding site. This can lead to polarity changes in the vicinity of the site, or may cause steric interference between co-monomer substituents and the complexing ion or, where ion pairs are involved, its associated counter ion. Co-monomer substituents or neighboring macrocyclic ligands may also furnish additional binding sites to further stabilize the macrocycle-ion complex. Incorporation into a polymer may also cause restricted mobility of the binding site, including interference in the conformational adaptation of the macrocyclic ligand needed for effective complex formation. Other effects on ion-binding to polymeric ligands may arise from the increase in charge density on the polymer as more complexes are formed and the resulting conformational changes generally observed in polyelectrolyte systems. With network polymers, factors such as crosslinking density and network swelling will influence the accessibility of binding sites, an important consideration in catalyst and chromatographic applications. An interesting application of certain *water-soluble* poly(crown ether)s is their ability to act as neutral polysoaps with a strong affinity to bind neutral or ionic hydrophobic solutes. The binding can be regulated by converting the polymeric ligand into a polycation on adding ligand-complexable cations.

The polymeric effects to be expected on binding solutes to polymers carrying macroheterocyclic ligands make it worthwhile to look more closely at these systems. In this chapter we will review the synthesis and solute binding properties of crown ether and cryptand-type ligands incorporated into linear or network-type polymers. Only a limited number of macrocycles have been used for this purpose. A few examples of polymers with acyclic ligands will be discussed for comparative purposes. Emphasis will be placed on contrasting the behavior of polymer-bound macrocycles with their monomeric analogues.

B Linear Polymeric Macroheterocyclic Ligands

Macroheterocyclic ligands can be incorporated into linear polymers as components of the backbone structure or as pendant moieties anchored to a macromolecular chain. The first type of polymer is generally made by condensation polymerization using a difunctional monomer containing the macrocycle. The second class of polymers can be synthesized from a vinyl-containing macrocycle or by reacting a polymer with a functionalized macrocycle.

B.I Macroheterocycles in Polymer Backbones

Linear condensation polymers containing a crown ether structure in the backbone were first reported by Feigenbaum and Michel [25]. They synthesized polycrown amides by reacting *cis* or *trans*-diaminodibenzo[18]crown-6 (*trans*-diamino-DB18C6)[1] in hexamethylphosphoramide (HMPA) with either iso-or terephthaloyl chloride. The *trans*-diamine and isophthaloyl chloride yielded the best polymer (*1*), the intrinsic viscosity in HMPA at 30 °C being 0.9- 1.1. Films cast from hexafluoroisopropanol

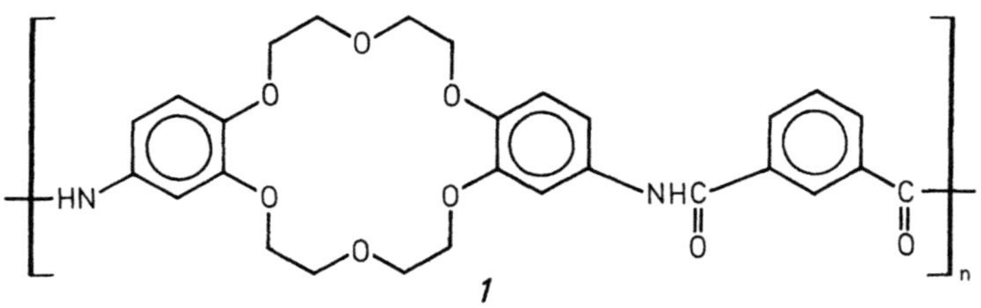

1

and exposed to a heated, 20% aqueous metal chloride solution adsorbed alkali ions. The fraction of crown ligands containing a cation increased in the order Cs^+ (0) < Li^+ (0.24) < Na^+ (0.54) < K^+ (0.64), while divalent cations did not dissolve in the films. Mechanical properties were not affected by the presence of cations.

Polymers of structure *1* with higher molecular weights ($[\eta]_{30}$ = 7.5 dl/g in HMPA) and yielding tougher films were made by Shchori and Jagur-Grodzinski who treated crystals of isophthaloyl chloride with an HMPA solution of *cis*-diamino DB18C6 [26,27]. Membranes of *1*, and its alloys with poly(vinylpyrrolidone) (PVP) cast from hexafluoroisopropanol were equilibrated with salt-containing solutions to determine their salt solubility and water-diffusion properties and to evaluate their potential application in desalination. Scanning electron microscopy indicated formation of true thermodynamic polymer blends of *1* with PVP, and only one transition was found by differential scanning calorimetry.

Some solubility data for films of polymer *1* and its PVP alloys are collected in Table 1. Exposure to 1 M NaSCN results in full saturation of the membranes by sodium ions although the salt saturation values decrease as the PVP content increases. Binding of Li^+ is weak, and cation solubility is lower for salts with less delocalized anions such as chlorides. Water solubility in the films is high, and is enhanced

1 Notations frequently used in this report are listed and explained on p. XI

by adding more PVP. Permeability of water (P_w) and salt (P_s) can both be described by Fick's first law for steady state diffusion. The observed water permeability ($P_w \approx 10^{-7}$ cm^2/sec) is much higher than of the salt ($P_s \approx 10^{-10}$–10^{-11} cm^2/sec). This demonstrates the strong interaction between crown ligand and cation. Salt and water permeabilities are affected by storage time and treatment histories of the membranes. However, the addition of PVP appears to minimize this dependence, apparently as a result of their role in maintaining the original membrane structure [26]. In some of the alloys water diffusion coefficients are only fifty times lower than that of pure water, but the presence of PVP also increases salt diffusion coefficients.

Table 1. Solubilities, C_s (mmole cc^{-1}), of Sodium and Lithium Salts in Membranes of *1* and its Alloys with Poly(vinylpyrrolidone)[a]

Equilibrating Solution	A(0)	A(20)	A(25)	A(30)
0.1 M NaCl	0.36	0.30	—	0.34
0.02 M NaSCN	0.48	0.38	—	0.34
0.1 M NaSCN	0.9	0.77	0.74	0.72
1 M NaSCN	1.92[b]	1.50	—	1.16
0.1 M LiSCN	0.095	0.14	0.15	0.14

[a] Taken from Ref. 26. The designation A(20) refers to a polymer *1* containing 20 wt.% poly(vinylpyrrolidone) in the dry alloy.

[b] The calculated C_s value for a 1:1 molar ratio of adsorbed salt to macrocycle is 1.97 for A(0), 1.47 for A(20), 1.33 for A(25) and 1.16 for A(30).

The observed concentration dependence of the diffusion coefficient is qualitatively consistent with a site jumping mechanism described by the expression $D_s = k\lambda^2$ where k is the cation decomplexation rate constant and λ the average jump distance between binding sites. The mechanism implies that the residence time of the cation on the crown site is long compared to the time of transfer. Semilog plots of D_s versus 1/T yield a salt diffusion activation energy of 50.2 kJ/mole. This value is the same as that obtained for the activation energy of decomplexation of a sodium ion from dibenzo[18]crown-6 (DB18C6) derivatives in several solvents [28–30].

Water permeabilities of the polycrown-amide membranes and their PVP alloys are at least an order of magnitude higher than those found with other aromatic polyamides, and some membranes exhibit better osmotic characteristics than cellulose acetate membranes. Salt rejection amounts to 95–97%, while water permeabilities range from 3.6×10^{-7}–7.3×10^{-7} cm^2/sec [26].

Membranes containing polymer-bound crown ethers useful for ultrafiltration and reverse osmosis have been described in a patent by Frost [31]. Most of these systems possess an aromatic crown ether such as DB18C6 as part of a polymer backbone with nitrogen linkages. Described in more detail are polymers derived from diamino-DB18C6 and pyromellitic anhydride or trimellitic anhydride-4-acid chloride. The structure obtained with the latter monomer is shown below. The nitrogen linkages, together with the pendant carboxyl groups, improve the hydrophilicity of these polymers and facilitate water transport through their membranes.

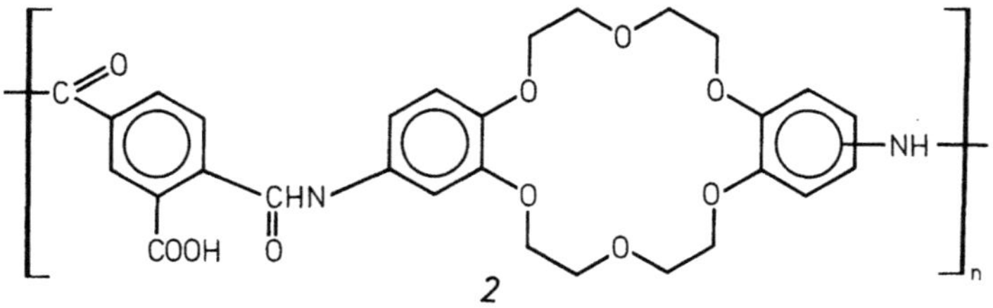

2

The polyamic acid *2* is converted at 150°–200 °C by chemical dehydration into the amide-imide derivative *3*. When this flexible and adherent film is exposed to a 0.5% KOH solution, structure *3* converts into the amid acid potassium salt of *2* with

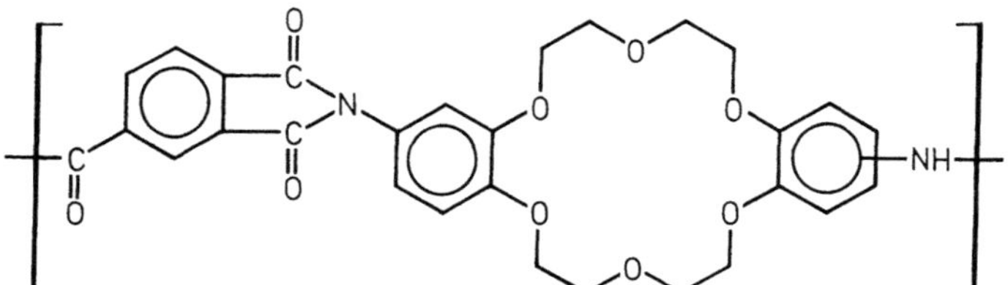

3

a high fraction of ionized carboxyl groups and the K^+ ion trapped in the crown cavity. In reverse osmosis tests with a 3.5% NaCl solution at 1500 psi this polymer salt gave 74% salt rejection and a pure water flux of 9.4 gal/ft^2/day. Macrocyclic polyethers with nitrogen linkages were found to be good film formers. Ion trapping ability is claimed to be superior for the carboxyl-containing polymers like *2* than for the polycrown imides (*3*) or amides. Few pertinent data are presented in the patent to evaluate the usefulness of these materials for reverse osmosis application, ultrafiltration or sewage treatment.

Several reports in the literature deal with the synthesis and ion-binding properties of linear condensation polymers with a *diazacrown ether* in the backbone. Gramain and his coworkers synthesized several of these materials by coupling diaza-15-crown-5 (DA15C5) or DA18C6 with difunctional epoxides [32–35] or diacid-chlorides [36]. Structure *4* depicts a condensation polymer of bisphenol-A with DA18C6 (kryptofix-22, Merck).

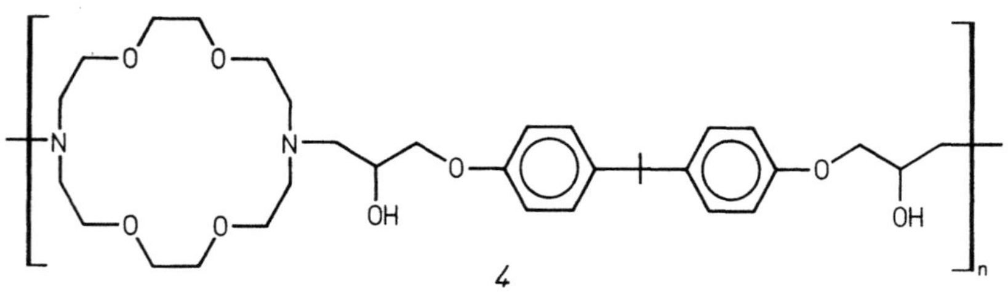

4

Most polymers have molecular weights between 2000 and 4000 and contain between 5 and 10 ligands per chain. Ion binding was studied by picrate salt extraction from water into a ligand-containing $CHCl_3$ phase. By determining solubilities of picrate salts in $CHCl_3$ in the absence of ligand, the association constant of the crown ether ligand with the picrate ion pair in the chloroform phase was obtained. To evaluate the effect of backbone linkages and comonomer substituents such as the neighboring hydroxyl groups, the results with the polymers were compared with those of monomeric diazacrown ether analogues containing $N-CH_2CH_2OH$ substituents. The importance of remote substituent effects was demonstrated by Cram et al. [38] from ion-binding studies with xylyl-type crown ethers endowed with a phenolic OH substituent. The OH group hydrogen-bonds with a crown ether oxygen binding site which adversely affects the ion-binding. The results of Gramain et al. show that association constants for polymer *4* are comparable or only slightly below those of its monomeric analogue, while cation selectivities remain essentially unchanged. Alkali and alkaline earth cations as well as ions such as Cu^{2+}, Cd^{2+} and Pb^{2+} all form 1:1 complexes with polymers containing the DA15C5 or DA18C6 ligand. An exception is Ba^{2+} which forms a sandwich-type 2:1 crown-cation complex with the copolymer of DA18C6 and diepoxyoctane [34]. Viscometric data revealed polymer contraction in the presence of Ba^{2+}, implying that the sandwich complex is intramolecular rather than intermolecular. The long flexible octane chain apparently folds back to bring two crown ligands close together. Studies of ^{13}C NMR chemical shifts and spin-lattice relaxation times of polyureas derived from diisocyanates and DA18C6 by Mathias and Al-Jumah also suggest considerable flexibility in the polymer backbone of the materials when hexamethylenediisocyanate is the comonomer [39].

Gramain and his coworkers also made the interesting observation that the picrate salt complex with *4* or with the monomeric analogue has an absorption maximum at 355 nm, suggesting some interaction between the picrate anion and the hydroxyl groups of the ligands [33]. The picrate complex with the acetylated derivative of *4*, or with the $N-CH_3$ substituted derivative of DA18C6, has an absorption maximum at 365 nm, consistent with the spectrum of picrate salt-crown ether complexes reported by Wong et al. [97].

DA18C6-containing polyethers and polyamides obtained by the synthesis of various diacid chlorides with the N,N^1-diethylene alcohol and N,N^1-diethylene amine derivatives of the diazacrown ligand were also reported by Gramain [36]. Studies with monomeric analogues containing a CH_3, CH_2CH_2OH, $CH_2CH_2NH_2$, $CH_2CH_2OCOCH_3$ or $CH_2CH_2NHCOCH_3$ substituent on each of the two nitrogen atoms show that ligand association constants with potassium picrate in $CHCl_3$ are very sensitive to the substituent structure near the ring. The respective association constants for the five ligands were found to be 10^5 K = 26, 30, 6, 0.7 and <0.01. The same was observed for the polyester and polyamide crown polymers in which the ester and amide linkage was removed from the nitrogen atom of the macrocycle by two methylene groups. The lower complex stabilities are believed to result from formation of five-membered ring structures such as those shown in *5*. No hydrogen bonding can occur with the ester functions and the decrease in the association constants is less.

Poor ion-binding activity was found by Gramain et al. [36] and by Cho and Chang [37] for polyamides or polyureas in which the carbonyl group was immediately adjacent to the nitrogen binding site of the diazacrown ether. Depressed cation-binding was also reported for the tetraaza polycryptate *6* recently synthesized by Gramain and

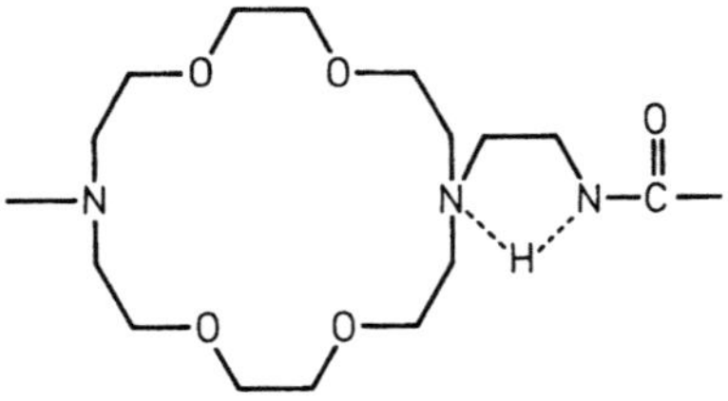

5

Frère [40]. Compared to monomeric analogues with a CH_3 or CH_2CH_2OH substituent, the potentiometrically determined association constants of alkali and alkaline earth cations in water with polymer *6* are considerably lower. The large polymer sub-

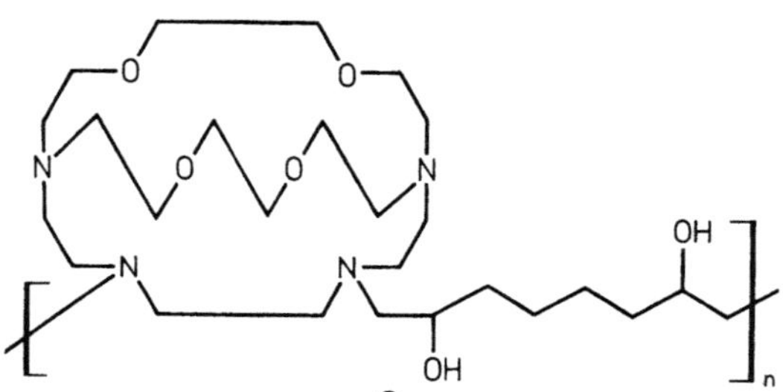

6

stituent increases the thickness of the ligand. This restricts the mobility of the cryptand chains and hinders the conformational adaptation of the cryptand cavity to form the cation-inclusion complex. The transition cations Cu^{2+}, Zn^{2+}, Co^{2+} and Ni^{2+} are not much affected by the increase in ligand thickness due to their small size, but the binding constant for Cd^{2+} is reduced by more than a factor 100. As a result, the Cu^{2+}/Cd^{2+} selectivity, while poor for the methyl derivative of the cryptand, is vastly improved for polymer *6*.

A final and very interesting example of a macrocycle incorporated into the backbone of a linear condensation polymer is the photoresponsive azobenzene-bridged DA18C6 polymer *7* [41]. Shinkai and his coworkers (cf. this series 67) have reported extensively on changes in the ion-binding behavior of monomeric azo-type crown ethers when photoirradiation converts the *E*-isomer into the thermodynamically less

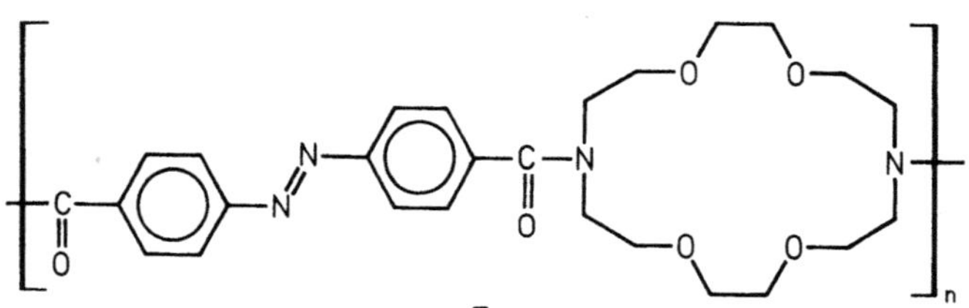

7

stable *Z*-compound. The isomerization may change the crown ring conformation to a structure which more effectively complexes a particular ion. Alternatively, conversion of azo-linked bis-(crown ethers) into their Z-isomer [42,43] may change the spacing between the two crown moieties into one that is more favorable in stabilizing 2:1 crown-cation complexes.

Shinkai et al. observed that photoirradiation of polymer *7* changes its intrinsic viscosity in *o*-dichlorobenzene from 0.057 dl/g to 0.047 dl/g, suggesting a more compact structure for the *Z*-isomer. Also, regeneration into the *E*-isomer is accompanied by an entropy *increase* of 24.7 eu, while for a comparable monomeric azo compound an entropy *decrease* was found of 39.3 eu. The 60 eu increase in the entropy for polymer *7* was compensated for by a 86.1 kJ/mole increase in the enthalpy of *E*-regeneration as compared to the monomeric analogue. The large changes in ΔH and ΔS were attributed to a polymeric effect, the steric crowding in the *Z*-isomer being alleviated by a conformational change in the polymer brought about by a crankshaft-like configurational change in the polymer. While entropically unstable, the compensating change in the enthalpy suggests that the compact *Z*-polymer conformation is stabilized by interactions between chain components [41].

For the *E*-polymer the extractability of methyl orange salts into *o*-dichlorobenzene decreases in the order $Na^+ > K^+ > Cs^+$, with no extraction observed for the lithium or rubidium salts [41]. Most DA18C6 compounds are reported to bind potassium salts considerably more effectively than sodium salts [35–37,41]. However, a drastic reduction in binding efficiency is observed when a carbonyl function is bound to the nitrogen binding sites of the diazacrown [36,37]. It apparently can also reverse the Na^+/K^+ selectivity [37] since the lower basicity of the carbonyl substituted N-atoms will adversely affect complexation of the larger K^+ ion more than it will Na^+. Comparison of *7* with other diazacrown compounds in terms of its cation binding ability is difficult since extraction data refer to different solvents, salts and conditions. However, it will be poor compared to polymers in which the nitrogen and carbonyl function are separated by methylene groups [37].

Isomerization of *7* to the *Z*-configuration increases the percent extraction of potassium methyl orange, but decreases that of the sodium salt [41]. In fact, the binding of Na^+ to the pure *Z*-isomer was calculated to be negligibly small (unfortunately, all extractions were carried out under only one set of conditions). The authors suggest that decreased Na^+ binding by the *Z*-isomer may result from electrostatic repulsion between the bound cations since the crown moieties are presumed to have a face to face orientation in this isomer [41]. However, a similar decrease would be expected for K^+, although an increase is observed. Moreover, electrostatic repulsions between bound cations are unlikely since under the experimental conditions polymer *7*, after salt extraction, contains less than one cation per hundred crown ether units [41]. While formation of the *Z*-isomer and the accompanying changes in the crown ether ring conformation are likely to have a different effect on Na^+ and K^+ binding, an equally important point to consider is the change in the microenvironment around the cation-binding site, especially since in dichlorobenzene a methyl orange salt is bound as a tight ion pair. The more compact *Z*-conformation may well hinder a close approach of the cation-complex and its associated methyl orange counterion, and force an increase in the interionic ion pair distance. Such a stretching of the ion pair distance would be more difficult to accomplish for the sodium salt than for the potassium compound.

B.II Macroheterocycles in the Side Chain of Linear Polymers

Ion-binding properties of macroheterocyclic ligands incorporated in the polymer backbone structure do not differ much from their monomeric analogues except in some special cases, for example, where backbone flexibility permits formation of 2:1 crown-cation complexes as discussed above for Ba^{2+} and diazacrown-containing polymers. The same holds for polymers with pendant macrocycles as long as the ligands are spaced sufficiently far apart. However, significant polymeric effects emerge in macromolecules with closely spaced ligands. The difference with similar, non-polymer-bound ligands is especially revealing in systems where crown and cation can form a stable complex of 2:1 stoichiometry. This is frequently encountered when the cation diameter is larger than that of the crown cavity, and the ion is sandwiched in between two ligands. Examples are [12]crown-4 type compounds with Na^+, [15]crown-5 type ligands with K^+, Rb^+, Cs^+ or Ba^{2+}, and Cs^+ with crown compounds containing an [18]crown-6 ring. In these systems both 1:1 and 2:1 crown-cation complexes can exist, but the stability of the latter is greatly enhanced by using compounds in which the two crown ether ligands are already in close proximity. Several studies have been reported on the complexation of cations to bis(crown ethers), compounds with two ligands linked by a short chain [14,15,42,43,45–52]. The data demonstrate that these compounds facilitate formation of 2:1 crown-cation complexes, depending on length, structure and flexibility of the connecting chain and, where ion pairs are concerned, on the nature of the counterion. The latter variable is especially important since a stable 2:1 crown-cation structure often requires increased charge separation between anion and cation. This process is generally easier with bulky or charge delocalized anions. Another interesting example of the cooperative effects of crown ligands was reported by Johnson and Sutherland who showed that stable intramolecular complexes can be formed between bis-primary alkylammonium salts and bis-monoazacrown ethers as long as the connecting chain is of the proper length [53]. Based on the results of bis(crown ether)s it is expected that poly(crown ether)s with pendant crown moieties will favor complexation of cations that form stable 2:1 complexes, and that such polymeric ligands can reverse the cation selectivity of the monomeric analogues.

A second effect encountered with closely spaced polymer-bound ligands concerns the electrostatic interaction between bound cations. Repulsion will decrease the binding constant as more cations become polymer-bound. This repulsion effect was also observed in the binding of cations to spiro-type bis(crown ether)s [54]. Using a crown ligand that forms a 1:1 complex with sodium ion, Laszlo and his coworkers showed that the binding constant for the second sodium ion is an order of magnitude smaller than that of the first ion. The electrostatic interactions between polymer-bound ions is expected to be less when ion pairs instead of free ions are bound. This is usually observed in low polarity solvents, and attractive intramolecular interactions between bound ion pairs could possibly stabilize polymeric complexes.

However, intermolecular ion pair-ion pair interactions frequently cause precipitation of charged polymeric ligands in such solvents. Also, in the binding of ion pairs, neighboring ligands can influence the interaction by preventing a close approach between the crown-bound cation and its associated anion.

In the following sections the ion-binding characteristics of a number of polymers with pendant crown ligands will be discussed as studied by viscometry, conductance, optical spectroscopy, potentiometry and other techniques. Of special interest to us have been the properties of some water-soluble poly(benzocrown ether)s. Usually, cation binding to poly(crown ether)s in aqueous solution is weak. However, with a proper balance of hydrophobic and hydrophilic regions in the polycrown, the macromolecule can assume a compact conformation in which it acts as a polysoap, capable of binding a variety of hydrophobic organic solutes. This aspect will be discussed in a separate section.

B.II.1 Synthetic Procedures and Polymer Structures

The first polymers with pendant crown ether ligands were reported by Smid and his coworkers [55,56] and their properties have been described in a number of review papers [57–61]. In their investigations they chiefly used homopolymers and styrene copolymers of vinylbenzo[15]crown-5 (V-B15C5) and V-B18C6 (*8*).

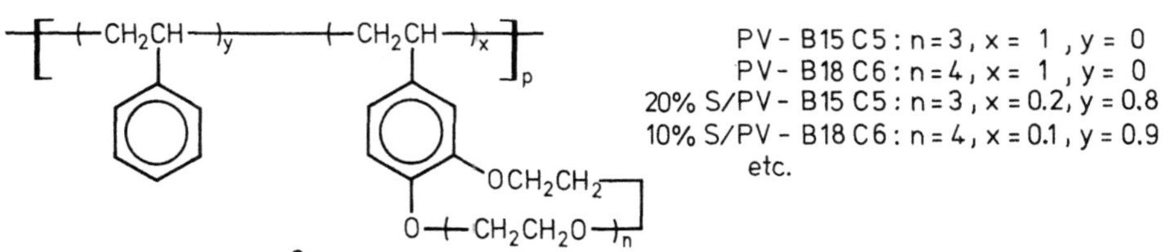

8

The monomers can be synthesized by converting 3,4-dihydroxyacetophenone (obtained by demethylation of the commercial product acetovanillone) into the 4′-acetobenzocrown ethers, followed by reduction to the alcohol and subsequent dehydration to the vinyl compounds [56]. An alternative route is via a 4′-formylbenzocrown ether which can be made from 3,4-dihydroxybenzaldehyde [62] or by introducing directly a formyl substituent into a benzocrown ether [63]. The 4′-formyl derivative is then converted into 4′-(1-hydroxyethyl)benzocrown ether by a Grignard reaction, followed by dehydration to the vinyl compound. The formyl derivative can also be reduced to 4′-hydroxymethylbenzocrown ether, an intermediate in the synthesis of crown ether-containing network polymers derived from chloromethylated polystyrenes, and in the preparation of crown ether esters of acrylic acids.

The vinylbenzocrown ethers were converted to high molecular weight polymers by radical polymerization in benzene or anionic polymerization in tetrahydrofuran [56]. Radical copolymerization with styrene as comonomer yields copolymers, the reactivity ratios being close to unity [64]. The crown monomers can also be polymerized by anionic initiators such as cumyl salts or salts of α-methylstyrene tetramer or dimer dicarbanions in THF at −80 °C [56]. The living polymers can be converted into diblock or triblock copolymers with monomers such as styrene, isoprene, vinylpyridine, methyl methacrylate or ethylene oxide. The abbreviated names used in the text for the homopolymers and styrene copolymers of the two vinylbenzocrown ethers are shown above with structure *8*. The notations x and y refer to the respective mole fractions of vinylbenzocrown ether and styrene in the random copolymers. The notation 20% S/PV-B15C5 refers to a copolymer of vinylbenzo[15]crown-5 and styrene containing 20 mole percent of the crown monomer. Other researchers have also reported on the synthesis and polymerization of vinylbenzocrown ethers [65–68].

Poly(acrylate)s and poly(methacrylate)s in which the ester substituent is a benzocrown ether (*9*) have been reported by Shono et al. [51,69,70] and by Smid and his coworkers [71]. The vinyl monomers can be obtained by reacting hydroxymethylbenzocrown ethers with acryloyl or methacryloyl chloride. The polymers can also be synthesized by converting the hydroxy derivative with hydrogen chloride into chloromethylbenzocrown ether which then is reacted with poly(methacrylic acid) in dimethylformamide in the presence of dicyclohexylmethylamine [69]. Structures and abbreviated names are given below, together with benzocrown ether derivatives of poly(acrylamide)s (*10*). The monomers for the latter polymers were synthesized

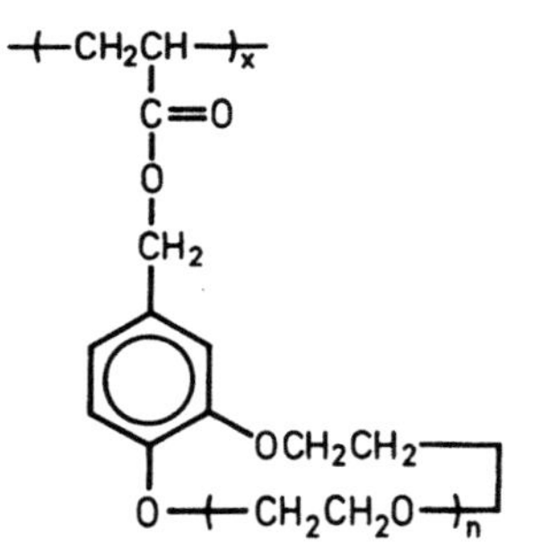

9 PMA-B15C5 : n = 3
PMA-B18C6 : n = 4

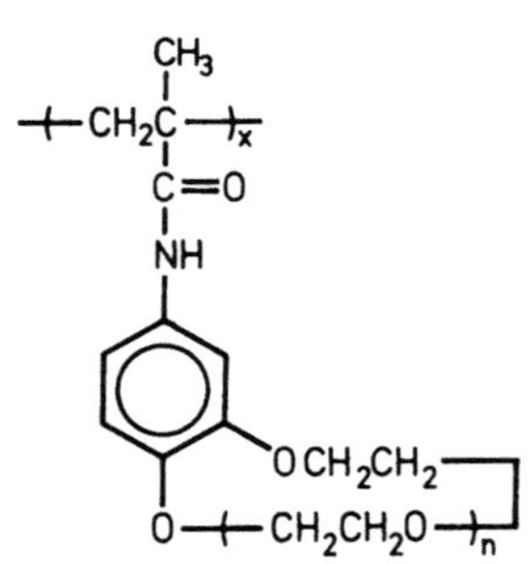

10 PAA-B15C5 : n = 3
PAA-B18C6 : n = 4

by Shono's group [46–49,72,73] and by Yagi et al. [74–76] by reacting aminobenzocrown ethers with acryloyl or methacryloyl chloride, or with acrylic anhydride [46]. A polyacrylate with an aliphatic hydroxymethyl[12]crown-4 ester group was recently reported by Maeda et al. [51].

A crown-modified poly(ethylene imine) (*11*) has been synthesized by Osa and his coworkers [77] using branched poly(ethylene imine) and *p*-nitrophenyl-B15C5-4′-carboxylate or 4′-B15C5-4′-propionate as acylating agents. They also described the properties of a polypeptide with pendant crown ether rings [78], i.e., the B15C5 derivative of poly(glutamate) (*12*). Finally, Quivoron and his coworkers have anchored DA18C6 ligands to a linear polyacrylamide (*13*) by means of the Mannich reaction with formaldehyde [79]. Structures of the polymers *11*, *12* and *13* are shown below.

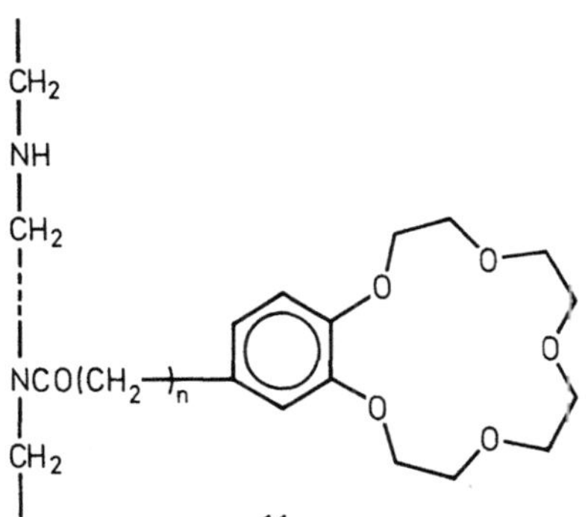

11

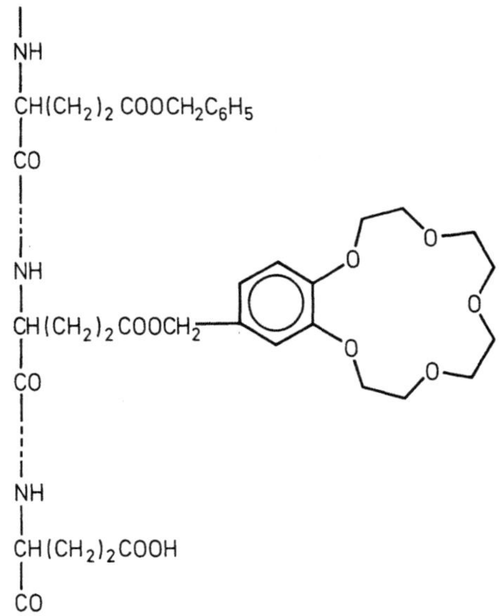

12

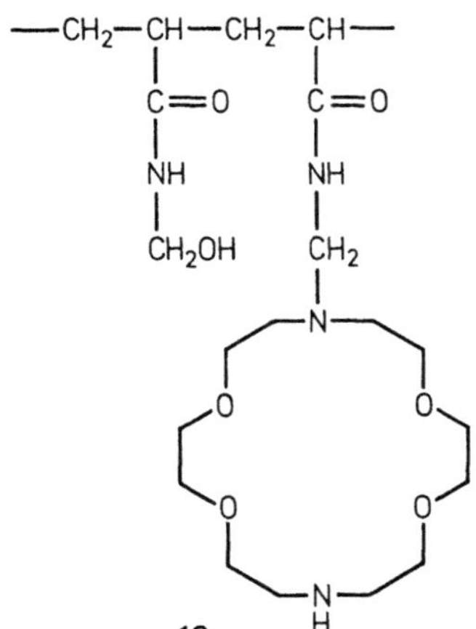

13

B.II.2 Intra- and Intermolecular Polycrown-Cation Complexes

Macromolecules with pendant crown ether ligands favour sandwich-type 2:1 crown:cation complexes if such complexes are likely to exist. Where both 1:1 and 2:1 complexes are formed, their relative probability of formation depends on the solvent, the nature of the anion, and the average distance between the chain-bound ligands. For example, 1:1 complexes are favoured when the ligands are spaced farther apart. Formation of 2:1 complexes may also result from cooperation of ligands bound to two different chains, especially at higher polycrown concentrations. The result would be an increase in the viscosity of such polycrown-salt solutions as these complexes act as physical crosslinks.

Intra- and intermolecular cooperation of crown ligands in binding cations can be demonstrated in a number of ways. *Optical spectroscopy* provides a simple and direct method to determine stoichiometries and formation constants of crown complexes. Smid and his coworkers utilized the observation that optical spectra of ion pairs of fluorenyl or picrate salts are sensitive to the interionic ion pair distance. Crown ether interaction with the tight ion pairs of these chromophores enlarges the distance between the ions which for these two salts results in a considerable bathochromic spectral shift [14,80]. In most cases, crown-complexed tight ion pairs (A^-M^+Cr) or loose ion pairs (A^-CrM^+ or A^-CrM^+Cr) are formed with distinctly different optical absorption maxima. The ratio of crown-complexed tight to loose ion pairs is a sensitive function of solvent and anion. For example, ion pair separation is more difficult in a low dielectric constant solvent or for a charge-localized anion. Formation of 2:1 complexes with two separate crown ligands or with a bis(crown ether) often results in ion pair separation. This process is easier for a fluorenyl carbanion salt than for the less delocalized picrate anion, and is more difficult to accomplish in dioxane than in tetrahydrofuran.

Binding constants of crown ethers and poly(crown ether)s to fluorenyl salts in ethereal solvents are often in the order of 10^6 to 10^7 M^{-1}. Therefore, a simple titration carried out under anaerobic conditions can be used to arrive at

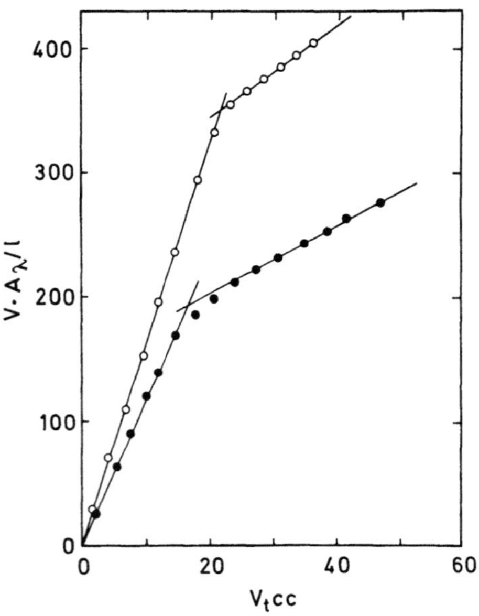

Fig. 1. Titration of (○) poly(vinylbenzo[15]-crown-5) and (●) poly(vinylbenzo[18]crown-6) with barium difluorenyl in THF at 25 °C. A_λ is the absorbance of the 373 nm absorption maximum of the crown-separated ion pair (Ref. [82])

the stoichiometry of the crown complexes of fluorenyl salts, or to obtain the maximum number of ion pairs which can be bound to a polycrown chain. Choosing conditions in which the added salt converts quantitatively to the complex, a sharp titration equivalence point is found on measuring the absorbance, A_λ, of the solution at the absorption maximum of the fluorenyl-crown complex. The data can be plotted as $VA_\lambda/1$ versus V_t, where 1 is the optical pathlength, V_t the volume of added carbanion salt and $V = V_t + V_0$, V_0 being the initial volume of the crown ether solution [81]. Figure 1 depicts such plots for the titration of barium difluorenyl with PV-B15C5 and PV-B18C6 [82]. The stoichiometry of the complex can be computed from the equivalence point. Some of the data are collected in Table 2. It was already established that 4′-methylbenzo-15-crown-5 (4-M-B15C5) forms a 1:1 and a 2:1 complex with potassium fluorenyl and a 2:1 complex with barium difluorenyl, while

4-M-B18C6 forms only 1:1 complexes with both these salts [83-85]. This is reflected in the data for the poly(crown ether)s. For example, nearly every crown ligand on PV-B18C6 can bind a potassium fluorenyl ion pair, while with PV-B15C5 two ligands are used to complex a potassium ion. For the two homopolymers the crown to cation ratio at saturation is slightly above 1.0 or 2.0, e.g., 1.16 for PV-B15C5/Na^+, 1.22 for PV-B18C6/K^+, 2.4 for PV-B15C5/K^+ and 2.2 for PV-B15C5/Ba^{2+}. Steric hindrance and possibly electrostatic repulsion caused by the larger number of bound cations and anions may prevent complete saturation. In 50% S/PV-B15C5, and even more so in 20% S/PV-B15C5, the crown ligands are spaced farther apart, diminishing steric interference or repulsion between bound ion pairs. Table 2 shows that an exact 2.0 stoichiometry is found for potassium fluorenyl with 20% S/PV-B15C5, implying a favorable arrangement of crown ligands along the chain of this copolymer for formation of 2:1 complexes. In 5% S/PV-B15C5, crown units are separated on the average by twenty styrene units, although statistical calculations show that even in this case a number of crown units are sufficiently close to form stable 2:1 complexes [64]. The crown to cation ration of 1.6 (Table 2) for this polymer implies a mixture of 1:1 and 2:1 crown-cation complexes. The cesium cation forms a 2:1 complex with the homopolymer PV-B18C6.

Table 2. Ratios of Crown to Cation at the Equivalence Titration Point for THF Solutions of Fluorenyl Salts and Crown Compounds (Ref. [82])[a]

Crown Compound	Na^+	K^+	Cs^+	Ba^{2+}
M-B15C5[b]	1.0	1.96		2.0
PV-B15C5	1.16	2.4		2.2
50% S/PV-B15C5		2.15		
20% S/PV-B15C5		2.0		
5% S/PV-B15C5		1.6		
M-B18C6[b]	1.0	1.0		0.98
PV-B18C6		1.22	2.2	~1.4
20% S/PV-B18C6		1.0		
10% S/PV-B18C6		0.97		

[a] The data for the copolymers S/PV-B15C5 and S/PV-B18C6 are from unpublished experiments by Dr. K. H. Wong of our laboratory
[b] 4′-Methylbenzo[15]crown-5 and 4′-Methylbenzo[18]crown-6 (cf. p. XI)

Additional information on the relationship between ligand spacing and complex stoichiometry was derived from spectra of picrate salts (Pi) in THF solutions of S/PV-B15C5 copolymers [64]. The Pi^-K^+Cr 1:1 tight ion pair complex with M-B15C5 has its absorption maximum in THF at 362 nm, while the 2:1 crown-separated ion pair Pi^-CrK^+Cr has a maximum at 378–380 nm [86]. At low polycrown concentration ($\approx 10^{-3}$ M) only intramolecular 2:1 complexes are formed, as no increase in viscosity can be detected [64]. Calculation of the sequence length distribution of the crown monomer units in S/PV-B15C5 copolymers yields an estimate of the fraction of "bis(crown ether)" units separated by a certain number (n) of styrene monomer units. Fractions for different values of n are then compared with the fraction of

2:1 crown-separated ion pairs determined experimentally by saturating an S/PV-B15C5 copolymer with potassium picrate. Calculated and experimental data agree best when ligands are separated by not more than three styrene units. Increasing the distance between crown ligands will decrease the fraction of 2:1 complexes and eventually only 1:1 complexes will be stable. The rather crude calculations do not account for specific effects such as rotational restrictions resulting from the styrene monomer units between ligands, but they do provide some information to rationalize spectral and viscometric data for polycrown-salt solutions.

Viscometric measurements provide some information on interactions of cations with poly(crown ether)s, especially when intermolecular 2:1 complexes are formed [56,64]. The neutral polycrown macromolecules give normal Huggins viscosity plots, but typical polyelectrolyte behavior is found in solutions with crown ether-complexable cations. For example, diluting a methanol/THF (v/v = 1/1) solution of PV-15C5 and KCl (ratio of crown/K^+ = 2) sharply increases the reduced viscosity η_r [56]. At constant polymer concentration, η_r increases on addition of salt as more cations become polymer bound, but its value passes through a maximum since at higher salt concentration increased counterion binding will oppose coil expansion and η_r eventually decreases again. The maximum η_r value at a particular PV-B15C5 or PV-B18C6 concentration depends on the binding constant of the cation to the polycrown. The viscosity maxima for PV-B18C6 decrease in the order $Cs^+ > K^+ > Na^+$ and for PV-15C5 in the order $K^+ > Cs^+ > Na^+$, consistent with the binding sequences obtained from extraction and conductance data. Note that the highest maxima are found for cations that form a stable 2:1 complex with the polymer.

The probability of forming *intermolecular* 2:1 complexes is proportional to the square of the total crown concentration, while that of *intramolecular* complexes depends on the first power. Hence, high polymer concentrations will favour the former complexes. This can be clearly demonstrated from viscosity plots of polycrown-salt solutions in systems where 2:1 complex formation is likely to occur. When concentrating methyl ethyl ketone solutions of alkali tetraphenylborate salts and S/PV-B15C5 or S/PV-B18C6 polymers only moderate viscosity increases are observed as long as a crown-cation system is used that only forms 1:1 complexes, for example, $NaBPh_4$ with PV-B15C5, PV-B18C6, S/PV-B15C5 or S/PV-B18C6, or $KBPh_4$ with PV-B18C6 or S/PV-B18C6. However, large increases in the reduced viscosity with polymer concentration are found when 2:1 complexes can be formed, e.g., with $KBPh_4$ and S/PV-B15C5 or $CsBPh_4$ and S/PV-B18C6. The increase is strongly dependent on the spacing between crown ligands along the macromolecular chain and on the ratio between cation and crown. For example, the Huggins plot of a MEK solution of $KBPh_4$/20% S/PV-B15C5 with a crown to K^+ ratio of 10 is not very different from that of systems with 1:1 complexes. Sufficient crown ligands with favorable spacing are available to bind all K^+ ions in the form of intramolecular 2:1 complexes. However, a sharp viscosity increase is observed in MEK solutions of $KBPh_4$/5.8% S/PV-B15C5 (crown/K^+ = 5 or 2). The crosslinking effect can be demonstrated in a different way from plots of η_r versus the ratio of $KBPh_4$ to crown ligand at constant polycrown concentration in concentrated solutions of S/PV-B15C5 copolymers. Figure 2 depicts the increase in η_r for MEK solutions (8 g/dl) of 5.8% S/PV-B15C5, 11.8% S/PV-B15C5 and 18% S/PV-B15C5. At a low K^+/crown ratio sufficient "bis(crown ether)" units along the chain are available to form intramolecular complexes, but

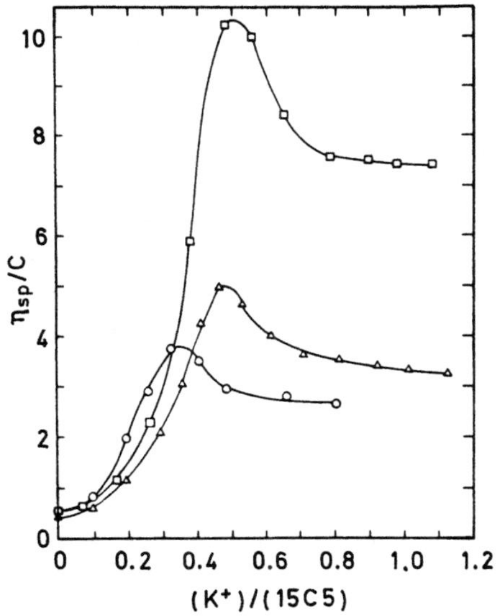

Fig. 2. Dependence of reduced viscosity of styrene-vinylbenzo[15]crown-5 copolymers (S/PV-B15C5) on K^+ concentration expressed as $K^+/15C5$ (molar ratio of $KBPh_4$ to benzo-[15]crown-5 units). Solvent: methyl ethyl ketone, $T = 25$ °C; [S/PV-B15C5] = 8 g dl^{-1}. (○) 5.8% S/PV-B15C5; (△) 11.8% S/PV-B15C5; (□) 18% S/PV-B15C5. (Ref. [64])

when more potassium ions are forced onto the macromolecule these units will be used up and intermolecular cooperation of crown ligands will become the predominant mode of 2:1 complex formation. The maximum value of η_r is reached close to a cation to crown ratio of 0.5, demonstrating again the tendency of these systems to form stable 2:1 crown-cation complexes. Since B15C5 can also form 1:1 complexes with potassium salts, addition of more K^+ can break up some of the 2:1 crown-cation complexes by converting them into 1:1 complexes. This reduces the number of crosslinks and is the probable cause of the decrease of η_r above a cation to crown ratio of 0.5. Although the fraction of "bis(crown ether)" units in 18% S/PV-B15C5 is higher than in 5.8% S/PV-B15C5, an 8 g/dl solution of the former polymer will have a substantially higher crown concentration. Therefore, at a cation to crown ratio of 0.5 the 18% S/PV-B15C5 polymer will contain more crosslinks and a higher reduced viscosity than a similar solution of 5.8% S/PV-B15C5.

Conductance has been used successfully to obtain information on the ion binding to ligands including crown ethers [2,14,87–94]. When only free ions are present, a conductance change is found when the mobility of the cation-ligand complex differs from that of the solvated cation [87–90]. Complexation to ion pairs can change the conductance as a result of a change in the interionic ion pair distance. Also, the dissociation equilibrium will shift when a ligand binds to the free ions [14,91–94]. Especially when combined with spectrophotometric measurements, ligand binding constants to both ion pairs and free ions can be obtained [14,92,94].

When a cation binds to a macromolecule, its mobility reduces to a low value. This is observed when a polycrown is added to a solution of crown-complexable cations. Moreover, when the number of bound cations per chain increases, the counter-ion activity also is reduced as a result of counterion binding to the charged macromolecule, similar to what is found for polyelectrolytes. Hence, while a bis(crown ether) decreases the limiting conductance of a 10^{-3} M $KBPh_4$ solution in MEK from 90 to close to 70 cm^2 ohm^{-1} $equiv^{-1}$, addition of PV-B15C5 and formation of the

K^+ ... PV-B15C5 complex reduces Λ to about 15 at a crown to K^+ ratio of about 3.0 [93] (Fig. 3). The conductance increases again on adding excess PV-B15C5 (crown to $K^+ > 3$), since the cations will be redistributed over the additional polycrown ether chains. This decreases the number of bound cations per chain, and the lower charge density causes the release of bound counterions. The same can be accomplished by increasing the distance between the crown ligands along a chain. For example, a 5% S/PV-B18C6 copolymer of DP $\approx$ 200 contains only a few bound cations per chain even when saturated with cations, and the fraction of polymer-bound counterions reduces to essentially zero. Calculations show that the limiting conductance for a K^+ ... 5% S/PV-B18C6 complex is approximately 4 cm^2 ohm^{-1} $equiv^{-1}$ as compared to 39.4 for the complex of K^+ with the monomeric analogue 4'-methyl-B18C6 [93].

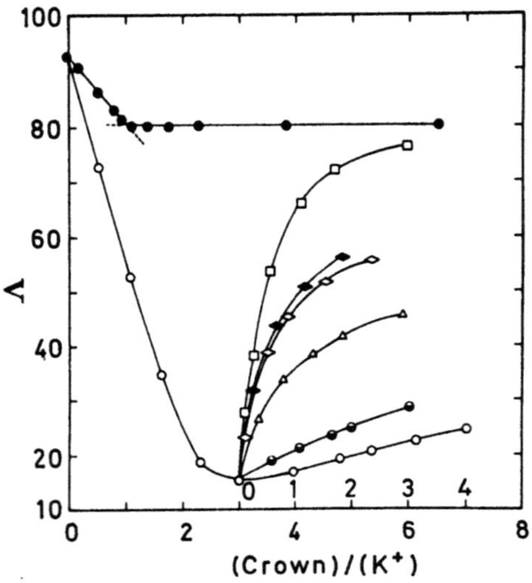

Fig. 3. Equivalent conductance Λ (cm^2 ohm^{-1} $equiv^{-1}$) versus the ratio crown to salt in methyl ethyl ketone solutions of $KBPh_4$ and poly(vinylbenzo[15]crown-5) (PV-B15C5), at 25 °C; PV-B15C5/$KBPh_4$ = 3.0; [$KBPh_4$] = 10^{-3} M.
(○) $KBPh_4$-PV-B15C5, no monomeric crown ligand added; (□) dicyclohexano[18]crown-6; (◆) 4'-methylbenzo[18]crown-6; (◇) bis(benzo [15]crown-5); (△) dibenzo[18]crown-6; (○) 4'-methylbenzo[15]crown-5; (●) $KBPh_4$-dicyclohexano[18]crown-6, no PV-B15C5 present. The values [crown]/[K^+] below the abscissa refer to the $KBPh_4$-PV-B15C5 system, and those above the abscissa refer to the molar ratio of added monomeric crown to salt. (Ref. [93])

The low conductance of the polycrown-cation complex can be used for a qualitative comparison of the binding efficiencies of low molecular weight crown compounds or other cation-binding ligands. An example is shown in Fig. 3. To a mixture of 10^{-3} M $KBPh_4$ in acetone or MEK is added PV-B15C5 until the conductance has reached its minimum value, i.e. at a crown to K^+ ratio of 3.0. Solid quantities of a low molecular weight ligand are now added and the conductance recorded after each addition. The added ligand will compete with PV-B15C5 for cations, and transfer of K^+ from PV-B15C5 to the new, smaller-sized crown will increase the mobility of the complex and also release polymer-bound counterions. Both processes will enhance the conductance. The rate of increase in Λ as a function of the ratio of low molecular weight crown to K^+ will depend on the binding constant of the cation to the crown ether. Eventually, when sufficient ligand is added and no cations are bound to polycrown anymore, Λ will reach the conductance of the salt-crown complex found in the absence of polycrown. Figure 3 shows that this point is reached for DCH18C6. Neglecting small differences in the ionic mobilities of the respective cation-crown complexes, the efficiency for binding K^+ in MEK is found to decrease in the order DCH18C6 > 4'-methyl-B18C6 $\approx$ bis(B15C5) > DB18C6 > methyl-B15C5.

B.II.3 Extraction Equilibria and Ion Transport

Extraction of picrate, methyl orange and salts of other chromophores from an aqueous solution into a macroheterocyclic ligand-containing organic phase (e.g., $CHCl_3$, CH_2Cl_2, benzene, octanol) has been extensively used to obtain quantitative information on the ion-binding properties of free and polymer-bound macrocycles. The results of some of the studies with macrocycles in the polymer backbone have already been discussed.

Some investigators only report the percent of salt (usually picrates) extracted in the organic phase when comparing cation-binding efficiencies of crown ether ligands. Frensdorff[95] determined the actual extraction equilibrium constant, K_e, describing the heterogeneous reaction $M^+ + Pi^- + Cr_{org} \rightleftarrows Pi^-M^+Cr_{org}$. He showed that K_e equals the product $K_sP_eP_c$, where K_s is the binding constant of the ligand Cr to the cation in water and P_e and P_c are the respective partition coefficients of the crown ether and its complex, the latter of which cannot be directly determined. The constant K_e can also be expressed as the product KK_c, where K_c is the formation constant of the complex in the organic phase, $Pi^+M^+_{org} + Cr_{org} \rightleftarrows Pi^-M^+Cr_{org}$, and K is the solubility equilibrium constant for the reaction $M^+ + Pi^- \rightleftarrows Pi^-M^+_{org}$ in the absence of ligand. It has been noted [56,95] that the experimental K_e values decrease considerably as more salt or cations are used in the aqueous phase (frequently, excess cations, e.g., metal hydroxides, are added to force more chromophore into the organic phase). The above expressions for K_e were derived on the assumption that the solubility equilibrium $M^+ + Pi^- \rightleftarrows Pi^-M^+_{org}$ in the absence of ligand is independent of salt concentration. Apparently this is not the case. Gramain and Frère [33] determined K values for picrate salts in $CHCl_3/H_2O$ mixtures at different picrate concentrations. For sodium picrate K (M^{-1}) = 1.78, 0.38, 0.062 and 0.0083 at picrate concentrations of 0.0003, 0.001, 0.03 and 0.01 M, respectively. Taking this large change in K into account, the calculated K_c values ($K_c = K_e/K$) deviated by not more than 30%. Using the same relationship to determine formation constants K_c for crown complexes with picrate salts in chloroform, Kimura et al. [49] report K values of 1.2×10^{-3} (Na^+), 2.67×10^{-3} (K^+), 3.66×10^{-3} (Rb^+) and 6.6×10^{-3} M (Cs^+). The authors do not state at what concentration these K values were determined, although most extractions were carried out at 10^{-3} M picrate. Even at 10^{-2} M picrate the K values would be much lower than those given by Gramain and Frère. It raises the question as to how accurately the solubilities of picrate salts can be determined since in the absence of ligand they are very low in chloroform.

To calculate K_c, Gramain and Frère have used the relationship $K_c = [f/(100\text{-}f)]/[Pi^-M^+]$, where f is defined as the percentage of total ligand complexed and $[Pi^-M^+]$ is the concentration of uncomplexed picrate salt in the organic phase. The latter is obtained from K values determined in the absence of ligand. However, Gramain and Frère also showed that in mixtures of sodium and potassium picrate the solubility of the latter salt in chloroform is decreased and that of the former increased relative to that for the pure salts. In view of this result it is somewhat questionable if in the calculation of K_c the same value for the concentration of uncomplexed Pi^-M^+ can be used regardless of the presence of a large excess of the complex. It is also important to realize that the complex formation constant,

K_e, for the reaction $Pi^-M^+_{org} + Cr_{org} \rightleftarrows Pi^-M^+Cr_{org}$ as determined from extraction equilibria actually describes interactions of crown ethers with hydrated picrate ion pairs, $Pi^-M^+ (H_2O)_n$, since the chloroform phase is water-saturated. Even the crown itself may be hydrated and its conformation different from that in dry chloroform [6]. The complex itself also may retain part of the water of hydration [96]. All these factors will affect the relative scale of binding efficiencies for a series of ligands with respect to a particular ion or ion pair. A better comparison would be obtained if no water were present in the organic phase. This can be accomplished by using crown ligands in competition with crown-containing networks, a method that will be discussed in Section C.II.

An additional problem in solvents of higher dielectric constant such as methylene chloride is dissociation of the extracted salt into free ions. Since binding constants of crown ligands to a free cation are frequently much higher than to its ion pair [14], a decrease in K_e is often observed as the amount of extracted picrate increases [56,95]. Correction for the dissociation into free ions can be made by means of a non-linear regression fitting procedure of the extraction data [95]. This yields the extraction equilibrium constant for the ion pair complex and its dissociation constant. The problem is more difficult to solve for polycrown complexes since the conductance behavior of polyelectrolytes is much more complex. The use of chloroform as the organic phase is, therefore, preferred since in this medium the dissociation of picrate salt complexes can usually be neglected. However, for weak-binding ligands (e.g. acyclic ligands) methylene chloride is often used because of the higher picrate solubility in this solvent.

While considerable problems and uncertainties are encountered in the treatment and interpretation of extraction data, the simple method still can serve as a useful tool to compare the ion-binding properties of macroheterocyclic ligands. This is especially so when comparing data for the same salt obtained under similar conditions of salt concentrations in the aqueous phase. We have collected in Table 3 most of the extraction equilibrium constants for poly(benzocrown ether)s and their monomeric analogues as reported by Shono, Osa, Yagi and Smid, and their coworkers [49,56,71,74,78,97].

Inspection of Table 3 leads to a number of observations. K_e for sodium picrate extraction in $CHCl_3$ with any of the polymeric or monomeric crown ethers does not vary by more than a factor five. In CH_2Cl_2 the variation is within a factor ten if the polyacrylamide benzocrown ethers are not counted. The same is true for complexes of potassium picrate with B18C6 and their polymers, where K_e in $CHCl_3$ ranges from 15 for 20% S/PV-B18C6 to 66 for PAA-B18C6. In all these systems the cations form 1:1 complexes with the crown ligands. Part of the variations in K_e can be attributed to the remote substituent effect at the 4′-benzo position. A change from 4′-amino-B15C5 to 4′-nitro-B15C5 decreases the binding constant to Na^+ in acetone by a factor 23! [87]. Note also that homopolymers appear to give slightly higher K_e values (compare PV-18C6 with 20% S/PV-B18C6 and PV-B15C5 with 20% S/PV-B15C5). It is not unlikely that oxygen atoms from adjacent ligands can provide additional binding sites to stabilize the complex [97]. However, the larger polymeric effects are found when the cation can form a 2:1 crown-cation complex. For example, compare K^+/M-B15C5 with K^+/PV-B15C5 or 20% S/PV-B15C5, K^+/AA-B15C5 with K^+/PAA-B15C5, the EI-B15C5 monomeric compounds with PEI-B15C5 ($n = 0, 2$)

Table 3. Extraction Equilibrium Constants, K_e, for Alkali Picrates

Crown[a]	$10^{-4}\,K_e$, M^{-2} ($H_2O/CHCl_3$)			Ref.	$10^{-4}\,K_e$, M^{-2} (H_2O/CH_2Cl_2)			Ref.
	Na^+	K^+	Cs^+		Na^+	K^+	Cs^+	
M-B15C5[b]	0.37	0.042[e]		97)	0.86	3.5	0.76	56)
AA-B15C5[c]	0.13	0.97		49)	2.5	8.4	3.1	74)
EI-B15C5 (n = 0)[d]					0.66	0.97	0.19	77)
EI-B15C5 (n = 2)[d]					1.1	1.7	0.24	77)
M-B18C6[b]	0.23	24		97)	1.5	53	9.7	56)
AA-B18C6[c]	0.14	17		49)	5.3	28	14	74)
PV-B15C5	0.55	0.23		97)	5.9	173	26	56)
20% S/PV-B15C5	0.41	4.5		97)				
PMA-B15C5	0.31	0.12		71)				
PAA-B15C5	0.63	5620	27.5	49)	25	1700	69	74)
PEI-B15C5 (*11*, n = 0)					6.7	410	11	77)
PEI-B15C5 (*11*, n = 2)					5.6	790	12	77)
PP-B15C5 (*12*)					3.3	75	5.3	78)
PV-B18C6	0.56	55		97)	5.0	116	215	56)
20% S/PV-B18C6	0.17	15		97)				
PMA-B18C6	0.40	29		71)				
PAA-B18C6	0.41	66	41.7	49)	16	120	1700	74)

[a] For the abbreviated names of polymers, see structures *8* through *12* in the preceding text, cf. also p. 115

[b] 4′-methyl-B15C5 and 4′-methyl-B18C6

[c] Benzocrown ether derivatives of acrylamide or methacrylamide

[d] B15C5 with a $(C_2H_5)_2{-}N{-}C(=O){-}(CH_2)_n$-substituent at the 4′ position

[e] Formation of the 1:1 complex

for both K^+ and Cs^+, and Cs^+/M-B18C6 or Cs^+/AA-B18C6 with their respective polymeric analogues PV-B18C6 and PAA-B18C6. Cooperative interactions of crown ligands are of course the main reason for the polymeric effect.

A striking result, reported by both Shono et al. and Yagi and his coworkers, is the very large extraction equilibrium constants for the poly(acrylamide)- or poly(methacrylamide)-bound benzocrown ethers with cations that form 2:1 complexes. For example, K_e for K^+/PAA-B15C5 is 46,000 times the value for K^+/PMA-B15C5, the only difference in the structure of the two polymers being that PAA-B15C5 contains an amide linkage and PMA-B15C5 an ester group connecting the crown ligand with the polymer backbone. The difference between Cs^+/PV-18C6 and Cs^+/PAA-B18C6 where also 2:1 complexes are formed is much less, but still a factor 8, while the 1:1 complexes of K^+/PV-18C6 and K^+/PAA-B18C6 give nearly the same K_e. Some of the reported data were not obtained under identical conditions, and others appear to have been derived from only one extraction (e.g., those for PAA-B15C5 and PAA-B18C6 in CH_2Cl_2/H_2O) with as much as 97% of salt extracted [74]. Nevertheless, both Shono and Yagi's group find consistently larger K_e values for the poly(acrylamide)-based benzocrown ethers, although no specific explanation is advanced by these authors. It is conceivable that the alignment of two adjacent benzocrown ligands in a

2:1 crown-cation sandwich-type complex may result in a favorable structure for hydrogen bond formation between the attached NH and CO groups. This in turn would further stabilize the 2:1 complex. It is interesting in this respect that by spacing the crown ligands farther apart (insertion of methylmethacrylate comonomer units) the extraction efficiencies of the poly(methacrylamidobenzocrown ether)s rapidly decrease [76] while that of a poly(vinylbenzo[15]crown-5) increases (compare K^+/PV-B15C5 with K^+/S/PV-B15C5 in Table 3). Hydrogen bonding, if it does occur, would be less favorable on increasing the distance between ligands. For PV-B15C5, it has been argued that insertion of a K^+ ion between the closely spaced ligands may be sterically hindered, while this would be alleviated in a 20% S/PV-B15C5 copolymer.

In addition to the extraction data of Table 3, a few other systems have been reported. Poly(acrylate) polymers with 12C4 ligands preferentially extract sodium picrate [51]. The distribution ratios in CH_2Cl_2/H_2O (metal concentration in the organic phase over that in the aqueous phase) were found to be 0.256 for sodium picrate and 0.067 for the potassium salt. Spectra of the picrate complexes show that the sodium salt is a separated 2:1 complex.

Sinta and Smid have reported the ion-binding properties of poly(vinylbenzoglyme)s in which the monomeric unit contains a benzene ring with two adjacent $CH_3(OCH_2CH_2)_nO$-substituents, n being 2 or 3 [98]. K_e values in CH_2Cl_2/H_2O for these acyclic ligands are two to three orders of magnitude lower than those of the corresponding macrocyclic ligands. However, the close proximity of the glyme chains in the macromolecules considerably improves the cation binding in comparison to the monomeric benzoglyme analogues by supplying more binding sites for the cation solvation shell. In one of these systems, a B18C6 ligand was incorporated in the polymer by copolymerization of a vinylbenzoglyme and the vinylbenzocrown ether monomer. The behavior of the 1:1 copolymers is very similar to that of the homopolymer PV-B18C6 except for sodium picrate where the copolymer is more effective by a factor five [98].

Shono and his coworkers have investigated the effect of polymer tacticity on the cation binding to polycrown compounds [69,70]. No difference exists between the extraction efficiencies of B15C5 bound to isotactic and to syndiotactic poly(methacrylic acid), as long as only 1:1 complexes are formed. For 2:1 complexes the isotactic polymer is slightly more effective. Under otherwise identical conditions, the isotactic polymer extracted 76% potassium picrate into $CHCl_3$, the syndiotactic one 59% and an atactic sample 65% [51]. No difference is found between picrate extractions for the corresponding *d,l* and *meso*-bis(crown ether)s. The authors suggest that chain flexibility may be the controlling factor when forming stable 2:1 crown-cation complexes in the two stereoregular polymers.

Poly(crown ether)s have also been studied in carrier-mediated salt *transport* through liquid membranes [97]. For monomeric crown compounds the cation selectivity in transport is essentially determined by the extraction equilibrium constants of the salts across the H_2O-organic solvent interphase, at least for the alkali ions [97,99]. For example, the K^+/Na^+ selectivity for picrates with 4′-methyl-B18C6 in $CHCl_3/H_2O$ is 104 when computed from the ratio of the extraction equilibrium constants and 50 from that of flux data. This implies an essentially diffusion-controlled carrier-mediated salt transport mechanism in the liquid membrane. However, there are exceptions, especially when the cation-macrocycle complex is very stable and the rate of salt transport is determined by the decomplexation rate [100,101].

A distinct polymeric effect on cation selectivity in salt transport is found when 2:1 crown-cation complexes are formed. For example, the K^+/Na^+ selectivity is 0.11 for picrate transport through a $CHCl_3$ membrane when the carrier is 4'-methyl-B15C5 (and increasing at higher crown concentration) and 11.0 for 20% S/PV-B15C5. The reversal results from the strong tendency of the S/PV-B15C5 copolymer to form stable 2:1 complexes with potassium. The rate of salt transport is slower for the polycrown carriers due to their lower diffusion coefficients [97]. Salt transport is also strongly anion dependent, and especially when the transport involves ion pairs, cation selectivities will also be dependent on the nature of the counterion [97,102]. In addition, the rate of transport and transport selectivity will depend on the liquid membrane itself, and on the presence of cation or anion-binding additives other then the crown or cryptand carriers. Such additives can be easily incorporated as comonomer substituents into a crown-containing macromolecule, and it should be feasible to obtain materials with considerable differences in cation selectivities.

B.II.4 Soluble Poly(crown ether) Catalysts

Macroheterocyclic and acyclic polyether and cryptand-type ligands have been extensively employed in anion-activated catalysis. Nearly all investigations have dealt with low molecular weight cation-binding ligands or with those immobilized on insoluble supports. The latter will be discussed in the last section of this review. Only a few studies have attempted to assess the polymeric effects of macroheterocyclic ligands bound to linear polymers. This is unfortunate since the catalytic action of network-immobilized ligands is usually more complex. In these systems, polymeric effects caused by ligand cooperativity or by microenvironmental changes arising from neighboring polymer substituents are often obscured by parameters that are characteristic of the polymer matrix itself, for example, catalytic site accessibility or resin porosity.

Polymeric effects encountered in polymer-bound anion-activating macrocycles often arise because of differences in the stretching of the ionic bonds in the bound ion pairs as compared to ion pairs complexed to low molecular weight ligands. This is especially so when the polymeric ligands bind the reactant in the form of a 2:1 ligand-separated ion pair. Smid and his coworkers studied the crown and polycrown-catalyzed decarboxylation of potassium 6-nitrobenzisoxazole-3-carboxylate in benzene [103] and that of the acid itself in dioxane [104]. The reaction, shown below, proceeds quantitatively via a charge-delocalized transition state. Complexation of a crown or cryptand stretches the $COO^-—K^+$ bond and destabilizes the initial state more than the transition state, causing a large increase in the decarboxylation rate constant.

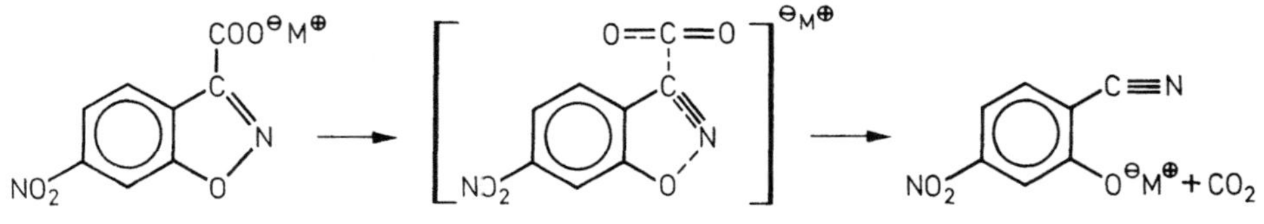

For S/PV-B18C6 polymers (structures *8*) the catalytic effect depends on the spacing between the B18C6 ligands along the chain. The first order rate constants, k (in sec^{-1}), at 25 °C in benzene are 0.9 (PV-B18C6), 0.54 (50% S/PV-B18C6), 0.21 (19% S/PV-B18C6), 0.17 (9% S/PV-B18C6) and 0.126 (5.5% S/PV-B18C6), while for the monomeric analogue M-B18C6, k equals 0.14 [103]. The polymeric effect probably results from a larger interionic COO^-—K^+ distance for ion pairs bound to polymers with a higher crown content, since close association of COO^- and the crown-complexed K^+ will become more difficult when other crown ligands are in the vicinity. In this respect it is interesting to find that the absorption maximum of the crown-complexed phenolate-type product shows a gradual bathochromic shift as a function of the crown content of the polycrown, i.e. from 444 nm for the complex with 1.6% S/PV-B18C6 and with M-B18C6 to 450 nm for the PV-B18C6 complex. Such a shift is usually caused by an increase in the interionic ion pair distance. The rate constant for the carboxylate-cryptand complex is five times that of the complex with PV-B18C6, the λ_m for the cryptand-phenolate product being 454 nm. Unfortunately, experiments with S/PV-B15C5 polymers which probably would have formed 2:1 crown-separated ion pairs with the potassium carboxylate were not successful due to the insolubility of the ion pair complexes.

Crown ethers are known to interact with protonic acids [92]. Therefore, it is not surprising that the decomposition of the acid itself is catalyzed by poly(crown ether)s [104]. In ethereal solvents, the decarboxylation of the acid in the absence of crown proceeds via the very reactive free carboxylate anions, as evidenced by the square root dependence of the observed rate constant on the inverse of the acid concentration, and by the inhibiting effect of a more dissociated acid like p-toluenesulfonic acid [105]. Crown ethers probably facilitate the ionization of the acid by forming the $RCOO^-H^+$ ... Cr complex [92]. Also in the presence of crown the observed rate constants are inversely proportional to the square root of the initial acid concentration. Their values for different crown ethers under otherwise identical conditions have been collected in Table 4. A considerable polymeric effect is observed. For example, 50% S/PV-B15C5 is nearly fifty times more active than M-B15C5. There are even significant differences between M-B15C5 and 5% S/PV-B15C5, and between M-B18C6 and 5% S/PV-B18C6. Apparently, neighboring comonomer substituents around the polymer-bound crown ligand alter the interionic ion pair distance in the $RCOO^-...H^+...Cr$ complex to such an extent that it significantly increases the fraction of free carboxylate anions.

Table 4. Observed Initial Rate Constants for the Crown-Catalyzed Decarboxylation of 6-Nitrobenzisoxazole-3-carboxylic Acid in Dioxane at 20 °C (Ref. [104])

Crown	$k_{obs,0} \times 10^4$ sec^{-1}	Crown	$k_{obs,0} \times 10^4$ sec^{-1}
M-B15C5	.59	M-B18C6	1.2
5% S/PV-B15C5	6.5	5% S/PV-B18C6	6.7
20% S/PV-B15C5	25	20% S/PV-B18C6	14
50% S/PV-B15C5	29	50% S/PV-B18C6	6.7
PV-B15C5	17.7	PV-B18C6	6.9

The homopolymers PV-B15C5 and PV-B18C6 together with their low molecular weight analogues B15C5 and B18C6 have been used by Akabori et al. in the allylation of sodium phenoxide [66]. This reaction can result both in oxygen (O) and carbon (C) alkylation, although the former is usually favored. However, C-alkylation is enhanced when phenoxide is stabilized by protonic solvents (e.g., water) or by tight ion pairing in solvents such as diethyl ether or benzene. Reaction yields are improved by adding crown compounds. This will promote transfer of the phenoxide to the organic allyl chloride phase, or enhance phenoxide solubility in benzene or ethyl ether while at the same time breaking down unreactive aggregates. The ligand addition also increases the O/O + C alkylation ratio in these solvents, by enlarging the interionic ion pair distance. PV-B18C6, and especially PV-B15C5, were found to be the most effective in this respect, an observation that is consistent with the trend in the optical absorption maxima of the polycrown phenolate complexes discussed in the previous paragraph.

Poly(crown ether)s have recently been used as catalysts in the conversion of salts of lower fatty acids (C_1–C_6) to the bromophenacyl- or pentafluorobenzyl esters for gas chromatographic determinations [73]. The esterification proceeds quantitatively under mild conditions in acetonitrile at room temperature. The use of low molecular weight crown ethers [106] has the disadvantage that they interfere with the gas chromatogram of the fatty acid esters unless they have been previously removed. The polycrown poly(methacryloylamino-B15C5) used by Shono et al. is of course non-volatile and does not require removal [73]. The strontium salts of the fatty acids have also been used recently in these reactions since strontium-coated glass beads are effective in preconcentrating fatty acids from ambient air [72]. The beads are then reacted with the derivatizing agent in the presence of polycrown, and the esters quantitatively analyzed by a flame ionization or electron capture detection system.

B.II.5 Polysoap-type Properties of Poly(crown ether)s in Aqueous Media

Dissolved in water, some poly(vinylbenzocrown ether)s possess polysoap-type properties due to a favorable balance between hydrophobic and hydrophilic regions in the polymer domain. As a result they can strongly interact with many different organic solutes. The best studied example is poly(vinylbenzo-18-crown-6) (PV-B18C6) which at 25 °C for $M_n = 106{,}000$ has a water solubility of 8 g/100 ml [107]. Like many other compounds with oligo-oxyethylene structures [108], PV-B18C6 exhibits inverse temperature solubility, the cloud point being 37 °C. The polymer of $M_n = 106{,}000$ has an intrinsic viscosity of only 0.107 dl/g, not much different from that of globular proteins. Apparently, the macromolecular chain in water assumes a compact conformation, with a hydrophobic interior consisting chiefly of the polystyrene backbone, and a more hydrophilic exterior where many of the macrocycles are hydrogen-bonded to water molecules to prevent precipitation. Poly(vinylbenzo glyme)s (PV-BG) exhibit a behavior similar to that of PV-B18C6 [98], and so does PV-B15C5, although its water solubility is much lower than that of PV-B18C6.

Binding of organic solutes to PV-B18C6 is largely hydrophobic in nature, but the neutral polymer has the interesting feature that it can easily be converted to a polycation of variable charge density by addition of crown ether-complexable cations such as K^+, Tl^+, Cs^+, Ba^{2+} and Pb^{2+}. This adds to the hydrophobic binding component an electrostatic force which makes it possible to regulate the binding of ionic and even some neutral organic solutes to PV-B18C6 by simply varying the concentration or type of added salt. This kind of control in solute binding is not found with PV-BGs since in aqueous solution cation-binding to the acyclic glyme ligands is very weak.

Over the past six years Smid and his coworkers have extensively investigated the interaction of optical and fluorescent dyes as well as other organic solutes with PV-B18C6 and PV-BG, and the effects of added salts. The work has recently been reviewed [61] and only some of the more interesting findings will be discussed here. The early studies dealt with the binding of anionic solutes like picrate and methyl orange to PV-B18C6 [107,109] and to a PV-BG polymer [98] consisting of a polystyrene backbone with $CH_3(OCH_2CH_2)_2O$-substituents at the 3 and 4 position of each benzene ring. Binding of picrate to the polymer domains is accompanied by a bathochromic shift from 354 to 384 nm, while a hypsochromic shift (465 to 435 nm) is found for methyl orange. A very large optical shift is found on binding the compound 2-(4'-hydroxybenzeneazo)benzoate to PV-B18C6. As also observed for binding to micelles and proteins, the structure of this dye changes from the azo form (λ_m 347 nm) into the more hydrophobic hydrazone form (λ_m 480–510 nm) [110]. The two neutral polymers PV-B18C6 and PV-BG also enhance the water solubilities of neutral aromatic hydrocarbons, especially that of pyrene and benzopyrene [111]. For pyrene, the 334 nm absorption band shifts to 341 nm for the polymer-bound species, and the equilibrium constant of binding can be determined spectrophotometrically.

More recent work has focused on the binding of neutral, anionic and cationic fluorescent dyes such as 1,8-anilinonaphthalene sulfonate (ANS) and 2,6-toluidinonaphthalene sulfonate (TNS) [112,113], auramine O (AuO, $[p\text{-}(CH_3)_2NC_6H_5]_2C = NH_2^+Cl^-$) [114,115], pyrene and ionic pyrene derivatives [111]. For ANS, TNS and AuO, all of which have been used extensively for probing the structures of micelles, proteins and synthetic polymers, the fluorescence quantum yield in water is very low, but the intensity is strongly enhanced when the dyes are bound to hydrophobic regions of a polymer. In the case of pyrene, the fluorescence spectrum of the polymer-bound molecule is different from that of free pyrene. Both optical and fluorescence binding data could be rationalized by assuming a simple Langmuir adsorption behavior, i.e., interactions between bound solutes can be neglected and binding is governed by statistical factors only. This is not surprising since even under saturation conditions the number of bound organic solute molecules per polymer chain is small in most systems. Deviations from Langmuir behavior is found when *cations* are complexed to the crown ligands of PV-B18C6 due to repulsion between the large number of bound cations [110].

The intrinsic binding constant (K), the first binding constant (K_1) and the minimum number (N) of monomer units per bound dye molecule can be determined from appropriate relationships derived on the basis of simple Langmuir adsorption behavior. The values of these constants, together with the thermodynamic binding parameters, are collected in Table 5. The intrinsic binding constants fall within the range $10^4 - 33 \times 10^4\ M^{-1}$, and compare favorably with those reported for the

Table 5. Intrinsic (K) and First (K_1) Binding Constants (25°) and Enthalpies and Entropies of Binding of Some Optical and Fluorescent Probes to PV-B18C6 and PV-BG

Polymer	Solute[a]	K $\times 10^{-4}$ M^{-1}	K_1 $\times 10^{-2}$ M^{-1}	N	ΔH Kcal/mol	ΔS eu	Ref.
PV-B18C6	Picrate	13.4	31.9	42	−11.9	−17	[107]
	MO	9.6	11.3	85	− 3.9	9.7	[107]
	TNS	6.18	24.1	26	− 6.7	− 0.64	[112,113]
	ANS	5.38	6.73	80	− 2.7	12.4	[112,113]
	AuO	2.2	3.17	67			[114,115]
	Pyrene	29	290	10	− 7.2	1.0	[111]
	PB	11.0	44.3	25	− 6.0	3.2	[111]
PV-BG	Picrate	23	60.5	38	−26.3	−64	[98]
	MO	6.9	23.8	29	− 7.8	− 3.8	[98]
	TNS	11.9	27.3	44	− 4.6	7.7	[113]
	AuO	1.2	0.6	200			[115]
	Pyrene	32.6	326	10			[111]
	PB	19	64	30			[111]

[a] MO = methyl orange; TNS = 2,6-toluidinonaphthalene sulfonate; ANS = 1,8-anilinonaphthalene sulfonate; AuO = auramine O; PB = 4-(1-pyrenyl)butyrate.

same solutes interacting with proteins or synthetic polymers like poly(vinylpyrrolidone). The N values vary considerably, and for PV-B18C6 range from 85 for methyl orange to 10 for pyrene. Hence, a PV-B18C6 chain of DP_n = 200 can bind a maximum of 20 pyrene molecules but only two methyl orange anions. It is not entirely clear what causes the differences in N values. Factors such as size, polarity and placements of ionic substituents in the solute will play an important role. These factors also determine whether the solute will be located more in the hydrophobic interior of the polymer domain or on its periphery. The negative binding enthalpies point to specific interactions between solute and the monomer units in the polymer, such as van der Waals and polar forces, in addition to hydrophobic contributions. The latter are dominant in processes where the entropy change is positive, e.g., in the binding of ANS to PV-B18C6.

The effect of *added salt* on the binding of the TNS anion to PV-B18C6 is shown in Fig. 4, while the effect on the cationic AuO is depicted in Fig. 5. In both figs. the ratio of observed fluorescence (F) to its maximum value F_m is plotted versus the

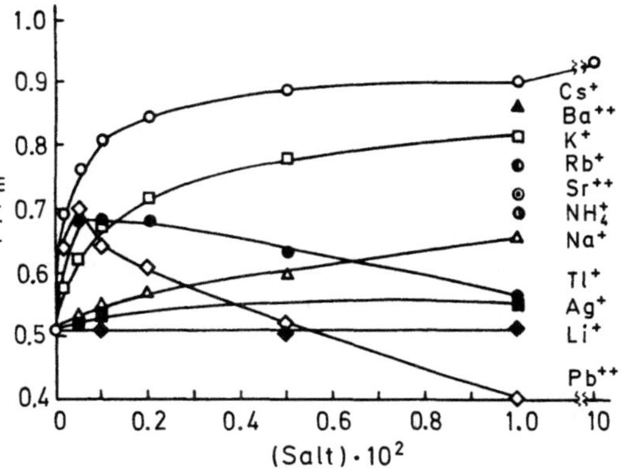

Fig. 4. Effect of added salts on the ratio F/F_m in the binding of 2,6-toluidinonaphthalene sulfonate (TNS) to poly-(vinylbenzo[18]crown-6) (PV-B18C6) in water at 25 °C. [TNS] = 2.72×10^{-7} M; [PV-B18C6] = 3.73×10^{-4} M. (Ref. [113])

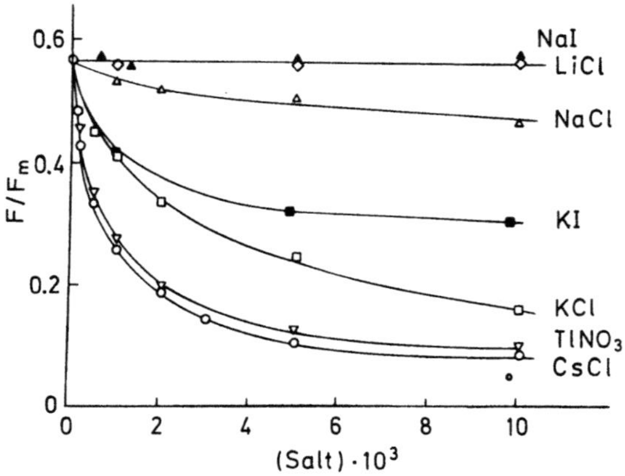

Fig. 5. Effect of added salt on the ratio F/F_m in the binding of auramine O to poly(vinylbenzo[18]crown-6) (PV-B18C6) in water at 25 °C. (Ref. [115])

added salt concentration. The value of F_m is obtained as the intercept of linear plots of 1/F versus 1/[polymer], and represents the fluorescence intensity under conditions when all solute is polymer-bound. Hence, F/F_m denotes the fraction of bound solute. When salts like KCl or CsCl are added, the cations will bind to the crown ligands of PV-B18C6. The number of crown-cation complexes per chain is a function of the type of salt and its concentration. Conversion of PV-B18C6 to a polycation will promote binding of anionic solutes like TNS, ANS, picrate or methyl orange, but it hinders the binding of cationic species like AuO. Figures 4 and 5 show that for alkali ions the salt effect decreases in the order $Cs^+ > K^+ > Na^+ \gg Li^+$.This order reflects the values of the binding constants of these ions to PV-B18C6, which are 300 M^{-1} for Cs^+, 110 M^{-1} for K^+ and 2 M^{-1} for Na^+ [110]. The effect of added LiCl is very small even at 0.5 M. Maxima in the TNS plots are found when Tl^+ or Pb^{2+} salts are added. Both cations are known to quench the fluorescence. Hence, as more TNS becomes bound, the expected increase in F is eventually more than offset by quenching when more and more Tl^+ or Pb^{2+} ions are bound to PV-B18C6 and the average distance between crown-bound Tl^+ or Pb^{2+} and the TNS^- in the polymer domain decreases.

The counterion of the added salt also affects the binding. This can be seen by comparing in Fig. 5 the results of alkali chlorides and iodides on the binding of AuO to PV-B18C6. More dramatic effects are found for the cationic solute 4-(1-pyrenyl)-butyltrimethylammonium bromide (PN^+) to PV-B18C6 [111], especially at high salt concentrations. The binding of this compound, like that of pyrene, can be studied by optical spectroscopy. Figure 6 shows that the fraction of bound PN^+ passes through a minimum, or immediately enhances on adding certain alkali salts in spite of the fact that conversion of PV-B18C6 to a polycation (e.g., with the potassium salts) is expected to hinder the adsorption of the PN^+ solute. Counterion binding to the polycrown-cation will of course increase in the order $Cl^- < Br^- < I^- < SCN^-$, and this can explain part of the behavior shown in Fig. 6. However, ion pairing of PN^+ with the salt anions cannot be neglected at the high salt concentrations, especially for softer anions like SCN^-. Formation of these ion pairs is likely to facilitate PN^+ binding to PV-B18C6. A very strong increase in the binding of PN^+

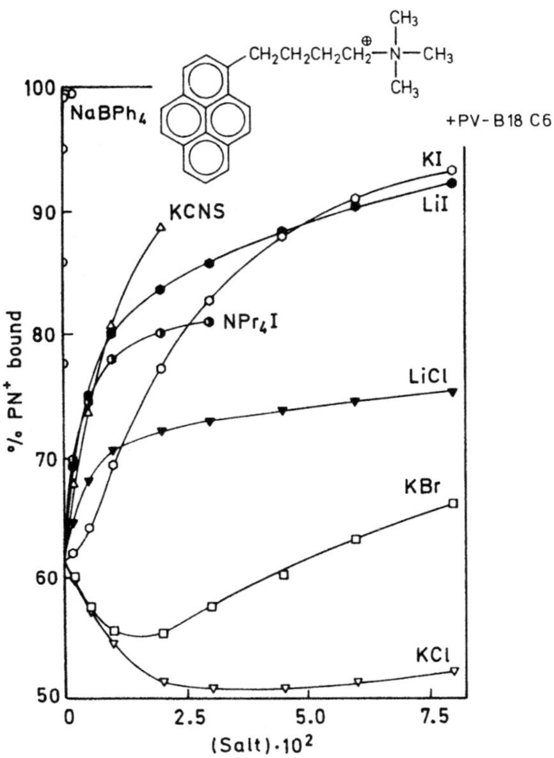

Fig. 6. Effect of added salt on the percent of 4-(1-pyrenyl)butyltrimethylammonium bromide (PN^+) bound to poly(vinylbenzo[18]crown-6) (PV-B18C6) in water at 25 °C. (Ref. [111])

is found on adding $NaBPh_4$. Optical and fluorescence studies in the absence of PV-B18C6 show that even at very low concentrations of PN^+ and BPh_4^- ($<10^{-4}$ M) a stable PN_2^{2+} $(BPh_4)_3^{3-}$ aggregate is formed. This aggregate apparently interacts very strongly with PV-B18C6. It is also found that while the N value for PN^+ with PV-B18C6 is 26 in the absence of $NaBPh_4$, it decreases to 2.5 in the presence of this salt, i.e., the capacity of PV-B18C6 to bind PN^+ dramatically increases on adding BPh_4^-. The fluorescence spectrum of the pyrene moiety clearly shows the excimer maximum, suggesting that the bound PN^+ ions are in close proximity to one another. They are probably located on the periphery of the PV-B18C6 chain, with BPh_4^- anions associated with quaternary ammonium ions of PN^+ molecules.

The property of PV-B18C6 to bind organic solutes to its hydrophobic domain has been applied to catalyze the decomposition of 6-nitrobenzisoxazole-3-carboxylate in aqueous media [116]. The reaction is the same as that given in Section B.II.4 where the decomposition of the potassium salt in benzene was shown to be accelerated by the anion-activating action of polycrown ligands. While in non-polar and aprotic media the free carboxylate anion is very reactive, it is stabilized by protonic solvents [117]. However, on adding PV-B18C6 to an aqueous solution of the carboxylate, the decarboxylation is accelerated as some of the reactant binds to the polymer, thereby changing its environment to a more reactive, hydrophobic one. The binding constant of this carboxylate to neutral PV-B18C6 at 25 °C is only 340 M^{-1}, much lower than for the dyes listed in Table 5. However, the fraction of bound solute can be increased by adding cations such as K^+ or Cs^+. Their addition to a PV-B18C6 solution of the carboxylate further increases the rate of decarboxylation until all carboxylate is polymer-bound. The studies show that the reactivity of carboxylate bound to neutral PV-B18C6 is 2300 times that of the free anion in water, while carboxylate bound to Cs^+-charged PV-B18C6 is 14,000 times more reactive. The increase in the decarboxylation rate constant essentially results from a decrease in the activation enthalpy of the reaction. Its values are 133 kJ/mole for the reaction in the absence of polymer, 88.2 kJ/mole for the reactant bound to neutral PV-B18C6 and 67.3 kJ/mole when the carboxylate is bound to the Cs^+-charged polymer. The respective activation entropies are 19, 2.3 and —10.7 e.u. [116].

Solutions of PV-B18C6 in water exhibit another interesting property. They form insoluble *complexes with polyacids* such as poly(acrylic acid) (PAA) [118,119] while cation-charged PV-B18C6 interacts with different polyanions to form insoluble *polyelectrolyte complexes* [71,120]. Figure 7 depicts plots of the fraction of precipitated

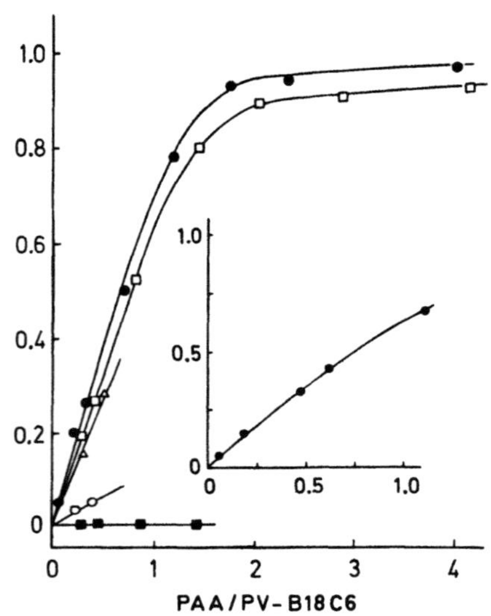

Fig. 7. Plot of the fraction of precipitated PV-B18C6 as a function of the ratio of poly(acrylic acid) (PAA) to poly(vinylbenzo[18]crown-6) (PV-B18C6) in water at 25 °C in the presence and absence of HCl. (□) 0.1 M HCl; (●) 0.01 M HCl; (△) 0.001 M HCl; (○) 10^{-5} M HCl; (■) no HCl added. $[PV\text{-}B18C6]_0 = 10^{-4}$ M. Insert: same plot as at 0.01 M HCl but for low ratios of PAA/PV-B18C6 (Ref. [119])

PV-B18C6 as a function of the ratio PAA/PV-B18C6 (polymer concentrations being expressed in moles/l of monomer base units) at different concentrations of HCl. The precipitation, which already occurs at polymer concentrations of 10^{-5} M, is most effective at pH = 2. The complexes are formed by cooperative hydrogen bonding between oxygen atoms of crown ligands and the carboxyl groups of the polyacid. At higher pH the sequences are interrupted by ionization of the carboxyl groups, and eventually the complexes solubilize. The precipitates at PAA/PV-B18C6 ratios of 0.5, 1.0 and 2.0 all have a composition of 1.8 COOH groups per crown ligand. Films of PAA-PV-B18C6 complexes can also be obtained at the interphase of a toluene solution of PV-B18C6 and a 0.01 N HCl solution of PAA [119]. The films can be drawn into a fiber yielding a complex that is much more stable than the precipitates formed in aqueous solutions. When dried they are insoluble in dioxane, a solvent that solubilizes the separate components PAA and PV-B18C6. The interfacial complexation permits the use of the water-insoluble S/PV-B15C5 and S/PV-B18C6 copolymers. The rate of formation of PAA complexes with these copolymers, and their strength as well, decreases with increasing styrene content.

On addition of crown-complexable cations, hydrogen bonding between PAA and PV-B18C6 can be prevented when enough crown ligands contain a cation [119]. The complexes, even in 0.01 N HCl, can be solubilized by adding 0.1 M KCl or 0.1 M CsCl. However, when these solutions are made basic, precipitation occurs in the form of polyelectrolyte complexes, a common phenomenon when polycations and polyanions are mixed. Insoluble complexes are formed between cation-charged PV-B18C6 and salts of carboxymethylcellulose [120], poly(styrene sulfonate) [71], poly(acrylate)s [71] and polynucleotides [121]. The precipitation is very sensitive to the number of crown ligands that contain a cation, and can be regulated by the type and concentration of salt. An interesting example of the formation of PV-B18C6 complexes with polynucleotides is the inhibitory effect of PV-B18C6 on the RNA-directed DNA polymerase activity from murine leukemia virus system, both in the absence and presence of crown-complexable cations like K^+ or Cs^+. Apparently, neutral PV-B18C6 interacts sufficiently strong with the polynucleotide poly(riboadenylate) to prevent its acting as a template in the polymerization reaction, although no polymer complex precipitates. Additional evidence for binding of PV-B18C6 to the polynucleotide was found by passing it over sepharose beads to which the polynucleotide was attached. Strong adsorption of PV-B18C6 was found both in the absence and presence of KCl, while no adherence is observed to sepharose itself [121].

Quivoron et al. [79] have used ^{13}C and ^{23}Na nmr to study complexation of alkali and alkaline earth cations to diaza-18-crown-6 (DA18C6) bound to soluble poly-(acrylamide)s in aqueous media (Structure *13*). A downfield chemical shift of about 3 ppm is found for the macrocycle carbon adjacent to the nitrogen binding site when Ag^+ is complexed to DA18C6. The other two types of carbon atoms show an upfield shift of about 0.5 ppm. Ba^{2+} and Ag^+ both give 1:1 complexes, and the viscosity of the polymer solutions in the presence of these ions increases.

^{23}Na relaxation time measurements provide information on the mobility and hydration of the macroheterocyclic ligand. The formation constant of the 1:1 Na^+ complex is 79 M^{-1}. Addition of Ba^{2+}, which competes effectively with Na^+ for binding to the macrocycle, modifies the Na^+ solvation sphere. The formation constant of the 1:1 Ba^{2+}-DA18C6 complex is obtained from the variation of the T_1 spin-lattice relaxation time as a function of the ratio Ba^{2+}/Na^+. Its value is 2.2×10^4 M^{-1}. The ^{23}Na nmr measurements provide insight into the electronic environment of the polymer-bound ion.

Finally, two studies have been reported on the use of polymer-bound benzocrown ethers in ion-selective electrodes. Based on their preference for K^+ ions, Kimura et al. [47] incorporated poly(acryloylamino-B15C5) polymers (Structure *10*, but with a poly(acrylamide) backbone) into a dibutylphthalate-plasticized PVC membrane. The selectivity coefficients, k_{KM}, defined as the ratio of the activities of K^+ and M^+ ion are 3×10^{-4} (Na^+), 2×10^{-1} (Rb^+) and 2×10^{-3} (Cs^+ and NH_4^+). Values for a styrene copolymer with 23% of crown monomer are slightly higher. All values are apparent values since the polycrown ethers, unlike bis(crown ethers) and other low molecular weight crown compounds respond slowly and do not give Nernstian or near-Nernstian behavior. However, the selectivity values are not very different from those of the corresponding bis(crown ether)s.

Fung and Wong [52] constructed K^+ and Cs^+-selective PVC membranes containing bis(crown ether)s and poly(vinylbenzocrown ether)s with B15C5 and B18C6 ligands. Selectivity coefficients comparable with that of the commercial valinomycin-based K^+-selective electrode were found for a B15C5 type bis(crown ether) in which the two benzene rings were linked by a CH_2OCH_2 chain. The same bis(crown ether) but with two B18C6 ligands was an effective Cs^+-selective electrode. Long response times were found with the corresponding S/PV-B15C5 and S/PV-B18C6 poly(crown ether)s, and were attributed to polymer incompatibility with PVC and to the low ion transport in the membrane [52].

C Macroheterocycles Anchored to Insoluble Supports

Macroheterocyclic ligands immobilized on insoluble supports offer distinct advantages over soluble compounds. In catalytic applications, the ease of recovery of the insoluble ligands by filtration facilitates product purification and prevents their contamination with toxic macrocycles. It also permits recycling of the rather expensive ligands, and the use of continuous flow reactors. Some of the materials are commercially available in small quantities. They have chiefly been applied for analytical purposes as chromatographic stationary phases in the separation of ionic and neutral inorganic and organic substances, and in preparative chemistry as heterogeneous phase-transfer catalysts.

C.I Applications as Exchange Resins in Chromatography

Because of their selective complexation of ions and neutral solutes, macroheterocyclic ligands have been extensively investigated for applications in chemical analysis [22]. Soluble macrocycles have been used as ion carriers in liquid membrane separations [122–124], and as complexing agents in the mobile phase in chromatographic separations [123]. It was especially Blasius and his coworkers [126] who explored in depth the analytical use of exchange resins with macrocyclic ionophores as anchor groups. Their work has been reviewed in a number of recent publications [67, 127–130], and only a few of the results will be discussed here.

Crown-and cryptand-containing exchange resins can be synthesized by common polycondensation processes, by copolymerization of vinyl-containing macrocycles in the presence of crosslinking agents, or by modification of existing resins or silica gel through substitution reactions with functionalized macrocycles. Some of the many different types of resin made by Blasius are shown below (*14–20*). Most condensation polymers were prepared by reaction of dibenzocrown ethers with formaldehyde in formic acid, yielding a network structure with methylene bridges (*14*). Also used were monobenzocrown ethers and monobenzocryptands with formaldehyde and additional crosslinking agents like toluene, xylene, phenol or resorcinol (*15*, *16*). Several crown ethers with oxygen and nitrogen binding sites have been employed, including macrocycles with urea structures (*17*). The same compounds can be attached to polystyrene surfaces via methylene bridges by reaction with formaldehyde. Another method involves coating of silica or polystyrene beads with polymeric ligands.

Blasius used hydroxymethyl, amino and ω-bromoalkylbenzocrown ethers to immobilize macrocycles on chloromethylated polystyrene resins and silica gel (*18*, *19*). They also made network polymers by polymerization of vinylbenzocrown ethers in the presence of divinylbenzene or divinyldibenzocrown ethers (*20*). This

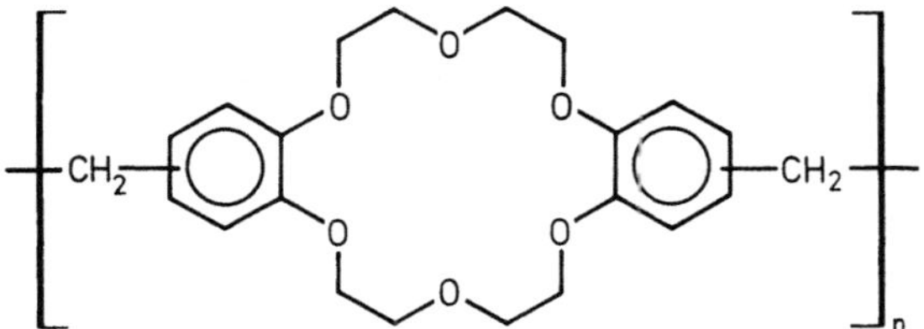

14

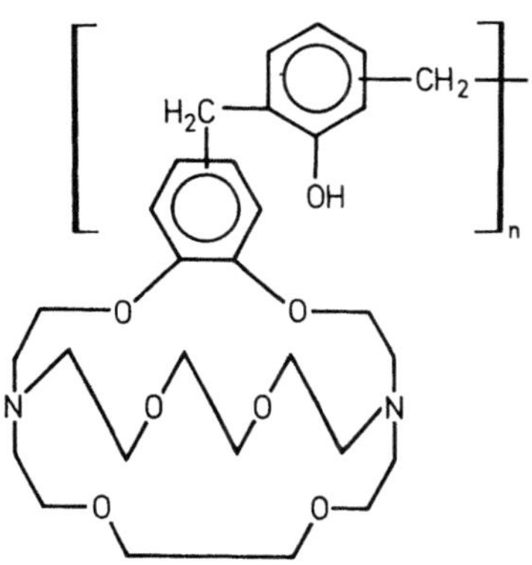

16

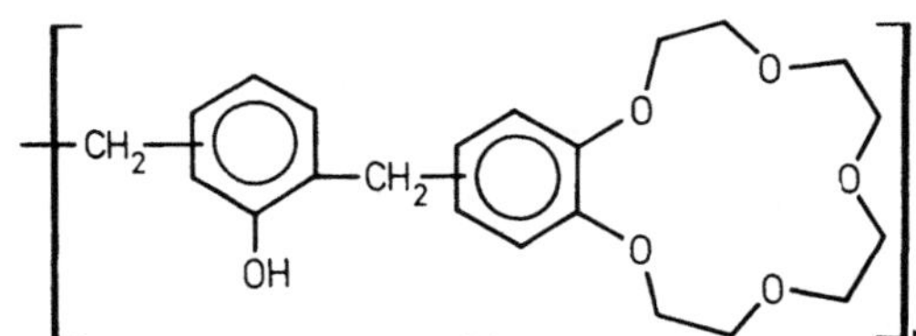

15

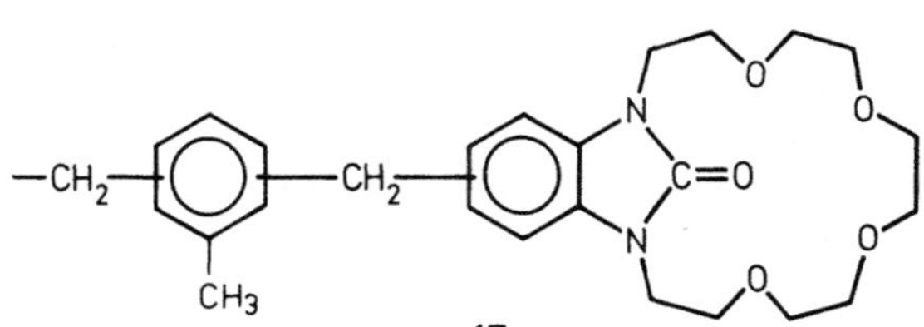

17

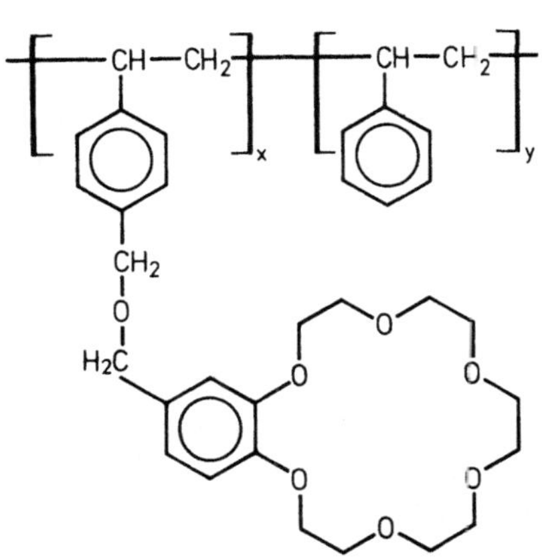

18

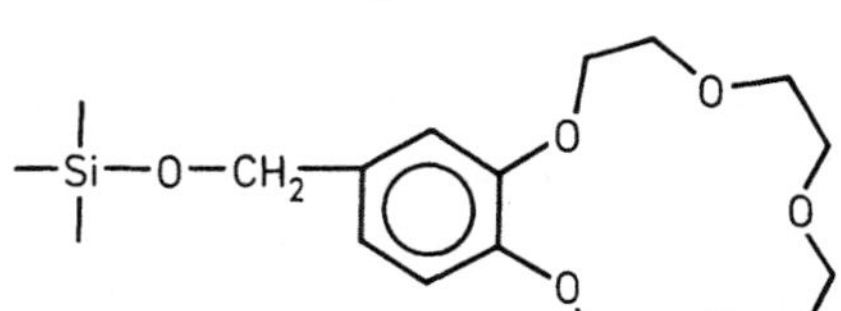

19

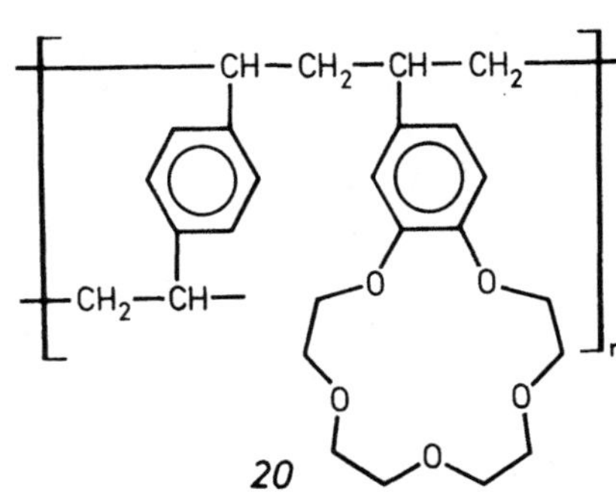

20

method provides an easy way to incorporate other monomers as well. All of the above exchange resins have been examined for use in thin layer chromatography and electrophoresis, in trace enrichment of radionuclides, in ion-selective membranes and as chromatographic stationary phases. Compared to commercial ion exchangers the resins are resistant to chemicals, heat and radiolysis. Since they contain neutral ligands, uptake of a cation is accompanied by anion binding to preserve electroneutrality. This means that the same material can separate either cations or anions.

Thermal stability of the network diminishes from about 250° to 180° with increased size of the polyether ring. Condensation resins are more stable than copolymerization materials. The former are also more resistant to organic solvents. Pure solvents such as water or methanol can function as eluents. No suppressor column is needed, and no time is lost in regenerating the stationary phase. It takes more time to reach equilibrium in water than in methanol, and is achieved more rapidly for condensation resins. Binding equilibria with silica-immobilized crown compounds are rapid enough to allow application of high-performance liquid chromatography. However, the silica-based materials hydrolyze rather easily and cannot be used in alkaline or acidic media. The other exchange resins are usable over the entire pH range, a distinct advantage over the sulfonated and aminated ion exchangers. Complex stabilities decrease at lower pH with nitrogen-containing macrocycles, and no complexation occurs below pH 2 due to extensive protonation of the N-binding sites.

Load capacities, stability of complexes and cation selectivities depend on solvent, anion, cation, temperature and type of ligand. Since binding is accompanied by removal of part or all of the cation solvation shell, complexes are more stable in methanol than water. In both solvents cation selectivities for all resins parallel those found for the corresponding free ligands. Salt binding generally decreases for less polarizable anions, i.e., $SCN^- > I^- > NO_3^- > Br^- > Cl^- > OH^- > F^-$, with the latter two ions giving sometimes a reversed order.

Clean solute separations can be accomplished with carefully packed columns and by choosing the proper resin-solvent combinations. Alkali halides can be separated with high-performance liquid chromatography using the silica resin *19* and methanol as eluent [129]. These halides can also be resolved with a DB24C8 condensation resin and water as eluent [128]. The latter crown is especially effective in binding Cs^+ and will separate this ion from medium-active waste solutions [127]. Shono et al. [132] reported recently an effective separation of alkali halides with silica-supported PV-B15C5 and a water-methanol (9:1) mixture as eluent (Fig. 8). The exchanger was synthesized by reacting aminopropyl-modified silica with methacrylic anhydride in dichloromethane followed by copolymerization with methacryloylamino-B15C5. The elution diagram clearly shows the earlier reported tendency of this polymeric ligand to form stable 2:1 complexes with Rb^+ and especially K^+, and the observed selectivity parallels that found for the soluble poly(B15C5). A silica-supported B15C5 resin did not yield a clean separation, although Blasius et al. used this support successfully in alkali halide separations [129].

Separation of multivalent ions can also be successfully accomplished. Figure 9 shows an elution diagram for a mixture of chlorides of Mg^{2+}, Ca^{2+}, Sr^{2+} and Ba^{2+} using the urea-type resin *17* and methanol as eluent [129]. Other separations

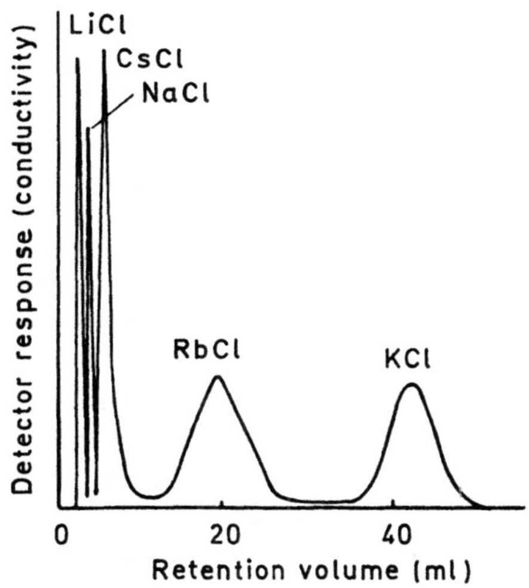

Fig. 8. Separation of alkali metal chlorides with poly(crown ether)-modified silica by elution with water/methanol (9/1). LiCl: 1.0 µg; NaCl: 1.4 µg; KCl: 2.8 µg, RbCl: 6.2 µg; CsCl: 7.3 µg (Ref. [132])

include $ZnCl_2/HgCl_2/CdCl_2$ [128], $FeCl_2/FeCl_3$ [67] and $CrCl_3/FeCl_3$. Blasius also reported the enrichment and separation of radioisotopes, e.g., $^{133}Ba/^{226}Ra$ [67], $^{85}Sr/^{152}Eu/^{133}Ba$ [67] and the removal of ^{90}Sr from large quantities of Ca (molar ratio 1:100,000). Water bound to salts or organic compounds can be determined quantitatively with DB21C7 and DB18C6 resins. Amounts of water can be calculated from calibration curves of known methanol-water mixtures. The exchange resins are also effective in separating anions. For example, resin *19* with water as eluent completely separates a mixture of 0.42 mg NaF, 0.35 mg NaCl, 0.62 mg NaBr, and 2.25 mg NaI [131]. The same resin also separated Na_2SO_4/NaCl up to a 1:10,000 molar ratio [131]. Clean separations also occur with ionic and neutral solutes. Examples are mixtures of alkylated ammonium chlorides [130], a mixture of furan,

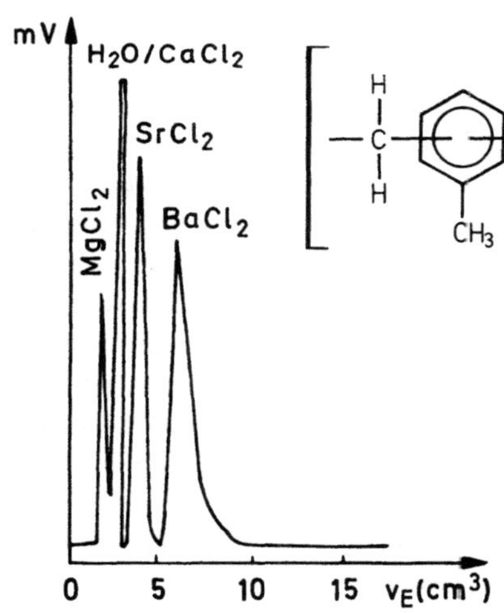

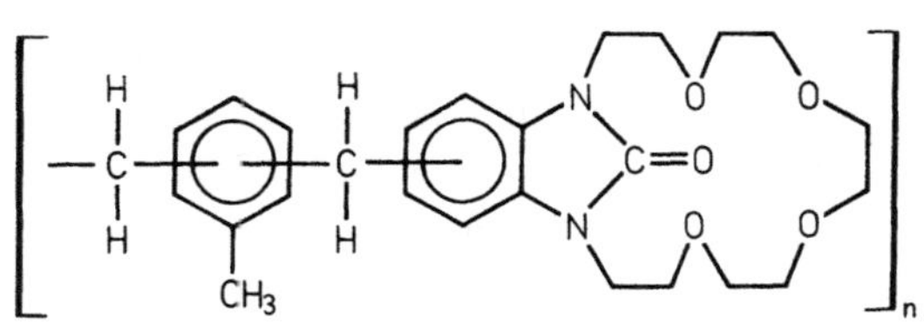

Fig. 9. Separation of alkaline earth metal chlorides with a crown ether containing exchanger (Structure *17*) by elution with methanol; 1.2 mg of $MgCl_2$, 0.1 mg of $CaCl_2$, 1.3 mg of $SrCl_2$ and 1.9 mg of $BaCl_2$. (Ref. [129])

thiophene and pyrrole chromatographed on a DB30C10 resin [127], mixtures of B-vitamins and of antibiotics [127], and the separation of the active substances (phenacetin, acetylsalicylic acid and caffeine) of an analgesic [127].

Exchange resins can be applied as adsorbents in thin layer chromatography and electrophoresis. For example, poly(ethylene terephthalate) sheets coated with suspensions of powdered resins in poly(vinyl alcohol) solutions are effective in separating neutral amino acids (isoleucine, valine, serine, α-aminobutyric acid) which on paper show non-resolved zones. Mixtures of the preservatives benzoic, salicylic and sorbic acid can be separated on a DB18C6 resin (*14*) with better results than when cellulose or Kieselgur layers are used.

Preparative applications of matrix-anchored macrocycles include salt conversions and anion-activated catalysis. The latter is discussed in the last section. Since cation complexation to the resin is anion-dependent, salt conversions can be quite easily accomplished. For example, gram quantities of $Sr(SCN)_2$, $Ba(SCN)_2$ and SrI_2 were prepared with resin *14* from mixtures of the chloride salts with either LiSCN or LiI [127]. Organic compounds have been purified on the same exchangers by separating out small amounts of water, salts and other impurities.

Recent efforts by other researchers to synthesize macroheterocycle-containing exchange resins include the anchoring of crown compounds to silica gel [133] and diatomaceous earth [134], and the coating of silica with the polycondensation product of diamino-DB18C6 and isophthaloyl chloride (*1*) [135]. The latter resin gave poor cation separation, but a reasonably clean resolution of a mixture of the anions SCN^-, I^-, NO_3^-, Br^-, Cl^-, F^- and SO_4^{2-} at 0.1 M salt concentration. Quivoron and his coworkers anchored DA18C6 to an acrylamide gel [136]. In static conditions, selectivity for alkaline earth cations is better than for alkali ions, and high extractions are achieved for Ag^+, Cd^{2+} and Hg^{2+}. Although retention volumes differ for the various salts, complete cation separation was not achieved. Frère and Gramain reported the immobilization of DA18C6 and cryptand-type ligands on chloromethylated polystyrene resins, including the polycryptand shown as structure *6* [137]. Selectivities parallel those found for the free ligands, but binding constants are higher for the resins. Resolutions are limited by slow cation-diffusion inside the resins and by low resin capacities (0.4–0.9 mmole M^{n+}/g). Yagi and Sanchez immobilized benzocrown ethers by thermal reaction of poly(2,3-epoxypropylmethacrylate) with amino-B15C5 and amino-B18C6 on the surface of chromosorb [75]. Separation of Na and K in methanol was especially effective with the B18C6-bound ligand. Shinkai recently bound one of his photoresponsive crown ligands (see Sect. B.I) with an azobenzene[24]crown-8-azobenzene structure in a bridge-type fashion to crosslinked polystyrene beads in an attempt to photocontrol the complexation of cesium to this cesium-selective crown ligand [138]. When the cesium-containing resin was photo-irradiated in dimethylformamide (which for the corresponding monomeric ligand converts 32% of the *E*- into the *Z*-isomer), most of the cesium (in the form of cesium p-nitrobenzoate) was released. It was suggested that the conversion of trans to cis changes the conformation of the crown ring into a more stretched one which is less effective in binding the cesium cation. No difference was found in the cesium extraction capabilities of the *monomeric* *Z*- and *E*-isomers. The rate of thermal isomerization of *Z*- into *E*-isomer for the polymer-bound ligand is the same as that for the free azo compound [138].

Of special significance has been the work of Cram and his coworkers who accomplished the resolution of racemic amino acid and ester perchlorate and hexafluorophosphate salts by means of chiral binaphthyl crown ethers [4,8] immobilized on silica gel [139] and macroreticular polystyrene [140,141]. Separation factors for racemic mixtures varied from 1.4 to 26, and resolution factors from 0.21 to 4.5. Increased loading of the chiral sites improved the resolution. This was interpreted to indicate that for sterically more confined sites where binding is more difficult the chiral recognition properties are better [141].

Useful separations of ionic solutes have also been accomplished with stationary phases of porous glass, silica and styrene-divinylbenzene copolymers to which electrically neutral, *acyclic* ionophores such as the podand-type ligands [17] have been anchored. While these immobilized ligands are not the subject of this review, they should be considered as possible alternatives for the often more expensive macroheterocyclic-containing resins, especially where it concerns catalytic applications. For a discussion of the use of such resins as stationary phases in liquid-solid chromatography the reader is referred to a review published recently by Grossman and Simon [142].

C.II Binding of Ion Pairs to Crown Ether Networks

Most work on immobilized macroheterocycles has been focused on their effectiveness in separating solutes and as phase-transfer catalysts in modifying reaction rates, yields and distribution of products. Few quantitative measurements have been reported on the binding of ionic solutes to immobilized macrocycles in low polarity media which are frequently employed in phase-transfer catalysis. Smid et al. recently determined formation constants of alkali picrate ion pair complexes with polystyrene-immobilized benzocrown ethers in ethereal solvents as a function of the crown content of the network and the spacer linking the crown ligand with the polymer backbone [143–146]. Use of picrate salts allows spectrophotometric determination of unbound solute at concentrations as low as 10^{-5} to 10^{-6} M. This minimizes their association to aggregates higher than ion pairs.

The networks used in these measurements have the same general structure *18*. Two low capacity resins R-B15C5 and R-B18C6 were synthesized from 0.9 meq Cl/g crosslinked chloromethylated polystyrene and possess on the average one benzo[15]crown-5 (R-B15C5) or benzo[18]crown-6 (R-B18C6) ligand per eleven styrene units. A 5 meq Cl/g resin was used in synthesizing the high capacity resins 5R-B15C5 and 5R-B18C6. They contain about one benzocrown ether per two monomer units. A fifth resin, RD-B15C5 has one B15C5 ligand per thirteen styrene units and a $CH_2O(CH_2)_{10}$-OCH_2 spacer between the benzocrown ligand and polystyrene. The four other networks have a CH_2OCH_2 spacer (structure *18*).

Equilibrium between network (a few mg) and a 5 ml picrate solution ($\approx 10^{-4}$ M) is reached in about 30 min. Picrate uptake in the network is determined spectrophotometrically at different initial picrate concentrations. Intrinsic binding constants, K_N, for the reaction $Pi^-M^+ + Cr^* \rightleftarrows Pi^-M^+Cr^*$ (Cr^* is a network-bound ligand) can be computed from the familiar Klotz relationship for solute binding to macromolecules: $1/R = 1/n + 1/nK_NA$. In this expression, 1/R denotes the ratio of total crown (in equivalents) to bound picrate, A is the measured free picrate concentration in solution and 1/n equals the number of crown ligands bound to one cation (1/n equals 1 or 2). Figure 10 depicts 1/R versus 1/A plots for binding of sodium, potassium and cesium picrate to R-B18C6 in THF at 25 °C. Linear plots with an intercept 1/n = 1 are found in all systems that form only 1:1 complexes, e.g. Na^+ with all five networks, and K^+ with R-B18C6 and 5R-B18C6. Linearity is maintained close to the point of maximum loading even for the high capacity networks. This implies that attractive or repulsive interactions between bound ion pairs do not play a significant role, at least in ether-type solvents.

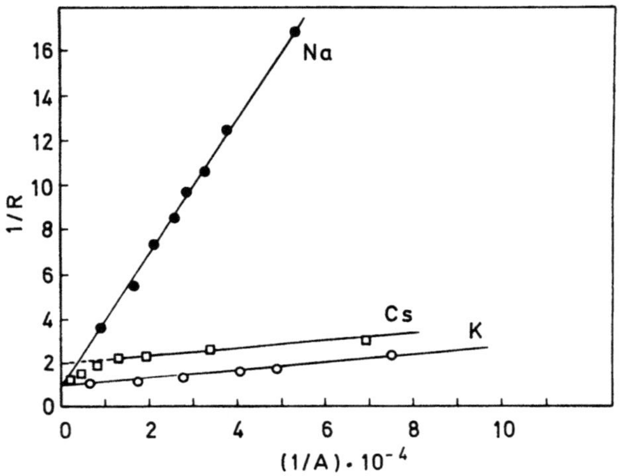

Fig. 10. Binding of alkali metal picrates to R-B18C6 in THF at 25 °C. (Ref. [143])

In a number of systems the binding plots are only linear above a certain value of 1/R, with an extrapolated intercept of 1/n = 2. However, close to the point of saturation (low 1/R values) the plots often curve downward to an intercept close to unity (Fig. 10 for R-B18C6 with cesium picrate). The intercept 1/n = 2 is found in systems that form stable 2:1 crown-cation complexes, for example, K^+ with R-B15C5 or 5R-B15C5, and Cs^+ with the two B18C6 networks. When more picrate is forced into these networks, the 2:1 crown-cation complexes can convert to 1:1 complexes: $Pi^-Cr^*M^+Cr^* + Pi^-M^+ \rightleftarrows 2\,Pi^-M^+Cr^*$, similar to what is observed with linear polymers (Sect. B.II.2). This causes the downward curvature at low 1/R values.

Complex formation constants for several network-salt systems in dioxane, tetrahydrofuran, 2-methyltetrahydrofuran and tetrahydropyran are collected in Table 6. As expected, binding is solvent dependent. The requirement to fully or partially desolvate the cation during complex formation lowers K_N in a better cation-coordinating solvent like THF. This effect is more pronounced for small cations, and the K_N differences between dioxane and THF, therefore, decrease in the order $Na^+ > K^+ > Cs^+$. Release of solvent molecules on picrate salt complexation to these resins is also implied by the positive entropy changes of about 8 e.u. derived from temperature dependent binding measurements.

Table 6. Intrinsic Binding Constants, K_N, of Alkali Picrates to Immobilized Crown Ethers in Ether-Type Solvents. T = 25 °C[a] (Ref. [144])

Network	$K_N \times 10^{-3}\ M^{-1}$										
	Sodium				Potassium				Cesium		
	THF	MeTHF	THP	DIOX	THF	MeTHF	THP	DIOX	THF	MeTHF	DIOX
R-B15C5	1.66	22.8	16.6	15.5	20.8	43.3	15.2	220*	1.3		(3.5*)
5R-B15C5	2.20			58.0	79*			350*	(55*)		66*
R-B18C6	3.41			25.8	54			172	125*		(150)
5R-B18C6	6.80			97.0	100			258	(96*)		410*
RD-B15C5	1.60	36.5	18.9			59.4	26.7			24.4	

[a] Values with an asterisk were calculated by using $1/n = 2$. For all other constants, $1/n$ equals unity. Bracketed values were derived from plots that were slightly curved.

The high capacity networks 5R-B15C5 and 5R-B18C6 invariably yield higher binding constants than the low capacity ones, irrespective of whether 1:1 or 2:1 crown-cation complexes are formed. Studies with linear poly(crown ether)s have shown that the stability of 2:1 complexes decreases on spacing the ligands farther apart, and eventually only 1:1 complexes may exist. This appears to be the case for RD-B15C5 and Pi^-K^+ where a $1/n$ value of 1 was found. The increase in K_N with crown content is especially large for cesium picrate when R-B15C5 is replaced by 5R-B15C5. The large interionic ion pair distance favors the formation of a 2:1 complex for this salt. Also, its 1:1 complex with a B15C5 ligand is not very stable. Where only 1:1 complexes are formed, K_N values for the high capacity networks are still higher by as much as a factor four (e.g., compare K_N values of Pi^-Na^+ with R-B18C6 and 5R-B18C6). The microenvironments around a crown binding site are different for low and high capacity networks. When the crown content is high, specific interactions may exist between the π cloud of a picrate anion and that of a benzocrown adjacent to the cation-crown complex. Or a crown-bound cation may acquire an extra oxygen binding site from an adjacent empty crown ligand. Insertion of a $(CH_2)_{10}$ spacer in RD-B15C5 appears to increase the K_N of 1:1 complexes by less than a factor two. Increased spacer length is known to improve the catalytic activity of immobilized macrocycles.

Binding constants of ionic solutes to networks can usually be measured in low polarity solvents if the anion is a chromophore. For salts such as alkali metal tetraphenylborates, competition measurements can be used. For example, $NaBPh_4$ competes effectively with sodium picrate for binding sites. It is possible to arrive at the binding constant of $NaBPh_4$ by measuring the amount of picrate released from the network on adding increased amounts of the borate salt [143,144]. The results show that the binding constant of $NaBPh_4$ to R18C6 in dioxane is at least fifteen times that of sodium picrate. The large interionic ion pair distance in $NaBPh_4$, and the resulting high electric field strength of Na^+, is the reason for the high binding constant of this salt.

Two observations pertinent to the use of immobilized crown ether ligands as anion-activating catalysts may be stressed. Firstly, high capacity networks may have higher binding constants with ionic reactants and also favour formation of 2:1 crown-complexed loose ion pairs in systems where their formation competes favorably with that of 1:1 crown-complexed tight ion pairs. Since the latter are often less reactive than loose ion pairs, a high capacity network may exhibit better catalytic properties than a corresponding low capacity one. Secondly, the catalytic activity may decrease at increased loading of active sites if this results in the conversion of 2:1 crown-separated ion pairs into 1:1 crown-complexed tight ion pairs. Other factors such as differences in the microenvironment around the bound reactant will of course also affect the catalytic properties of the network.

Immobilized macroheterocycles can be used effectively in studying the interactions of soluble ligands with ionic solutes in apolar media[142,146]. When adding a soluble ligand (L) to a solvent containing a solute A^-M^+ bound to a crown ether-containing network, A^-M^+ will be released into the solvent if L can compete effectively with the immobilized crown (Cr*) for the solute:

$$A^-M^+Cr^* + L \overset{K}{\rightleftarrows} A^-M^+L + Cr^* \tag{1}$$

Assuming that only 1:1 complexes are formed, it can be easily shown that

$$[(1/F_2) - 1] = K[(1/F_1 - 1)] \tag{2}$$

where F_1 denotes the fraction of ligand L bound to picrate salt and F_2 the fraction of immobilized crown ether complexed to A^-M^+. It can also be easily verified that $K = K_L/K_N$, where K_N is the binding constant of A^-M^+ to the network and K_L is the formation constant of the complex A^-M^+L, i.e., $A^-M^+ + L \rightleftarrows A^-M^+L$. Hence, if K is measured for a network-ionic solute system for which K_N is known (e.g., any of the systems of Table 6), K_L can be determined. In a solvent such as toluene, K_N cannot be measured due to the insolubility of salts like alkali metal picrates. However, K can still be determined by solubilizing the ionic solute with a small amount of ligand L, followed by addition of sufficient network to bind most of the salt, then adding more ligand to release the salt again as A^-M^+L.

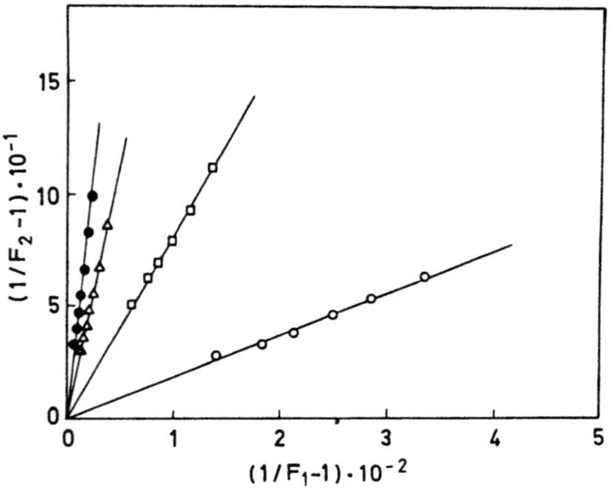

Fig. 11. Plots of equation (2) for the release of R-B18C6-bound potassium picrate in toluene on addition of [18]crown-6 (●), dicyclohexano[18]crown-6 (△), 4′-methylbenzo[18]crown-6 (□) and dimethyldibenzo[18]crown-6 (○). (Ref. [145])

Table 7. Equilibrium Constants, K, and Formation Constants, K_L, for Complexes of Potassium Picrate with Crown Ether Ligands in Dioxane at 25 °C with R-B18C6 and 5R-B18C6 as Network Polymers (Ref. [146])

Ligand[a]	R-B18C6[b]		5R-B18C6[b]		R-B18C6 (Toluene)
	K	$K_L \times 10^{-4}$, M^{-1}	K	$K_L \times 10^{-4}$, M^{-1}	K
18C6	5.62	96.7	3.76	97.0	4.09
DCH18C6	3.13	53.8	2.19	56.5	2.08
M-B18C6	0.91	15.6	0.60	15.5	0.60
DM-B18C6	0.595	10.2	0.437	11.3	0.206
PV-B15C5	1.05	18.0			
50% S/PV-B15C5	0.77	13.2			
10% S/PV-B15C5	0.453	7.8			

[a] 18C6 = [18]crown-6; DCH18C6 = dicyclohexano[18]crown-6; M-B18C6 = 4′-methylbenzo-[18]-crown-6; DM-B18C6 = dimethyldibenzo[18]crown-6 (cf. p. 11); PV-B15C5, 50% S/PV-B15C5 and 10% S/PV-B15C5 (see Section B II 1, Structures *8*).

[b] K_N's for R-B18C6 and 5R-B18C6 with Pi^-K^+ in dioxane are 1.72×10^5 M^{-1} and 2.58×10^5 M^{-1}, respectively.

Plots of $1/F_2 - 1$ versus $1/F_1 - 1$ for the system potassium picrate/R-B18C6/toluene with four different crown ligands are shown in Fig. 11. Good linear plots are obtained with other networks and ligands as well. Curvatures are found when the ligands form both 1:1 and 2:1 complexes, but by using appropriate equations the formation constants of both complexes can be computed [146]. Curvatures can also be expected when the solute-network combination does not give linear Klotz plots, e.g., when 1:1 and 2:1 complexes are formed simultaneously in the network.

K and K_L values for potassium picrate with several monomeric and polymeric crown ligands in competition with R-B18C6 or 5R-B18C6 are listed in Table 7. Similar data were obtained by this method with other soluble ligands, and also with sodium picrate. A comparison of columns two and four in Table 7 demonstrates the reliability of the procedure. When two networks differing in K_N are used for a particular ligand, the K values will differ but K_L should be the same. The data obtained for four ligands with either R-B18C6 or 5R-B18C6 differ by less than 4%.

The competition equilibrium constant K for R-B18C6/B18C6 in dioxane is 0.91 (Table 7). The structures of the soluble 4′-methyl-B18C6 and the immobilized crown ligand are nearly identical, and K = 0.91 implies, therefore, that with picrate salts in dioxane the binding constant to a soluble crown ligand is not much changed when it is anchored to a polystyrene resin, at least for a low capacity network and as long as only 1:1 complexes are formed. Whenever the ligand and cation can form a 2:1 complex, ligand immobilization will make it a more effective complexer, especially in high capacity networks where cooperativity of ligands is favored. Also, for 1:1 complexes the immobilized crown appears to be a better complexer in high capacity networks, e.g., K for Pi^-K^+ with 5R-B18C6/B18C6 is only 0.60 (Table 7). In toluene, only K values can be obtained. A comparison of the first and last column of Table 7 reveals that the complexation order 18C6 > DCH18C6 > 4′-methyl-B18C6 > dimethyl-DB18C6 for potassium picrate is the same in dioxane as in toluene. However, the absolute values are lower in toluene by as much as a factor three, implying that in this solvent the cation affinity of an immobilized ligand is increased relative to that of the free ligand. Differences between microenvironmental effects in the dioxane and toluene swollen networks are possible causes for the changes in K.

The competition method is useful for either monomeric or polymeric ligands (Table 7). In fact, any anion or cation-binding ligand can be used as long as it competes effectively with the crown ligand for salt. We recently applied the method to soluble oligo-oxyethylene (glyme) compounds, sulfoxides and polyamines in conjunction with oligo-oxyethylene-containing resins. Complications can arise in more polar solvents due to increased dissociation of the ion pair-ligand complexes into free ions. This can lead to curved plots since binding to free ions is often much stronger than to ion pairs. Difficulties are also encountered in poorly-swollen networks where binding equilibria establish slowly.

C.III Catalytic Properties

Polymer networks and other insoluble supports have been extensively used for anchoring reactive groups and catalysts [147,148]. Considerable interest has recently been focused on polymer-supported phase-transfer catalysts for application in either liquid/liquid/solid or solid/solid/liquid triphase catalysis [149,150]. The most widely investigated catalytic systems are immobilized organophilic quaternary ammonium and phosphonium salts. Microporous, 1–4% crosslinked divinylbenzene-styrene resins have been the most common supports used [147,148,150]. Macroporous [151–153] and popcorn [151] polymers have also been utilized, but the experimental evidence appears to indicate that these materials do not offer any special advantages over gel-type networks [152]. A highly porous "macronet"-type support derived from the reaction of linear polystyrene with 1,6-dibromohexane in the presence of $AlCl_3$ and toluene, and carrying phosphonium salt exhibits higher catalytic activity than an analogous microporous resin [154]. Silica-based onium salts have also been reported [155,156], but they are not useful in high alkaline media.

Polymer-supported onium salts derived from chloromethylated polystyrene resins and substituted phosphines and amines have been employed as catalysts in a number of nucleophilic displacement reactions [150,157–159]. In most cases, the polymer-supported catalysts are as effective as their soluble analogues in terms of percent conversion and product distribution, although reaction rates are usually lower due to decreased accessibility of catalytic sites. Rates appear to be linearly dependent on substrate concentration [151,160,161] and the amount of catalyst [151,160]. Increased network capacity in terms of the ratio of catalytic sites to styrene monomer units usually enhances the reaction rate per unit concentration of sites [159,162]. Higher crosslink densities generally lower reaction rates [153,163] due to poorer site accessibility, and in some cases the catalysts exhibit substrate selectivity depending on size and crosslink density [164]. Immobilized ammonium salts are usually less stable and not as efficient as phosphonium salts [159,165]. They can lose activity upon recycling as a result of dequaternization [166].

Crown ethers and cryptands have been examined extensively for use as phase-transfer catalysts [167,168]. They are thermally more stable when immobilized than the onium-based catalysts. The synthesis of the different types of immobilized macroheterocycles applied as catalysts follows the same procedures as described in Section C.I for chromatographic stationary phases. The most frequently used method is the reaction of an appropriately functionalized macrocycle with a preformed network, usually a chloromethylated polystyrene resin. The disadvantage of this procedure is that complete reaction is often difficult to achieve, and that side reactions also may occur. These factors can introduce unwanted functionalities into the network which may influence its catalytic behavior.

Montanari and Tundo have used both hydroxy and amino derivatives of aliphatic crown ethers and cryptands for anchoring macrocycles to polystyrene resins of different crosslink densities [160,165,169–171]. Most reactions concern the conversion of octyl bromide into iodide, cyanide or thiophenolate by means of the corresponding alkali metal salt under phase-transfer conditions, using a toluene/water mixture. The results show that immobilized cryptands are more effective than the crown ligands or quaternary ammonium salts. Resins with higher crosslink densities are less effective. Using the crown network shown as structure *21*, Montanari and Tundo found that an increase

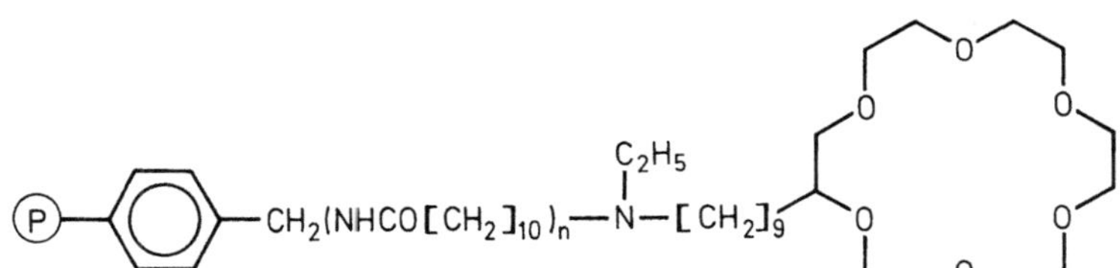

21

in spacer chain length from 11 to 23 to 35 atoms increased the observed pseudo-first order rate constant of the reaction between octylbromide and KI in toluene by not more than 30% [170]. This spacer length effect tends to level off in more polar solvents like chlorobenzene. In a poor swelling solvent like n-heptane the 11 chain catalyst (n = 0) gives a four times lower rate constant than in toluene. The rate constant increases by a factor three in heptane when n is increased to 2 (35 atoms). Similar spacer length effects have been reported for ammonium [172] and phosphonium salts [158]. Improved yields in the Merrifield synthesis of polypeptides upon lengthening the connecting chain is another example of this effect [176]. The spacer chain allows the catalytic site to protrude into the solution and become more accessible and solvated. Higher binding constants of picrate salts to immobilized crown ethers on increasing the spacer length was mentioned earlier (see Sect. C.II). Montanari and Tundo conclude that diffusive factors are minimized by a combination of low crosslink density and a spacer chain of at least ten atoms. The activity of such immobilized catalysts approach those of comparable soluble phase-transfer catalysts [170]. Low crosslink densities unfortunately decrease the mechanical stability of the resins and breakdown of the material is often noticeable under prolonged stirring.

Tomoi and his coworkers catalyzed reactions similar as those reported by Montanari and Tundo by using aliphatic crown ethers and cryptands (*22*, *23*) bound to macroreticular chloromethylated resins (10% divinylbenzene) [174,175]. They also

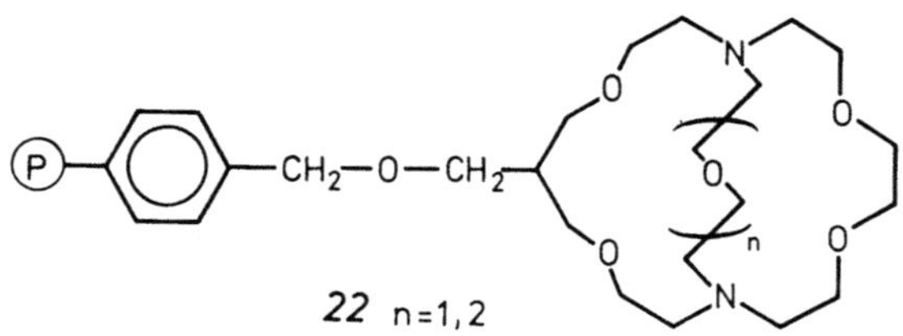

22 n=1,2

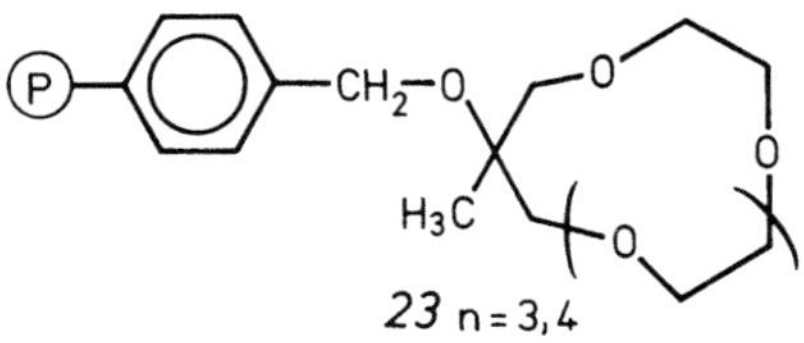

23 n=3,4

found the cryptand to be a more active catalyst than the crown ligand, but it lost its activity in the octylbromide-KI exchange reaction due to formation of triiodide. The activity can be restored by treatment with $Na_2S_2O_3$ [175]. Reactions with sodium salts are more effective when catalyzed with immobilized cryptand [2.2.1], while those with the potassium salt are better with cryptand [2.2.2]. This agrees with the fact that Na^+ better fits the [2.2.1] cavity and K^+ the [2.2.2] cavity.

Van Zon et al. carefully studied the exchange reaction of octyl bromide in toluene and aqueous KCN catalyzed by hydroxymethyl-B18C6 bound to a 2% crosslinked polystyrene resin [176]. The reaction rate is proportional to the concentration of 1-bromooctane, the amount of catalyst and the concentration of KCN. At high KCN/octyl bromide ratios the reaction follows pseudo first order kinetics, but marked deviations occur at smaller excess of KCN due to concurrent KCN hydrolysis. A network with one benzocrown ether per two styrene units is by one order of magnitude more effective than a resin with one crown per twenty styrene units. Binding constants of alkali metal picrates to similar crown networks (Sect. C.II) and reactivities of carboxylates bound to linear poly(crown ether)s (Sect. B.II.4) also increase with crown content even in cases where only 1:1 complexes are formed. A somewhat larger interionic distance in the KCN ion pair bound to the higher capacity network may in part explain its enhanced activity. The higher concentration of ionic solute in this network is also believed to cause more aqueous KCN to interact with the solid catalyst and to enlarge the active liquid-liquid-solid surface.

Manecke prepared immobilized crown and cryptand catalysts by direct polymerization of vinyl-containing macrocycles with divinylbenzene or divinylbiphenyl as crosslinking agents [68,177,178]. Styrene or 2-vinylpyrrolidone were added to vary the content of catalyst and the polarity of the resin. One of the network structures is given below (*24*). The reaction of benzyl chloride with solid potassium acetate in acetonitrile was catalyzed by either gel-type or macroreticular resins. Surprisingly, the catalytic activity of the cryptand resins is not affected by the copolymer composition or the polarity, degree of crosslinking and porosity of the network [178]. The same variables do affect the cation complexation properties and capacities of these resins, but these data were obtained in aqueous media. A macroporous (20% divinylbenzene, 80% cryptand monomer) and a 4% crosslinked 1:1 copolymer of vinylbenzo-cryptand and vinylpyrrolidone were nearly saturated with alkali metal

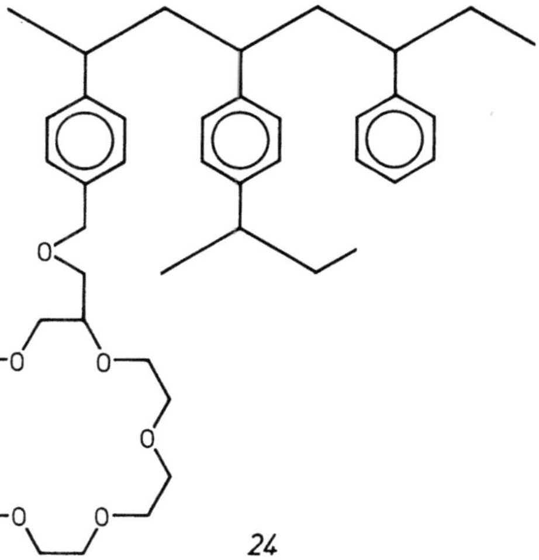

24

picrate in about two hours, while in the same time a 1:1 copolymer with styrene as comonomer complexed only 20% of the theoretical maximum capacity. Site accessibility in the hydrophobic styrene-containing network is of course more difficult when the salts are in aqueous media. For crown ether-containing networks the 20% crosslinked resins are rather unreactive, but the difference in activity between a 5% or 10% crosslinked network is small. A more porous and, therefore, active catalyst is obtained with divinylbiphenyl as crosslinking agent. In the conversion of dichlorobutane to adiponitrile by KCN no enhanced catalyst activity is observed when the spacer between aliphatic crown and styrene unit is lengthened by one ethylene oxide unit. The monomeric 18C6 as catalyst in this reaction is about six times more active than a 5 or 10% crosslinked resin with the same crown. The network catalysts also catalyze the conversion of dinitrochlorobenzene into the fluoro derivatives by potassium fluoride.

Blasius and his coworkers [181] catalyzed a variety of reactions with the formaldehyde-DB18C6 resin *14* (P-DB18C6). Although the cryptand catalysts are more active, the ease of preparation of P-DB18C6 makes this an attractive alternative. A number of reactions catalyzed by this resin are listed in Table 8. The same catalyst is active in dichlorocarbene reactions with α-methylstyrene, cyclohexene and styrene by means of $CHCl_3$/KOH. It also promotes the gas phase reaction of 1-chlorobutane with potassium iodide which at 135 °C in one hour gives a 60–70% yield of the iodide product, while none is found in the absence of P-DB18C6. The resin catalyzes the anionic living polymer-type polymerization of ethylene oxide and propylene sulfide initiated by carbazyl potassium in THF [129]. It also is able to convert 2-pyrrolidone into the polyamide after the monomer has been contacted with an alkali metal mirror to form small quantities of potassium pyrrolidonate [181].

The same network (P-DB18C6) and a similar one containing a 10:1 molar ratio of anisole and dibenzo[18]crown-6 (P-DB18C6-A) were used by Regen et al. [182] in reactions of octyl bromide with aqueous potassium salts of iodide, cyanide and phenolate. For comparison, 1% crosslinked polystyrene resins with pendant hydroxymethyl-18C6 (A) and polyglyme ligands (B, eight oxygen atoms, and C, fifteen oxygen atoms) as well as the monomeric phenoxymethyl-18C6 (D) and some other catalysts

Table 8. Yields for the Reaction $RCH_2Y + KX \rightarrow RCH_2X + KY$ in the Presence and Absence of Crown Catalyst P-DB18C6 (Ref. [181])[a]

Reactants		Time (h)	Yield (%)	
			with P-DB18C6	No catalyst
Benzyl-Br	KCl	1	94	6
	KI	0.5	94	18
Benzyl-Cl	KBr	3	83	1
Octyl-Br	KCN	2	83	0
	$KOCOCH_3$	18	93	0
	KSCN	3	95	0
	KNO_2	6	60	0
Butyl-Br	KOC_6H_5[b]	1	95	5
	KSC_6H_5[b]	1	57	10
Butyl-Br, Benzyl-$COCH_3$	KOH	2	70	15

[a] T = 110 °C; Solvent: Toluene/H_2O; [b] Solvent: Toluene

were also examined. In the reaction with potassium iodide, the pseudo-first order rate constants (10^5 k_{obs}, sec^{-1} g^{-1}) were 632 (D), 14 (A, 5% ring substitution), 207 (A, 20% ring substitution), 16 (B), 116 (C), 5 (P-DB18C6) and 37 (P-DB18C6-A). Note the nearly fifteen times increase in activity of the crown resin A when substitution is increased from 5 to 20%, confirming the observation of Van Zon et al. [176]. Completely different results are found with potassium cyanide or phenolate [182]. For the cyanide reaction the respective rate constants, in decreasing order, are 55 (P-DB18C6), 5 (A, 20% ring substitution), 5 (D), 2 (P-DB18C6-A), 2 (A, 5% ring substitution), 1 (B) and 1 (C). In this reaction even the monomeric crown is much less active than P-DB18C6 and the same was found with the phenolate, although in the latter case the immobilized high capacity crown catalyst A was slightly more active than either of the DB18C6 resins. The contradictory results for the three reactions are hard to interpret, and they stress the need for careful mechanistic studies under more simplified conditions (e.g., homogeneous systems) to determine the relationship between catalyst structure and activity. The author's conclude that resins with pendant crown ligands are not essential for polyether-based triphase catalysis, and that cheaper resins such as P-DB18C6 or networks with pendant polyglyme chains may be attractive alternatives.

It should be stressed that the effectiveness of polystyrene resin-based oligo-oxyethylenes (polyglymes) as phase transfer catalysts in reactions discussed in this Section has been well established [183–189], and that in some instances they are comparable with crown-containing resins in their catalytic activity. Several mechanistic studies have been reported with these catalysts and their action resembles that of the macroheterocyclic systems. However, since they fall outside the scope of this review, the reader is referred to the quoted references.

In all systems discussed so far, the macroheterocyclic ligands were synthesized prior to their incorporation in the polymer matrix. Since from a commercial point of view crown ethers and cryptands are expensive chelating agents, attempts have been made to develop synthetic procedures in which macroheterocyclic structures would be formed directly on the resin. The most successful approach appears to be that of Warshawsky et al. who reacted polyethylene glycols (PEGs) with highly chloromethylated polystyrenes [190]. Exhaustively chloromethylated polystyrene is used to ensure a homogeneous chlorine profile throughout the bead cross section [191]. Interchain crosslinking during the reaction is suppressed by taking an equimolar amount of diol and polymer, by templating with sodium ions, and by carrying out the reaction in a good swelling solvent to ensure maximum chain separation. The resulting resins coined "*polymeric pseudocrown ethers*" contain ring structures and cavities in which part of the macromolecular network is incorporated into the macrocyclic polyether ring. The polymer derived from a PEG of molecular weight 600 (14 oxygen atoms) proved to be a strong complexer of anionic transition metal halide complexes of the type MX_4^- or MX_4^{2-} (M = Au, Zn, Fe; X = Cl, Br, I) without competition from large excesses of hydrogen or X^- ions. Complexation occurs in the form of insoluble polymeric oxonium salts forming a 1:1 type complex. Only an equivalent amount of H^+ is required. Coordination is most effective when the anions fit well in the polyether cavity. This is the case with the 14-oxygen ring system which has a cavity diameter of 10.5–13 Å versus 9–11 Å for the MX_4^- or MX_4^{2-} anions.

Warshawsky and Kahana also synthesized crown ethers directly on a polymer network by using a chloromethylated polystyrene resin with $SnCl_4$ as an alkylating agent to bind first catechol, the precursor of benzocrown ethers [192]. The catechol-containing resin is then reacted with polyethylene glycol dihalides to form the polymeric benzo-substituted 12C4, 15C5, 18C6 and 24C8. The complexation pattern of the immobilized crown ligands parallel those of the corresponding monomeric analogues. The polymers bind alkali ions by an ion-exchange mechanism via residual catechol moieties and through complexation to crown ligands. The latter process was found to be strongly temperature dependent. A column packed with polymeric B18C6 in methanol can be saturated at room temperature with KCl, but the salt is spontaneously eluted at 60 °C.

Polymeric crown structures have also been obtained by cationic polymerization of divinyl ethers of tetraethylene glycol [193]. A clear, light brown gel is formed which is insoluble in common solvents but easy to swell in toluene or chloroform. The reaction resembles the cyclopolymerization of monomers such as 1,2 bis (2-ethenyloxyethoxy)benzene into linear polymers with crown ether-type structures [194]. The network polymer derived from tetraethylene glycol divinyl ether is claimed to have catalytic properties comparable to that of monomeric crown compounds like 15C5 in the exchange reaction of octyl bromide and NaCN [193]. The reaction with KCN is equally fast as with NaCN. This lack of selectivity is believed to be the result of uncommon macrocyclic polyether structures such as 16C5 or possibly 17C5 cycles. The exact structure of the macrocycles and the presence of linear polyether chains will be difficult to determine. However, some of these networks may provide an inexpensive but effective alternative as heterogeneous phase transfer catalysts.

D Conclusions

Macroheterocyclic ligands of the crown ether and cryptand type have proven to be versatile complexing agents for a variety of ionic and neutral solutes. When bound to a macromolecular chain, their binding properties can change significantly as a result of cooperative binding effects involving adjacent ligands or comonomer substituents, microenvironmental effects caused by the polymer backbone or by neighboring groups, conformational restrictions imposed upon the macrocyclic ring due to the close proximity of other polymer-bound entities, interaction between the bound solutes, etc. In this respect, soluble polymeric crown ethers and cryptands can serve as excellent model compounds for studying solute binding to neutral macromolecules in both aqueous and non-aqueous systems. The effect of variables such as distance between bound macrocycles, length and structure of spacer between bound ligand and polymer backbone, flexibility of the polymer backbone and ionic or neutral comonomer substituents can be determined by a variety of techniques, both in catalytic and solute binding applications. The behavior of certain poly(crown ether)s in water is especially of interest for a better understanding of the binding of monomeric and polymeric solutes to neutral and charged polysoap-type macromolecules.

The knowledge acquired from the soluble polymeric systems is necessary in unraveling the often more complex phenomena encountered with immobilized macroheterocyclic ligands. The latter, of course, are of interest in chromatographic separations, phase-transfer catalysis and other applications where their insolubility or ease of recovery is required. The effect of network swelling, crosslink density, particle size, site accessibility, porosity and other network parameters are not always well understood, and more quantitative work with both linear and network polymeric macroheterocyclic ligands is needed. Although commercial application may at present be limited by economic considerations, the immobilized macrocycles nevertheless can function as very valuable compounds to study the characteristics of solute binding to network structures, and are also useful in studying interactions between solutes and soluble ligands in low polarity media.

E Acknowledgements

The authors are grateful for the many contributions of their coworkers whose names are listed in the appropriate references. They also acknowledge with thanks the generous support of the National Science Foundation for research carried out in the authors' laboratory.

F References

1. Pedersen, C. J.: J. Am. Chem. Soc. *89*, 7017 (1967)
2. Pedersen, C. J., Frensdorff, H. K.: Angew. Chem. *84*, 16 (1972); Angew. Chem., Int. Ed. Engl. *11*, 16 (1972)
3. Lehn, J. M.: Pure Appl. Chem. *50*, 871 (1978)
4. Cram, D. J., Trueblood, K. N.: Top. Curr. Chem. *98*, 43 (1981)
5. Lehn, J. M.: Pure Appl. Chem. *52*, 2441 (1980)
6. De Jong, F., Reinhoudt, D. N.: Adv. Phys. Org. Chem. *17*, 279 (1980)
7. Stoddart, J. F.: Quart. Rev. *1979*, 85
8. Cram, D. J., Cram, J. M.: Acc. Chem. Res. *11*, 8 (1978)
9. Bell, T. W. et al.: J. Am. Chem. Soc. *104*, 5185 (1982)

10. Schultz, R. A., Dishong, D. M., Gokel, G. W.: ibid. *104*, 625 (1982)
11. Cram, D. J., et al.: ibid. *104*, 6827, 6828 (1982)
12. Weber, E.: Angew. Chem. *91*, 230 (1979); Angew. Chem., Int. Ed. Engl. *18*, 219 (1979)
13. Weber, E.: J. Org. Chem. *47*, 3478 (1982)
14. Bourgoin, M., et al.: J. Am. Chem. Soc. *97*, 3462 (1975)
15. Wong, K. H., Ng, H. L.: Tetrahedron Lett. 4295 (1979)
16. Lehn, J. M. et al.: J. Am. Chem. Soc. *99*, 6766 (1977)
17. Vögtle, F., Weber, E.: Angew. Chem. *91*, 813 (1979); Angew. Chem., Int. Ed. Engl. *18*, 753 (1979); Vögtle, F., Weber, E.: Kontakte (Merck) *1977* (1), 11
18. Raßhofer, W., Müller, W. M., Vögtle, F.: Chem. Ber. *112*, 2095 (1979)
19. Top. Curr. Chem. *98* (1981)
20. Izatt, M., Christensen, J. J. (Eds.): Synthetic Multidentate Macrocyclic Compounds, Acad. Press, New York 1978
21. Izatt, M., Christensen, J. J. (Eds.): Progress in Macrocyclic Chemistry, Wiley, Interscience, New York, Vol. I, 1979; Vol. II, 1981
22. Kolthoff, I. M.: Anal. Chem. Rev. *51*, 1R–22R (1979)
23. Poonia, N. S., Bajaj, A. V.: Chem. Rev. *79*, 389 (1979)
24. Bradshaw, J. S., Stott, P. E.: Tetrahedron *36*, 461 (1980)
25. Feigenbaum, W. M., Michel, R. H.: J. Polym. Sci. *A-1*, 9, 817 (1971)
26. Shchori, E., Jagur-Grodzinski, J.: J. Appl. Polym. Sci. *20*, 773 (1976)
27. Shchori, E., Jagur-Grodzinski, J.: ibid. *20*, 1665 (1976)
28. Wong, K. H., Konizer, G., Smid, J.: J. Am. Chem. Soc. *92*, 666 (1970)
29. Shchori, E., Jagur-Grodzinski, J., Shporer, J.: ibid. *95*, 3842 (1973)
30. Shchori, E. et al.: ibid. *93*, 7133 (1971)
31. Frost, L. W.: U.S. Pat. 3, 956, 136 (1976)
32. Bormann, S. et al.: Tetrahedron *31*, 2791 (1975)
33. Gramain, Ph., Frère, Y.: Macromolecules *12*, 1038 (1979)
34. Gramain, Ph., Frère, Y.: Polymer *21*, 921 (1980)
35. Gramain, Ph., Frère, Y.: Ind. Eng. Chem. Prod. Res. Dev. *20*, 524 (1981)
36. Gramain, Ph., Kleiber, M., Frère, Y.: Polymer *21*, 915 (1980)
37. Cho, I., Chang, S. K.: Makromol. Chem., Rapid Commun. *2*, 155 (1981)
38. Koenig, K. E., Helgeson, R. C., Cram, D. J.: J. Am. Chem. Soc. *98*, 4018 (1978)
39. Mathias, L. J., Al Jumah, K.: J. Polym. Sci., Polym. Chem. Ed. *18*, 2911 (1980)
40. Gramain, Ph., Frère, Y.: Makromol. Chem., Rapid. Commun. *2*, 161 (1981)
41. Shinkai, S. et al.: J. Am. Chem. Soc. *102*, 5860 (1980)
42. Shinkai, S. et al.: J. Chem. Soc., Perkin Trans. 1, *1981*, 3279
43. Shinkai, S. et al.: J. Am. Chem. Soc. *104*, 1960 (1982)
44. Cho, I., Chang, S. K.: Chem. Lett. *1981*, 515
45. Wong, K. H., Ng, H. L.: Synth. Comm. *10*, 385 (1980)
46. Kimura, K., Maeda, T., Shono, T.: Anal. Lett. *A11*, 821 (1978)
47. Kimura, K., et al.: J. Electroanal. Chem. *95*, 91 (1979)
48. Maeda, T., Kimura, K., Shono, T.: Fres. Z. Anal. Chem. *298*, 363 (1979)
49. Kimura, K., Maeda, T., Shono, T.: Talanta *26*, 945 (1979)
50. Kimura, K. et al.: ibid. *27*, 801 (1980)
51. Maeda, T. et al.: Chem. Lett. *1981*, 1573
52. Fung, K. W., Wong, K. H.: J. Electroanal. Chem. *111*, 359 (1980)
53. Johnson, M. R., Sutherland, I. O.: J. Chem. Soc., Chem. Commun. 306 (1979)
54. Bouquant, J. et al.: J. Am. Chem. Soc. *104*, 686 (1982)
55. Kopolow, S., Hogen Esch, T. E., Smid, J.: Macromolecules *4*, 359 (1971)
56. Kopolow, S., Hogen Esch, T. E., Smid, J.: ibid. *6*, 133 (1973)
57. Smid, J.: Chapter 3 of ref. 21, Vol. II, p. 169 (1981)
58. Smid, J.: Pure Appl. Chem. *51*, 111 (1979)
59. Smid, J.: Ind. Eng. Chem. Prod. Res. Dev. *19*, 364 (1980)
60. Smid, J.: Makromol. Chem., Suppl. 5, *1981*, 203
61. Smid, J.: Pure Appl. Chem. *54*, 2129 (1982)
62. Smid, J. et al.: Org. Prep. Proceed. Int. *8*, 193 (1978)
63. Wada, F., Hirayama, H., Namiki, H.: Bull. Chem. Soc. Jpn. *53*, 1473 (1980)
64. Shah, S. C., Kopolow, S., Smid, J.: Polymer *21*, 189 (1980)

65. Kikukawa, K., Nagira, K., Matsuda, T.: Bull. Chem. Soc. Jpn. *50*, 2270 (1977)
66. Akabori, S., Miyamoto, S., Tanabe, H.: J. Polym. Sci., Polym. Chem. Ed. *17*, 3933 (1979)
67. Blasius, E. et al.: Z. Anal. Chem. *284*, 337 (1977)
68. Manecke, G., Krämer, A.: Makromol. Chem. *182*, 3017 (1981)
69. Kimura, K. et al.: Polymer Bull. *1*, 403 (1979)
70. Kimura, K., Maeda, T., Shono, T.: Makromol. Chem. *182*, 1579 (1981)
71. Varma, A. J., Majewicz, T., Smid, J.: J. Polym. Sci., Polym. Chem. Ed. *15*, 1189 (1977)
72. Kimura, K., Sawada, M., Shono, T.: J. Chromatogr. *240*, 361 (1982)
73. Kimura, K., Sawada, M., Shono, T.: Anal. Lett. *A12*, 1095 (1979)
74. Yagi, K., Ruiz, J. A., Sanchez, M. C.: Makromol. Chem., Rapid Commun. *1*, 263 (1980)
75. Yagi, K., Sanchez, M. C.: Makromol. Chem. *2*, 311 (1981)
76. Yagi, K. et al.: Polym, Prepr. *23*, 169 (1982)
77. Anzai, J. et al.: Makromol. Chem., Rapid Commun. *3*, 399 (1982)
78. Anzai, J. et al.: ibid. *3*, 55 (1982)
79. Ricard, A., Lafuma, F., Quivoron, C.: Polymer *23*, 907 (1982)
80. Smid, J.: Angew. Chem. *84*, 127 (1972); Angew. Chem., Intern. Ed. Engl. *11*, 112 (1972)
81. Ambroz, L., Wong, K. H., Smid, J.: Anal. Chem. *44*, 872 (1972)
82. Kopolow, S. et al.: J. Macromol. Sci. *A7*, 1015 (1973)
83. Takaki, U., Hogen Esch, T. E., Smid, J.: J. Am. Chem. Soc. *93*, 6760 (1971)
84. Takaki, U., Smid, J.: ibid. *96*, 2588 (1974)
85. Hogen Esch, T. E., Smid, J.: ibid. *91*, 4580 (1969)
86. Wong, K. H., Bourgoin, M., Smid, J.: J. Chem. Soc., Chem. Commun. *1974*, 715
87. Ungaro, R., El Haj, B., Smid, J.: J. Am. Chem. Soc. *98*, 5198 (1976)
88. Kulstad, S., Malmsten, L. A.: J. Inorg. Nucl. Chem. *42*, 573 (1980)
89. Evans, D. F. et al.: J. Phys. Chem. *75*, 1714 (1971)
90. Shchori, E., Jagur-Grodzinski, J.: Isr. J. Chem. *11*, 243 (1973)
91. Gilkerson, W. R., Jackson, M. D.: J. Am. Chem. Soc. *101*, 4096 (1979)
92. Nae, N., Jagur-Grodzinski, J.: ibid. *99*, 489 (1977)
93. Shah, S. C., Kopolow, S., Smid, J.: J. Polym. Sci., Polym. Chem. Ed. *14*, 2023 (1976)
94. Hogen Esch, T. E., Smid, J.: J. Phys. Chem. *79*, 233 (1975)
95. Frensdorff, H. K.: J. Am. Chem. Soc. *92*, 391 (1970)
96. Iwachido, T., Kimura, M., Toei, K.: Chem. Lett., *1976*, 1101
97. Wong, K. H., Yagi, K., Smid, J.: J. Membr. Biol. *18*, 379 (1974)
98. Sinta, R., Smid, J.: Macromolecules *13*, 339 (1980)
99. Reusch, C. F., Cussler, E. L.: Amer. Inst. Chem. Eng. J. *19*, 736 (1973)
100. Kobuke, Y. et al.: J. Am. Chem. Soc. *98*, 7414 (1976)
101. Kirch, M., Lehn, J. M.: Angew. Chem. *87*, 542 (1975); Angew. Chem., Int. Ed. Engl. *14*, 555 (1975)
102. Lamb, J. D. et al.: J. Am. Chem. Soc. *102*, 3399 (1980)
103. Smid, J., Varma, A. J., Shah, S. C.: ibid. *101*, 5764 (1979)
104. Shirai, M., Smid, J.: J. Polym. Sci., Polym. Chem. Ed. *18*, 659 (1980)
105. Shirai, M., Smid, J.: J. Am. Chem. Soc. *102*, 2863 (1980)
106. Lindner, W., Truger, W.: J. Chromatogr. *168*, 171 (1979)
107. Wong, L., Smid, J.: J. Am. Chem. Soc. *99*, 5637 (1977)
108. Taylor, L. D., Cerankowski, L. D.: J. Polym. Sci., Polym. Chem. Ed. *13*, 2551 (1975)
109. Smid, J. et al.: J. Polym. Sci., Polym. Symp. *64*, 267 (1978)
110. Wong, L., Smid, J.: Polymer *21*, 195 (1980)
111. Roland, B., Smid, J.: J. Am. Chem. Soc. *105*, 5269 (1983)
112. Kimura, K., Smid, J.: Makromol. Chem., Rapid. Commun. 2, 235 (1981)
113. Kimura, K., Smid, J.: Macromolecules *15*, 966 (1982)
114. Kimura, K., Smid, J.: Polym. Bull. *6*, 395 (1982)
115. Wong, K. H., Kimura, K., Smid, J.: J. Polym. Sci., Polym. Chem. Ed. *21*, 579 (1983)
116. Shah, S. C., Smid, J.: J. Am. Chem. Soc. *100*, 1426 (1978)
117. Kemp, D. S., Paul, K. G.: ibid. *97*, 7305 (1975)
118. Jaycox, G. D., Smid, J.: Makromol. Chem., Rapid Commun. *2*, 299 (1981)
119. Jaycox, G. D., Sinta, R., Smid, J.: J. Polym. Sci., Polym. Chem. Ed. *20*, 1629 (1982)
120. Varma, A. J., Smid, J.: ibid. *15*, 1189 (1977)
121. Pitha, J., Smid, J.: Biochem. Biophys. Acta *425*, 287 (1976)

122. Cussler, E. L., Evans, D. F.: Sep. Purif. Meth. *3*, 399 (1974)
123. Izatt, R. M. et al.: J. Membrane Sci. *9*, 263 (1981)
124. Newcomb, M. et al.: J. Am. Chem. Soc. *101*, 4941 (1979)
125. Sousa, L. R. et al.: ibid. *96*, 7100 (1974)
126. Blasius, E. et al.: J. Chromatogr. *96*, 89 (1974); and subsequent papers
127. Blasius, E. et al.: ibid. *167*, 307 (1978)
128. Blasius, E. et al.: Talanta *27*, 127 (1980)
129. Blasius, E. et al.: J. Chromatogr. *201*, 147 (1980)
130. Blasius, E., Janzen, K. P.: Top. Curr. Chem. *98*, 165 (1981)
131. Blasius, E., Janzen, K. P.: Pure Appl. Chem. *54*, 2115 (1982)
132. Kimura, K., Nakajima, M., Shono, T.: Anal. Lett. *13*, 741 (1980)
133. Bogatskii, A. B., Luk'yanenko, N. G., Pastushok, V. N.: Dokl. Akad. Nauk, SSSR *247*, 379 (1979)
134. Sjulek, W., Lada, W.: Radiochem. Radioanal. Lett. *30*, 199 (1979)
135. Igawa, M. et al.: Anal. Chem. *53*, 1942 (1981)
136. Kutchukow, P., Ricard, A., Quivoron, C.: Eur. Polym. J. *16*, 753 (1980)
137. Frère, Y., Gramain, Ph.: Makromol. Chem. *183*, 2163 (1982)
138. Shinkai, S., Kinda, H., Manabe, O.: J. Am. Chem. Soc. *104*, 2933 (1982)
139. Sogah, G. D. Y., Cram, D. J.: ibid. *97*, 1259 (1975)
140. Sogah, G. D. Y., Cram, D. J.: ibid. *98*, 3038 (1976)
141. Sogah, G. D. Y., Cram, D. J.: ibid. *101*, 3035 (1979)
142. Grossmann, P., Simon, W.: J. Chromatogr. *235*, 351 (1982)
143. Sinta, R., Smid, J.: J. Am. Chem. Soc. *103*, 6962 (1981)
144. Sinta, R., Lamb, B., Smid, J.: Macromolecules *16*, 1382 (1983)
145. Smid, J., Sinta, R.: Crown Ethers and Phase Transfer Catalysis in Polymer Science, Carraher, C. E., Mathias, L. (Eds.) Plenum, New York 1983
146. Sinta, R., Rose, P. S., Smid, J.: J. Am. Chem. Soc. *105*, 4337 (1983)
147. Hodge, P., Sherrington, D. C.: Polymer-supported Reactions in Organic Synthesis. Wiley Interscience, New York 1980
148. Mathur, N. K., Narang, C. K., Williams, R. E.: Polymers as Aids in Organic Chemistry, Academic Press., New York 1980
149. Regen, S. L.: J. Am. Chem. Soc. *97*, 5956 (1975)
150. Regen, S. L.: Angew. Chem. *91*, 464 (1979); Angew. Chem., Int. Ed. Engl. *18*, 421 (1979)
151. Regen, S. L.: J. Am. Chem. Soc. *98*, 6270 (1976)
152. Ford, W. T., Lee, J., Tomoi, M.: Macromolecules *15*, 1246 (1982)
153. Regen, S. L., Bolikal, D., Barcelon, C.: J. Org. Chem. *46*, 2511 (1981)
154. Tundo, P.: Synthesis *1978*, 315
155. Rolla, F., Roth, W., Horner, L.: Naturwissenschaften *64*, 377 (1977)
156. Tundo, P., Venturello, P.: Tetrahedron Lett. *21*, 2581 (1980)
157. Regen, S. L., Heh, J. C. K., McLick, J.: J. Org. Chem. *44*, 1961 (1979)
158. Molinari, H., et al.: J. Am. Chem. Soc. *101*, 3920 (1979)
159. Chiles, M. S., Reeves, P. C.: Tetrahedron Lett. *36*, 3367 (1979)
160. Molinari, H., Montanari, F., Tundo, P.: J. Chem. Soc., Chem. Commun. *1977*, 639
161. Tundo, P.: J. Chem. Soc., Chem. Commun. *1977*, 641
162. Ohtani, N. et al.: Macromolecules *14*, 516 (1981)
163. Ford, W. T., Tomoi, M.: J. Am. Chem. Soc. *102*, 7141 (1980)
164. Nigam, A., Regen, S. L.: ibid. *100*, 7773 (1978)
165. Cinquini, M. et al.: J. Chem. Soc., Chem. Commun. *1976*, 394
166. Dou, J. M. et al.: J. Org. Chem. *42*, 4275 (1977)
167. Starks, C. M., Liotta, C. (Eds.): Phase Transfer Catalysis; Academic Press, New York 1978
168. Weber, W. P., Gokel, G. W. (Eds.): Phase Transfer Catalysis in Organic Synthesis; Springer, New York 1977
169. Montanari, F., Tundo, P.: Tetrahedron Lett. *52*, 5055 (1979)
170. Montanari, F., Tundo, P.: J. Org. Chem. *46*, 2125 (1981)
171. Montanari, F., Tundo, P.: ibid. *47*, 1298 (1982)
172. Brown, J. M., Jenkins, J. A.: J. Chem. Soc., Chem. Commun. *1976*, 458
173. Sparrow, J. T.: J. Org. Chem., *41*, 1350 (1976)
174. Tomoi, M. et al.: Tetrahedron Lett. *33*, 3031 (1978)

175. Tomoi, M., Kihara, K., Kakiuchi, H.: ibid. *36*, 3485 (1979)
176. Van Zon, A., De Jong, F., Onwezen, Y.: Rec. Trav. Chim. Pays-Bas *100*, 429 (1981)
177. Manecke, G., Reuter, P.: Makromol. Chem. *182*, 1973 (1981)
178. Manecke, G., Reuter, P.: J. Mol. Cat. *13*, 355 (1981)
179. Manecke, G., Reuter, P.: Pure Appl. Chem. *51*, 2313 (1979)
180. Manecke, G., Stärk, M., Krämer, A.: Plaste und Kautschuk, *28*, 489 (1981)
181. Blasius, E. et al.: Makromol. Chem. *183*, 1401 (1982)
182. Fukunishi, K., Czech, B., Regen, S. L.: J. Org. Chem. *46*, 1218 (1981)
183. Yanagida, S., Takahashi, K., Okahara, M.: ibid. *44*, 1099 (1979)
184. Mackenzie, W. M., Sherrington, D. C.: J. Chem. Soc., Chem. Commun. *1978*, 541
185. Mackenzie, W. M., Sherrington, D. C.: Polymer *21*, 791 (1980)
186. Heffernan, J. G., Mackenzie, W. M., Sherrington, D. C.: J. Chem. Soc. Perkin Trans. 2, *2*, 514 (1981)
187. Mackenzie, W. M., Sherrington, D. C.: Polymer *22*, 431 (1981)
188. Hiratani, K., Reuter, P., Manecke, G.: J. Mol. Cat. *5*, 241 (1979)
189. Hiratani, K., Reuter, P., Manecke, G.: Isr. J. Chem. *18*, 208 (1979)
190. Warshawsky, A., et al.: J. Am. Chem. Soc. *101*, 4249 (1979)
191. Warshawsky, A., Kalir, R., Patchornik, A.: J. Org. Chem. *43*, 3151 (1978)
192. Warshawsky, A., Kahana, N.: J. Am. Chem. Soc. *104*, 2663 (1982)
193. Mathias, L. J., Canterberry, J. D.: Polym. Prepr. *22*, 38 (1981)
194. Butler, G. B., Lien, Q. S.: ACS Sym. Ser. *195*, 149 (1982)

Synthesis and Metal Complexes of Aza-Macrocycles with Pendant Arms having Additional Ligating Groups

Thomas A. Kaden

Institute of Inorganic Chemistry, University of Basel, Spitalstraße 51, CH-4056 Basel, Switzerland

Table of Contents

1 Introduction 158

2 Synthesis of Functionalized Azamacrocycles 159
2.1 N-Functionalized Derivatives 159
2.1.1 Tetrasubstituted Tetraazacycloalkanes 160
2.1.2 Disubstituted Tetraazacycloalkanes 161
2.1.3 Monosubstituted Tetraazacycloalkanes 161
2.2 C-Functionalized Derivatives 166
2.3 Bis-Macrocycles 168

3 Metal Complexes 168
3.1 Structures 168
3.2 Equilibria in Solution 170
3.2.1 Protonation and Complexation 170
3.2.2 Equilibria Involving the Side-Chain 172

4 Kinetical Aspects 173

5 Conclusions 177

6 Acknowledgements 177

7 References 177

1 Introduction

After the discovery of the crown ethers by Pedersen [1] on one side and of the template reactions by Curtis [2] on the other side, a large number of synthetic macrocycles have appeared in the literature [3]. The complexes formed by the macrocycles often exhibit properties which are distinctly different from those of analogous open chain ligands. Macrocycles being not very flexible can impose a specific coordination geometry to the metal ion, whereas open chain chelators generally can more easily adapt themselves to the geometrical requirements of the metal ion. In addition many labile metal ions form kinetically stable complexes with macrocycles, thus allowing to study their reactivity either in strongly acidic or alkaline solutions. Both these aspects are found to an even larger extent in the chemistry of the cryptates as described in the many papers of Lehn et al. [4].

In the last years efforts have been made to combine the properties of the relatively rigid and kinetically stable structural element of the macrocycles with those of the more flexible and kinetically labile open chain ligands. The reasons are various. By introducing additional ligating groups into a macrocycle, its properties can be modified so that more stable complexes are formed or its specificity in metal ion binding is increased or its solubility is changed. For example the 12-membered tetraazamacrocycle *1* to which four acetate groups have been attached gives thermo-

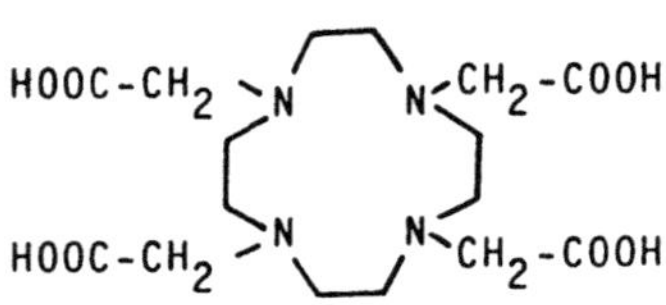

1

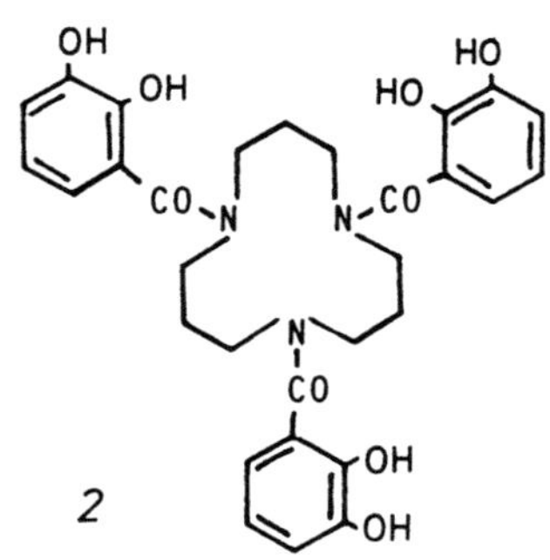

2

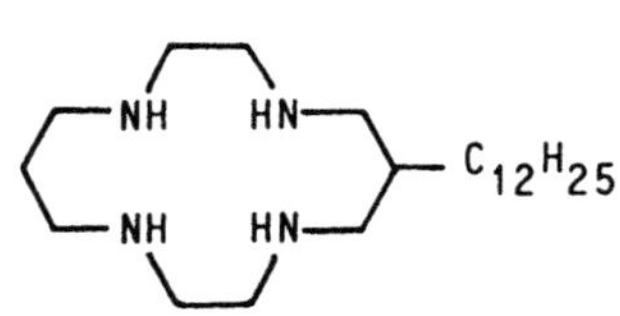

3

dynamically stable alkali and alkaline earth complexes [5-7] whereas the unsubstituted ligand only binds transition metal ions. Another interesting example is tris(2,3-dihydroxybenzoyl)-1,5,9-triazacyclotridecane (*2*) which was synthesized by Raymond et al. [8] to specifically bind Fe^{3+}. Finally the macrocycle *3* to which a long lipophilic side chain has been attached to make the ligand and its metal complexes soluble in organic solvents can be used to extract metal ions from aqueous solution into an organic phase [9].

Beside these aspects which are very important for analytical applications macrocycles with additional ligating groups can also be used to study the selective influences of such a group on the properties of metal ions. It is concievable that for example the redox potential of a coordinated metal ion can be changed and adjusted introducing an additional ligand in his axial position. It is also possible that the selective binding of a further ligand will depend on the nature of what is already coordinated in the axial position of the metal ion. For example it is thought that the oxygen binding capability of iron in haemoglobin and its reversibility are, among other effects, a consequence of the axially coordinated imidazole of the distal histidine [10]. To mimic and model this Traylor et al. [11] have prepared an iron porphyrin to which a long chain with a histidine terminal is attached and have shown that this derivative reversibly binds oxygen at low temperature.

Another possible application of functionalized azamacrocycles could be to bring close together a metal ion and a weakly binding or even a non-coordinating organic group so that an interaction between the two can take place. Thereby the properties and reactivity of the organic group might be changed. Thus, by covalent coupling of the organic group to the macrocycle which binds the metal ion, metal ion promoted reactions can be studied. The metal ion can act as a Lewis acid or as a redox catalyst and the organic group as the substrate. This allows to follow such processes on a molecular basis and to obtain information and insight into the mechanism of metal ion catalyzed or promoted reactions.

The aim of this review is to cover these different aspects but also to outline the principles in syntheses, structures, stabilities and reactivities of functionalized azacycloalkanes and of their metal ion complexes.

2 Synthesis of Functionalized Azamacrocycles

2.1 N-Functionalized Derivatives

Beside many examples of N-functionalized tetraazamacrocycles only two triazacycloalkanes *2* and *4* carrying pendant side-chains have been described. Their synthesis is simple and consists in the alkylation or acylation of the three amino

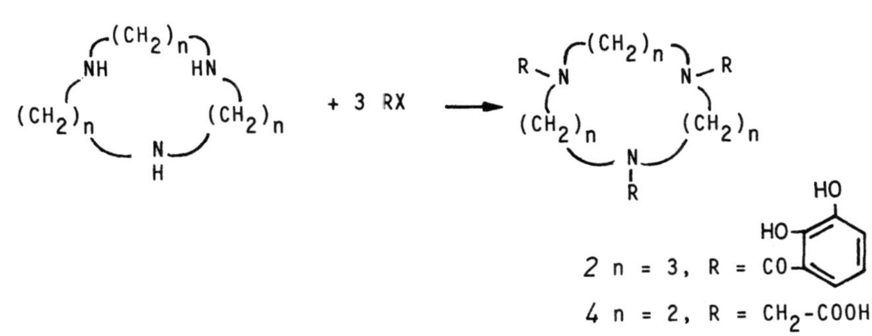

Fig. 1. Synthesis of functionalized triazacycloalkanes through acylation and alkylation

nitrogens by chloroacetic acid [12] and 2,3-dioxymethylenebenzoyl chloride [8], respectively (Fig. 1); in the case of compound *2* followed by BCl_3-cleavage of the methylene ether bonds.

For the tetraazamacrocycles a larger variety of substituted products have been prepared. Since four nitrogens are present up to four side-chains can be introduced. As shown in the following sections the syntheses of these compounds markedly differ depending on the degree of substitution. Whereas no large problems are encountered when tetra-N-substituted derivatives are prepared, more specific routes have to be devised for the synthesis of mono-N-substituted derivatives.

2.1.1 Tetrasubstituted Tetraazacycloalkanes

The synthesis of tetra-N-substituted tetraazamacrocycles is straight-forward. The unsubstituted macrocycle is reacted with an excess of alkylating agent in the presence of a base to bind the acid thereby formed (Fig. 2).

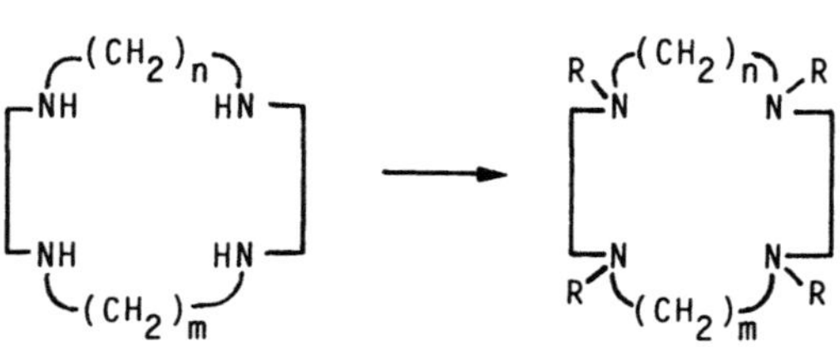

m = n = 2

m = 2, n = 3

m = n = 3

1 R = CH_2-COOH, n = m = 2

5 R = CH_2-COOH, n = 2, m = 3

6 R = CH_2-COOH, n = m = 3

7 R = CH_2-CH_2-OH, n = m = 2

8 R = CH_2-CH_2-CN, n = m = 3

9 R = CH_2-CN, n = m = 2

10 R = CH_2-CN, n = m = 3

Fig. 2. Synthesis of tetraazacycloalkanes with four pendant arms

Stetter et al. [6] have described the preparation of the macrocycles *1*, *5* and *6* with different ring sizes (n, m = 2 or 3) using chloroacetic acid as alkylating agent in alkaline solution. The compounds were purified by converting them to the esters which then were hydrolysed with water without the addition of acid or base. Similarly, Häfliger et al. [13] have reacted *meso*-5,13-dimethyl-1,4,8,11-tetraazacyclotetradecane (meso-5,13-dimethyl-TA14C4)[1] with chloroacetic acid to obtain the tetra-N-substituted macrocycle *11* with four acetate groups.

1 Notations frequently used in this report are listed and explained on p. XI

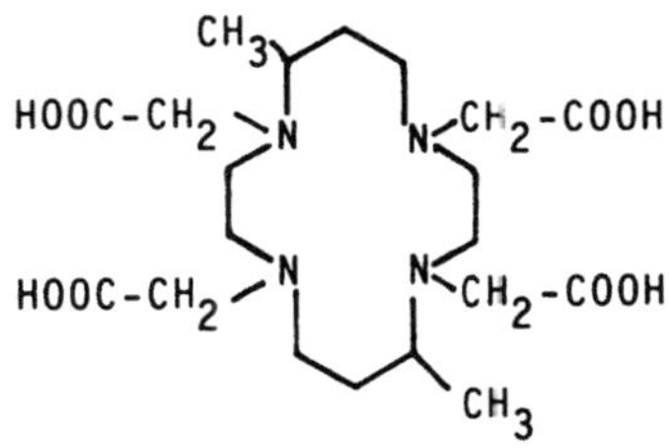

11

Dale et al. [14] have used ethylene oxide in the preparation of a 12-membered (n = m = 2) macrocycle *7* containing four 2-hydroxyethyl side-chains. Four 2-cyanoethyl groups were introduced into 1,4,8,11-tetraazacyclotetradecane (TA14C4) by Wainwright [15] using acrylonitrile as alkylating agent. Compound *8* represents a versatile starting material for the preparation of other tetra-N-substituted derivatives, since the nitrile functions can easily be hydrolyzed to amides [15] or reduced to amines [16], both derivatives being interesting new ligand systems. Following Strecker's synthesis which uses formaldehyde and KCN it is also possible to introduce four cyanomethyl groups into tetraazacycloalkanes giving *9* and *10* [17]. These macrocycles offer the same reaction modification possibilities as mentioned for the cyanoethyl derivative.

2.1.2 Disubstituted Tetraazacycloalkanes

Only a few disubstituted tetraazamacrocycles have been prepared. Schibler [18] describes the cyanomethylation of 1,4-dimethyl-TA14C4 to give *12*, which can be selectively hydrolyzed to the monoamide-mononitrile in the presence of Cu^{2+} (see

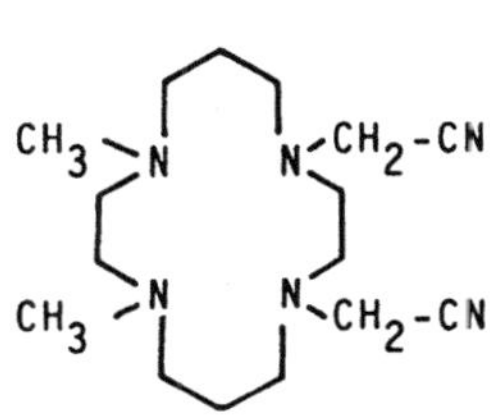

12

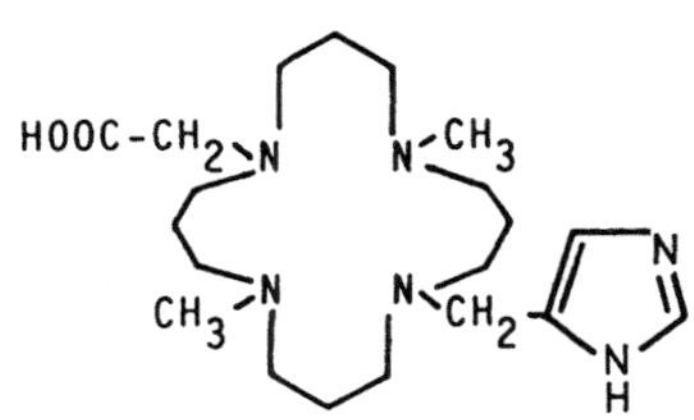

13

section 4) [19]. A difunctionalized 16-membered macrocycle having an acetate and an imidazolylmethyl side-chain *13* was prepared by Vitali [20]. Its Cu^{2+} complex was studied and from the VIS-absorption spectrum a coordination of all ligating groups was postulated.

2.1.3 Monosubstituted Tetraazacycloalkanes

For the synthesis of monosubstituted tetraazamacrocycles with different side-chains new routes had to be devised. These are schematically shown in Fig. 3: the selective alkylation (a), the cyclisation with side-chain (b), the modification of a side-chain which is already present (c) and the alkylation of a selectively protected macrocycle (d).

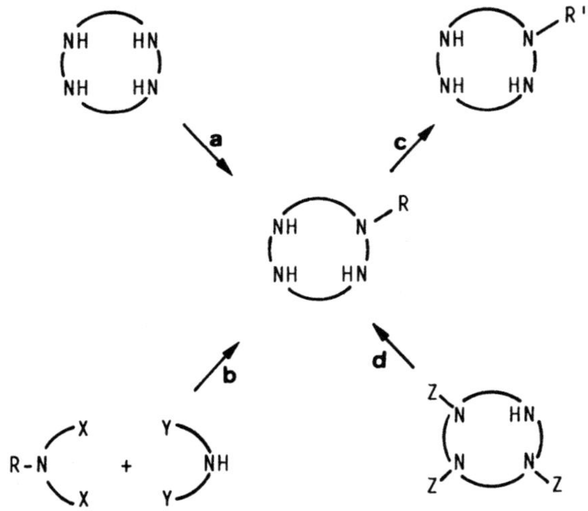

Fig. 3. Synthetic routes for the preparation of mono-N-substituted tetraazacycloalkanes: a) selective alkylation, b) cyclisation in the presence of a side-chain, c) modification of the side-chain and d) alkylation of a selectively protected tetraazamacrocycle

a) *Selective alkylation*

The simplest and most naive approach would be the reaction of an alkylating agent with a large excess of the corresponding tetraazamacrocycle, so that the monosubstituted product will be favoured over the other ones. If the separation of the starting material and of the products can easily be performed, one would then obtain the desired compound. To our knowledge such reactions have not been tried yet or have not been reported because of failure. Wainwright [15] claims that he has not detected any intermediate in the reaction of TA14C4 with acrylonitrile in the preparation of *8*, but he has used acrylonitrile as solvent. The large excess of reagent makes it difficult to obtain products less strongly substituted than the tetraderivative. Also in the preparation of the tetraacetates *1*, *5* and *6* there is no indication that a monosubstituted species is formed [6]. Perhaps the product of the reaction of TA14C4 with a functionalized polymer in DMF using NaH as base is an example for the selective alkylation [21], although the product has not been fully identified. Selective alkylation, however, was used in the case of two N,N'-dimethyl-tetraazacycloalkanes to give trisubstituted compounds. Vitali [20] has reacted 1,9-dimethyl-1,5,9,13-tetraazacyclohexadecane (1,9-dimethyl-TA16C4) with bromoacetic acid methyl ester to give compound *14* with 14% yield. Barefield et al. [22] have shown that one can depro-

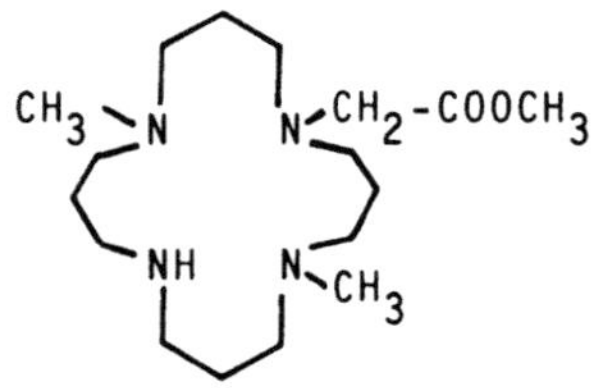

14

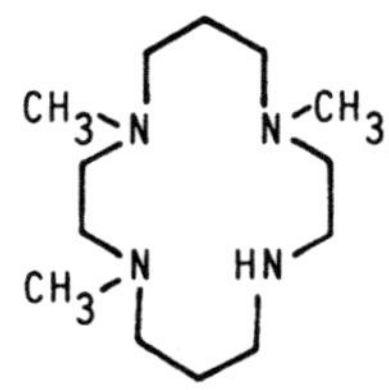

15

tonate and selectively methylate the Ni^{2+}-complex of 1,4-dimethyl-TA14C4 by reacting it in DMSO with methyl iodide in the presence of a strong base. The product *15* is obtained in good yield. These examples indicate that under the suited experimental conditions selective alkylation could become a realistic and effective synthetic path which has the advantage of being relatively simple.

b) *Cyclisation in the presence of side-chain*

A second possibility for the synthesis of mono-N-substituted derivatives is the cyclisation of two components, one of which has the side-chain with a functional group or a precursor of it already attached. Any reaction such as template condensation [2,3], high dilution cyclisation [23] or the procedure of Richman and Atkins [24] could be used. Although this synthetic approach allows to introduce a side-chain into a macrocycle at a specific nitrogen atom, it requires a new cyclisation for each new side-chain. Since this is the most difficult step of the whole synthesis, this method is time consuming and labour intensive. It is not the method of choice if one wants to study the effect of different functionalized side-chains on the properties of the metal ion.

One example for this type of reaction is the condensation of 2,6-diacetylpyridine with 4-(2-dimethylaminoethyl)-1,7-diamino-4-azaheptane or 4-(2-hydroxyethyl)-1,7-diamino-4-azaheptane in the presence of Ni^{2+} as template [25] (Fig. 4). One obtains the Ni^{2+} complexes *16* and *17*, respectively, with a 2-dimethylaminoethyl or a 2-hydroxyethyl pendant arm.

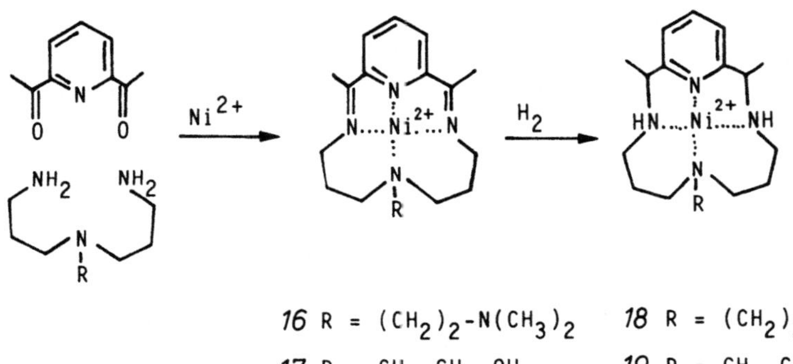

16 R = $(CH_2)_2$-$N(CH_3)_2$ *18* R = $(CH_2)_2$-$N(CH_3)_2$

17 R = CH_2-CH_2-OH *19* R = CH_2-CH_2-OH

Fig. 4. Synthesis of the monosubstituted macrocycles *16–19* using a Ni^{2+} template reaction and subsequent reduction of the Schiff bases

The Schiff bases *16* and *17* can be reduced with Pt/H_2 to the corresponding saturated Ni^{2+} complexes *18* and *19*, respectively. By treatment with excess of NaCN they are decomposed to give the free ligands, from which other metal complexes can be prepared. Pierpoint et al. [26] have used the same amine 4-(2-dimethylaminoethyl)-1,7-diamino-4-azaheptane to prepare a 14- and a 16-membered macrocycle using

Richman and Atkins cyclisation procedure [24]. After deprotection, the free ligands *20* and *21* (Fig. 5) were obtained and their Ni^{2+}, Cu^{2+} and Zn^{2+} complexes were prepared.

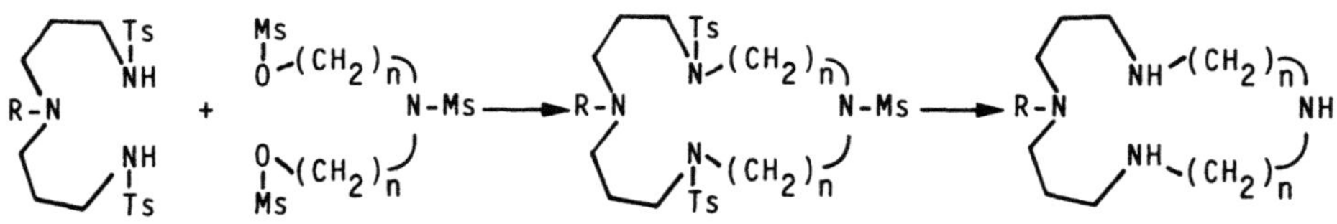

R = $(CH_2)_2$-$N(CH_3)_2$

20 n = 2

21 n = 3

Fig. 5. Synthesis of the monosubstituted macrocycles *20* and *21* using Richman and Atkins cyclisation procedure

c) *Modification of the side-chain*

If an appropriate side chain has once been introduced into the macrocycle by one of the other methods, it should be possible to modify it to obtain other derivatives. Therefore it is interesting to prepare a macrocycle with a pendant arm at the end of which an easily modifiable group is present. However, very little has been done in this respect probably because, up to now, no good starting materials are available in large quantities. In a series of reactions Furter [27] has modified the 2-hydroxyethyl side-chain of ligand *19* to give a 2-chloroethyl-, a 2-cyanoethyl- and a 2-carboxyethyl-derivative (Fig. 6), of which the corresponding Ni^{2+} complexes were also prepared.

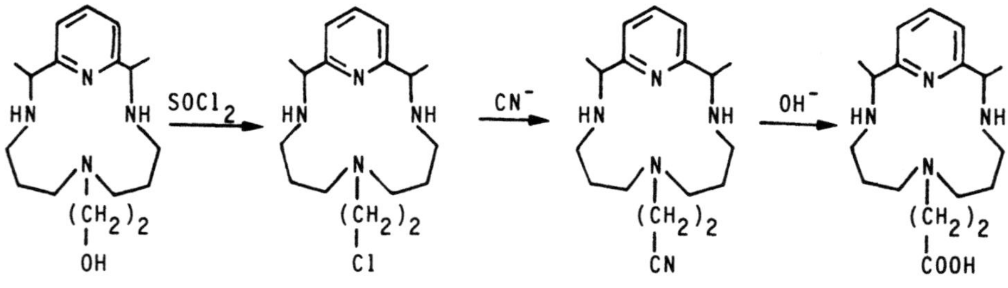

19

Fig. 6. Side-chain modifications of macrocycle *19*

d) *Alkylation of a selectively protected macrocycle*

The last synthetic approach which has been undertaken is shown as route d) in Fig. 3. Assumed that it is possible to prepare a tetraazamacrocycle which is selectively protected at three nitrogens, so that these are blocked for further reactions, but has one nitrogen accessible for alkylation, then one could easily introduce into the macrocycle any side-chain, which is robust enough to survive the deprotection step. This route has the great advantage that the cyclisation has to be done only once and thus allows optimization of the yield.

The two synthetic approaches described by Hediger et al. [28] are a modification of the cyclisation procedure of Richman and Atkins [24]. Instead of using only tosyl as protecting and activating groups, one nitrogen atom was substituted by a benzyl- or by a triphenylmethyl-group. Such large protecting groups were chosen to bring the reacting components in a conformation which is suitable for cyclisation. Probably these groups induce the same conformation as do the tosyl groups in the Richman and Atkins cyclisation, thus allowing to work at relatively high concentrations and obtaining good yields even without high dilution technique. Both paths (Figs. 7 and 8) give after cleavage of the benzyl- and trityl group, respectively, a tritosylated tetraazamacrocycle, which has one nitrogen accessible for alkylation. From these two key products *22* and *23* it is easy to prepare a large number of derivatives, especially such with a functionalized side-chain. Up to now the following pendant arms were introduced: 2-cyanoethyl [28], 2-carbamoylethyl [28], carbamoylmethyl [28], 2-tosylamidoethyl [28], 2-hydroxyethyl [29], 3-hydroxypropyl [29], 2-carbethoxyethyl [29] and carbethoxymethyl [29].

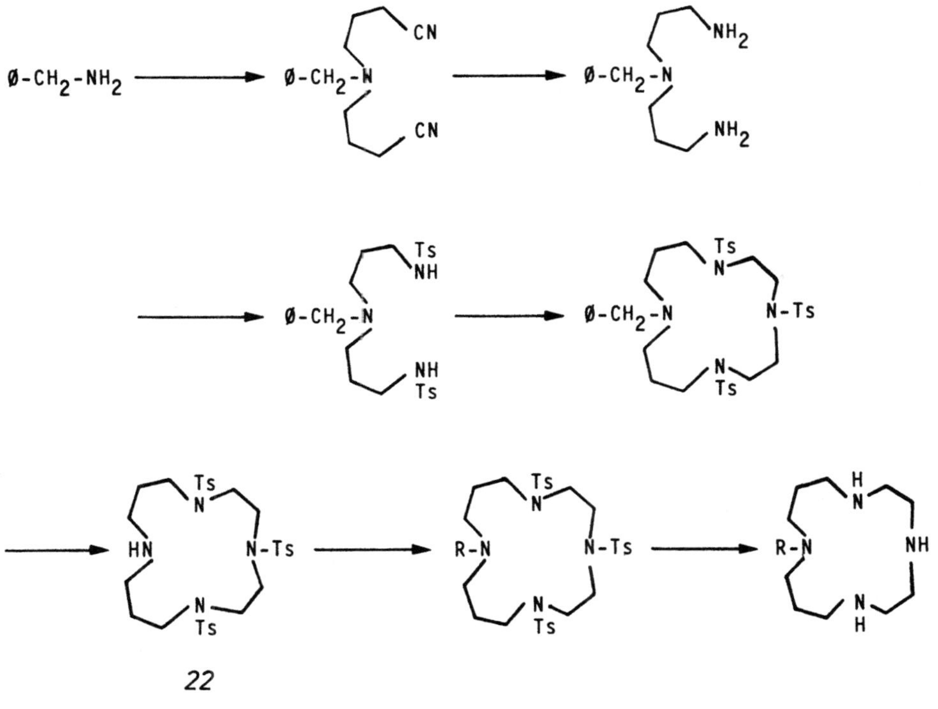

22

Fig. 7. Reaction scheme for the preparation of the tritosylated macrocycle *22* and its alkylation to give mono-substituted derivatives

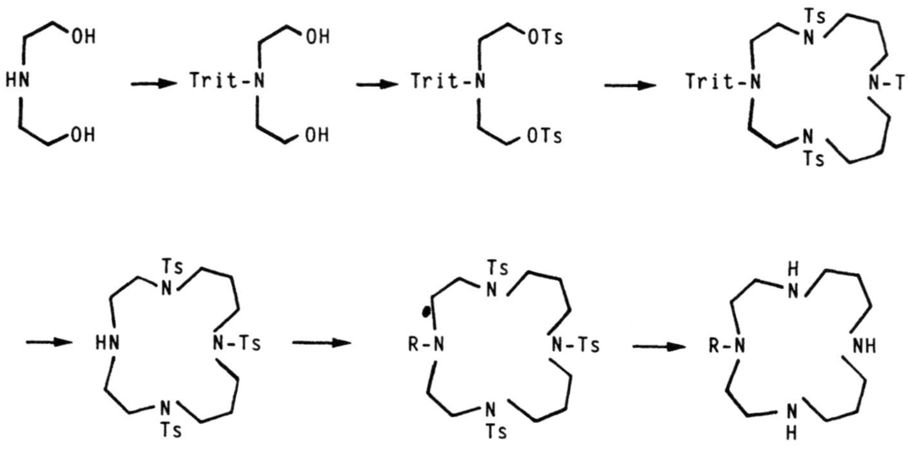

23

Fig. 8. Reaction scheme for the preparation of the tritosylated macrocycle *23* and its alkylation to give mono-substituted derivatives

The last step of this synthesis consists in a detosylation reaction which should be done under possibly mild conditions so that the functional group of the side chain can withstand the reaction. Hydrogen bromide in glacial acetic acid in the presence of phenol or an electrochemical reduction method according to Horner et al. [30] have been used successfully for this conversion [28]. The nitrile [28] and the ester [29] functions were attacked whereas the other ones remained intact.

Thus, the route d), although a multistep synthesis, seems most promising and versatile, especially if one wants to systematically vary the nature of the ligating group in the pendant arm.

2.2 C-Functionalized Derivatives

Pendant side-chains have also been introduced into tetraazamacrocycles by attaching them to one of the carbon atoms of the ring. In general the synthesis of all these compounds starts with the side chain already present in one of the cyclisation components and therefore resembles path b) of Fig. 3 for the corresponding N-substituted derivatives. A large number of C-functionalized macrocycles of the type *24* were prepared by Jäger et al. [31] by reacting diamines such as ethylene diamine, 1,3-diaminopropane and o-phenylene diamine with β-ketoimidato complexes. These compounds have in the rest R a carbonyl or an ester function but there is no evidence that these groups can coordinate to the metal ion since the carbon chain is to short.

Tabushi et al. [32] describe a series of compounds obtained by condensation of 1,9-diamino-3,7-diazanonane with different malonic esters substituted at the α-carbon with a side-chain (Fig. 9). Interesting is that this cyclisation does not require high

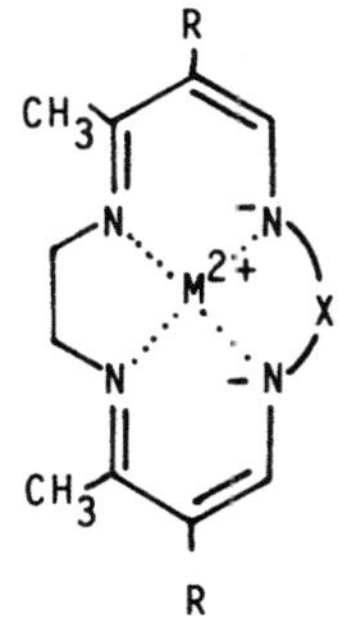

$X = (CH_2)_2,\ (CH_2)_3,\ o\text{-}C_6H_4$

$R = CO\text{-}CH_3,\ CO\text{-}C_6H_5,\ COOEt$

24

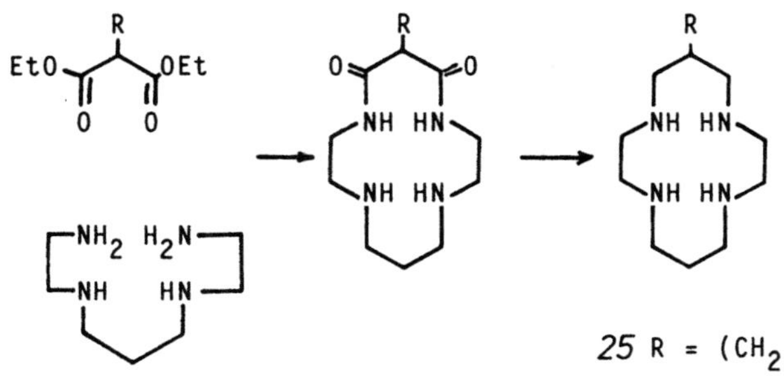

25 $R = (CH_2)_3\text{-}OH$

26 $R = (CH_2)_4\text{-}NH_2$

Fig. 9. Synthesis of the C-functionalized macrocycles *25* and *26*

dilution conditions and nevertheless gives good yields of the cyclic product. The resulting amides were reduced with diborane to the corresponding tetraazamacrocycles *25* and *26* with a pendant 3-hydroxypropyl- and 4-aminobutyl group, respectively. Up to now, no report of metal complexes with these compounds has been given.

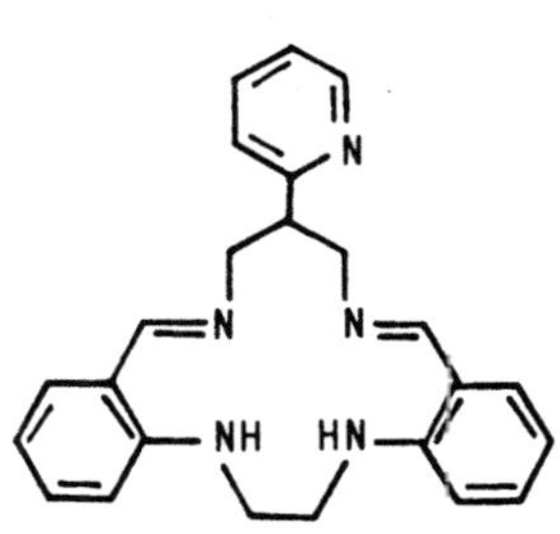

27

Another C-substituted macrocycle has been prepared by Tasker et al. [33]. The pyridyl group of *27*, however, cannot coordinate to the metal ion, since the chain is to short. The synthesis is based on a template reaction between a dialdehyde and 2-(2-pyridyl)-1,3-diaminopropane, in the presence of Ni^{2+} or Cu^{2+}.

2.3 Bis-Macrocycles

A special class of functionalized tetraazamacrocycles are compounds obtained by connecting two macrocycles together so that binuclear metal complexes can be prepared.

Murase et al.[34] describe the synthesis of a binuclear Cu^{2+} complex which contains two N_4-macrocyclic units linked by an ethylene bridge (*28*). The synthesis follows path b) of Figure 3: 2,6-diacetylpyridine and N,N,N',N'-tetrakis-(3-amino-

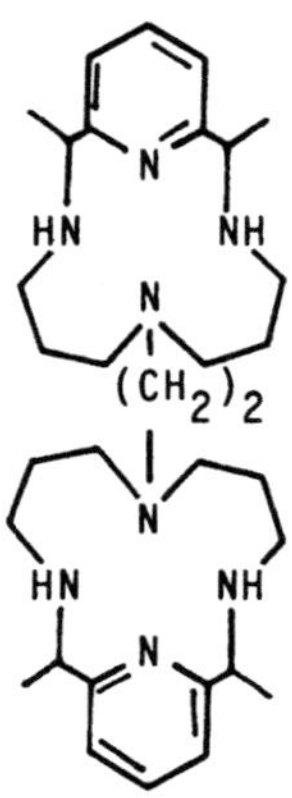

28

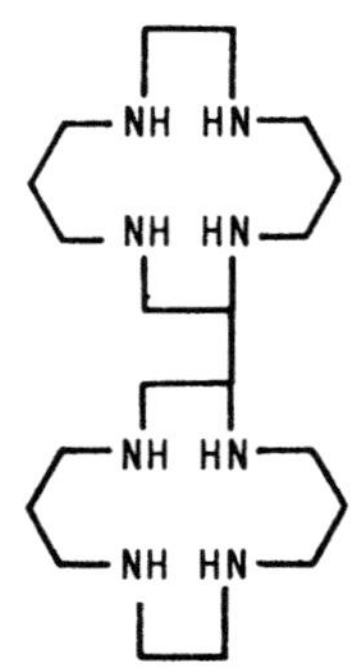

29

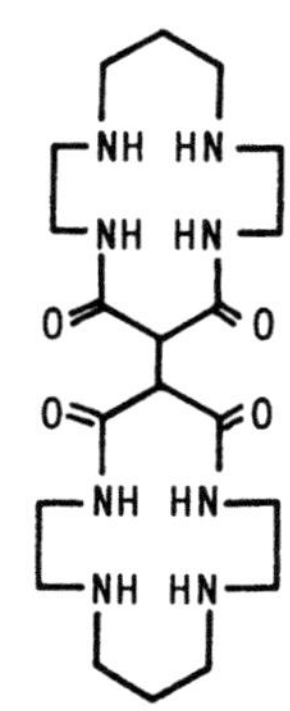

30

propyl)-1,2-diaminoethane are reacted in the presence of Cu^{2+} as template. The binuclear Cu^{2+} complex exhibits a weak antiferromagnetic interaction as indicated by the temperature dependence of the magnetic moment and by the EPR spectrum.

Another macrocycle (*29*) was obtained in yields of 1.5–3% by Barefield et al.[35] as a byproduct of the synthesis of 1,4,8,11-tetraazacyclotetradecane. This compound probably forms during the reduction by condensation of the active methylene group of the half reduced Schiff-base complex with the imine group of the second complex. The ligand forms binuclear Cu^{2+} and Ni^{2+} complexes. Fabbrizzi et al.[36] have used the condensation of 1,9-diamino-3,7-diazanonane with tetraethyl-1,1,2,2-ethane-tetracarboxylate to prepare the bis-macrocycle *30*. The binuclear Cu^{2+} complex was studied by EPR and electrochemically. It was found that the two Cu^{2+} ions are oxidized to Cu^{3+} in two steps separated by 110 mV.

3 Metal Complexes

3.1 Structures

There are only a few structural studies of the metal complexes with azamacrocycles having pendant arms. In an interesting paper Wieghardt et al.[37] compare the structures of the Cr^{3+}, Fe^{3+} and Cu^{2+} complexes of *4*. In these compounds all six ligating atoms bind to the metal ion with a facial N_3O_3 donor set. The structures, however, vary from distorted octahedral for Cr^{3+}, to pseudo trigonal-prismatic

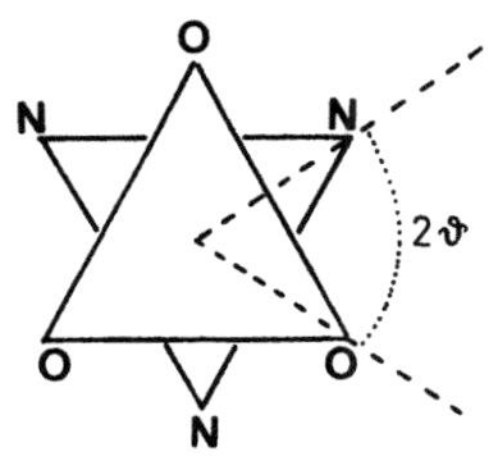

Fig. 10. Twist angle ν of the N_3O_3-donor set of macrocycle *4* in different metal complexes

for Fe^{3+} and to distorted pseudo-prismatic for Cu^{2+}. This is shown by the values of the twist angle ϑ (Fig. 10) which is 24.5° for Cr^{3+}, 12.6° for Fe^{3+} and 13.3° for Cu^{2+}. The authors explain this differences by comparing the ligand field stabilisation for an octahedral arrangement to that for a prismatic one: for a d^3 system (Cr^{3+}) it is −1.3 Dq, for a d^9 ion (Cu^{2+}) only −0.5 Dq whereas for a d^5 system (Fe^{3+}) no ligand field stabilisation is expected.

Zehnder et al. [38] have solved the structure of the Ni^{2+} complex *18*. The metal ion is encircled in a square planar arrangement by the four nitrogens of the macrocycle, whereas the dimethylamino group at the end of the side-chain remains uncoordinated (Fig. 11).

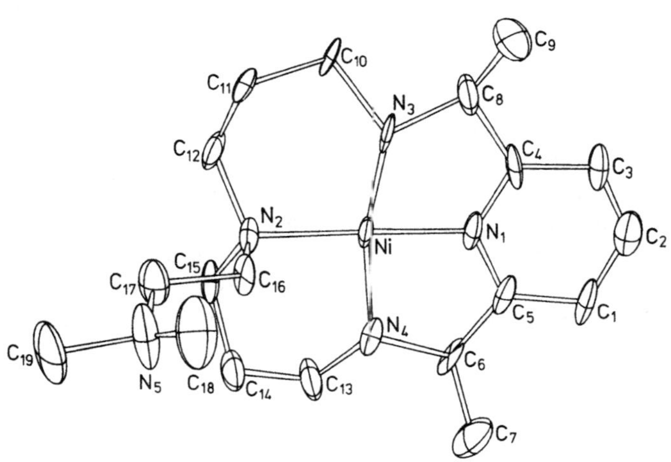

Fig. 11. X-ray structure of the Ni^{2+} complex *18*

This is because the complex was obtained from acidic solution at a pH at which the dimethylamino group is still protonated (see 3.2.2). For the Ni^{2+}-complex with the unprotonated dimethylamino group up to now no good quality crystals were obtained. It is interesting to note that the complex is a racemic isomer with the α-methyl groups, the hydrogens of the secondary amines and the pendant arm all on the same side of the Ni—N_4-plane.

Desreux et al. [39] have solved the structure of the Eu^{3+} complex with *1*. The lanthanide ion is coordinated by the four nitrogens of the macrocycle, by the four

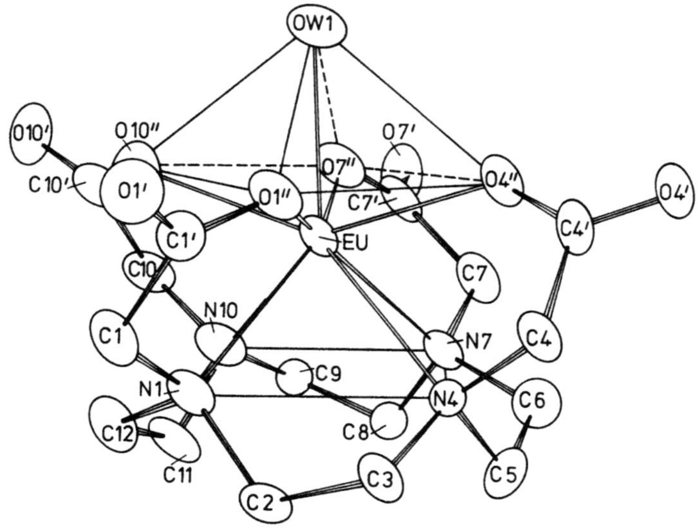

Fig. 12. X-ray structure of the Eu^{3+} complex of *1*

carboxylates and by one water molecule giving as coordination polyhedron a distorted square capped antiprism (Fig. 12). The metal ion lies between the two planes of the four nitrogens and of the four oxygens, which are nearly parallel to each other. Another study of Desreux et al. [40] in which the chemical shifts and the line broadening of a series of lanthanide complexes with *1* were measured, gives some indication for the conformation of the macrocycle in solution. The approximate distance between the lanthanide and the nitrogens is about 2.85 Å, calculated from the values of the shift ratio of the axial protons. The metal ion sits on a fourfold symmetry axis above the plane of the four nitrogens. Below room temperature the ethylene groups have all the same conformation. The energy barrier for ring inversion is high ($\Delta G = 60$ kJ/Mol) for all the complexes and relatively insensitive to the size of the encapsulated metal ion.

3.2 Equilibria in Solution

3.2.1 Protonation and Complexation

The protonation and complexation equilibria of the tetraazamacrocycles with four acetate groups *1*, *5*, *6* have attracted much attention [5–7,41]. The pH-titrations show that all these ligands have two pK_H above 9.5, two around 4 and two below 2 (Table 1). In an interesting NMR study the microconstants were obtained by following the chemical shift of the methylene protons adjacent to the nitrogens as a function of pD. Desreux et al. [41] were able to show that the protonation first occurs at two amino nitrogens of the macrocycle which are trans to each other, then at two carboxylates attached at the two nitrogens which are not yet protonated and finally at the other two carboxylates. The remaining nitrogens must have extremely low pK_H-values, which is consistent with the finding that 12–14 membered tetraazamacrocycles

Table 1. pK_H values of the macrocyclic tetraacetates *1*, *5* and *6*

		pK_1	pK_2	pK_3	pK_4	pK_5	pK_6
1	a)	11.36	9.73	4.54	4.41		
	b)	11.08	9.23	4.24	4.18	1.88	1.71
	c)	11.22	9.75	4.37	4.36		
	d)	12.09	9.68	4.55	4.13		
5	a)	11.22	9.18	4.59	3.28		
	c)	11.35	9.73	4.16	3.23		
6	a)	11.07	9.75	4.31	3.43		
	b)	11.56	10.18	4.05	3.38	2.17	1.42
	c)	10.68	10.14	4.09	3.34		

a) T = 20°, I = 0.1 M (KCl) from Ref. [5]
b) T = 25°, I = 1 M (NaCl), corrected, from Ref. [41]
c) T = 25°, I = 0.1 M (KNO_3) from Ref. [7]
d) T = 25°, I = 0.1 M ($(CH_3)_4N\ NO_3$) from Ref. [7]

without pendant arms have two high and two low pK_H-values, because of the strong electrostatic repulsion of the positive charged ammonium groups [42]. The discrepancy in the $pK_{H,1}$ values obtained by different authors were explained by Delgado et al. [7] by the fact that these ligands form relatively stable Na^+ and K^+ complexes so that it is necessary to use tetramethylammonium nitrate as non-complexing medium and tetramethylammonium hydroxide as base to obtain reliable results.

The stability constants of the different metal complexes were first measured by Stetter et al. [5] who found that ligand *1* forms the most stable Ca^{2+} complex in aqueous solution known to date. They also reported that in the case of ligand *5* the stability of the Sr^{2+} complex was higher than that of the Ca^{2+} complex. The same authors, however, were not able to confirm their initial result [6].

Newer measurements by Delgado et al. [7] clearly indicate that in addition to the 1:1 complexes (ML), which were the only ones taken into account by Stetter et al. [5,6], protonated species such as MLH_2 and MLH must also exist. This is in accordance with preparative work of Häfliger et al. [13] who were able to isolate solid compounds with the stoichiometry MLH_2 (M = Cu^{2+}, Ni^{2+}, Zn^{2+} and L = *11*). The IR spectra of these complexes have strong NH^+ intercombination bands indicating that at least one but probably two nitrogens are protonated and not involved in complexing the metal ion. Thus the metal ion probably is not encircled but is bound in a relatively open structure. The stabilities of the earth-alkali ions follow the normal order $Ca^{2+} > Sr^{2+} > Ba^{2+}$ and the small ring *1* gives the most stable complexes. For the transition metal ions, however, the ring size has only little effect on the stability constants. Hama et al. [43] have measured the stabilities of the Mn^{2+}, Co^{2+}, Cu^{2+}, Zn^{2+}, Cd^{2+} and Pb^{2+} complexes with *4* by polarographic techniques. Since the complexation reactions were slow, it was necessary to heat the solutions at 60° to reach equilibrium conditions.

The protonation and complexation equilibria of the Fe^{3+} complexes with *2* and its sulfonated derivative were measured by Raymond et al. [44]. At high pH the ligands coordinate to the Fe^{3+} through the six phenolic oxygens giving an extremly stable complex, which can remove essentially all Fe^{3+} from transferrin. At lower pH, protonated complexes $FeLH^{2-}$, $FeLH_2^-$ and $FeLH_3$ were also observed for which the authors propose a different binding mode through the carbonyl oxygen and one phenolate oxygen of each unit.

3.2.2 Equilibria Involving the Side-Chain

An interesting aspect of macrocycles with a pendant arm is the internal equilibrium between the open and chelated form of the side-chain (Fig. 13), which depends on the nature of the ligating atom and the length of the side-chain.

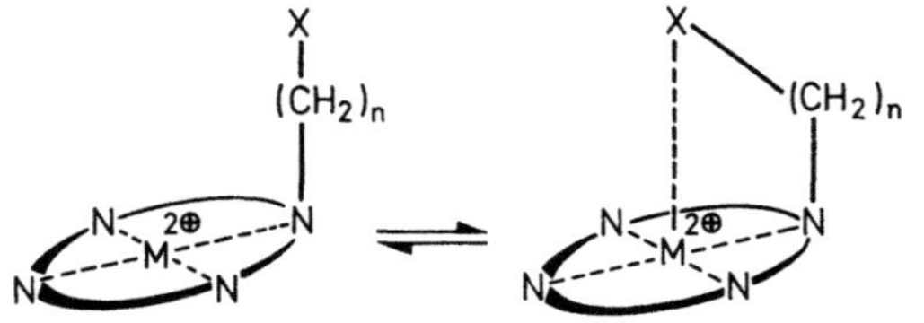

Fig. 13. Equilibrium between the open and apically coordinated form in metal complexes of macrocycles with a side-chain

For an aliphatic side-chain with a high degree of flexibility one would expect that if the chain length is suitable for the formation of a chelate ring, the equilibrium will be shifted in favour to the closed form. The first example of this kind was observed by Lotz et al. [25,45] for the Ni^{2+} complex *18*. In this case two solids can be isolated: from acidic solution one obtains a yellow diperchlorate, whereas from alkaline solution a violett triperchlorate crystallizes. Complexes of the same colour are also observed in aqueous solution. Below pH 6.3 the Ni^{2+} complex is diamagnetic and yellow, with an absorption at 470 nm typical for square planar geometry. This is in accordance with the results of the X-ray structure analysis of the solid (see sect. 3.1). At pH above 6.3 the dimethylammonium group looses its proton, becomes a potential ligand and binds in the axial position of the metal ion, whereby a hexa-coordinated paramagnetic violett Ni^{2+} complex is formed. The acid-base equilibrium and the concomitant reversible geometry change is a consequence of the combination of the kinetically stable and rigid macrocycle and the kinetically labile and flexible structural unit of the side-chain. Whereas the complexation capability of the macrocycle is not influenced by the pH change, the coordination potential of the dimethylaminoethyl group is strongly dependent on the pH. Similar equilibria have also been observed for the Ni^{2+} complex *19* [25].

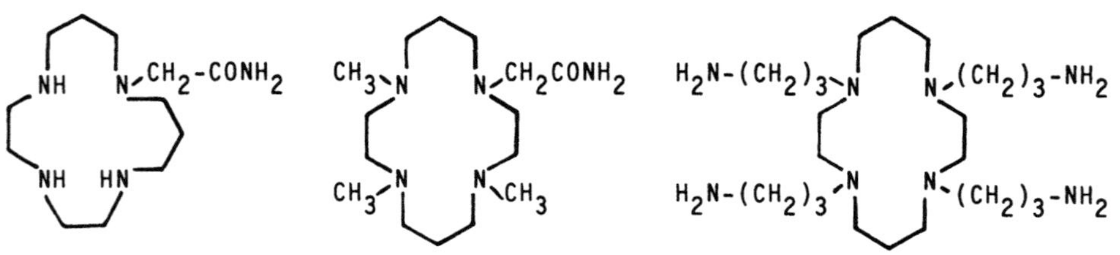

31 *32* *33*

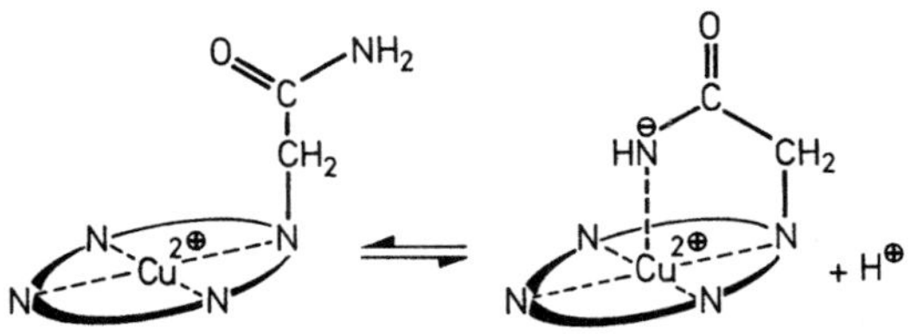

Fig. 14. Deprotonation and coordination of the amide group in metal complexes

Another example of this kind of axial coordination is found in the Cu^{2+} complexes of *31* [46] and *32* [18]. At pH below 10 the amide group probably does not coordinate. If the pH is increased above 10, however, deprotonation and coordination of the amide group through its nitrogen atom takes place (Fig. 14). That an axial interaction with the amide group is present can be seen from the IR-spectrum in D_2O. The amide band shifts from 1660 cm^{-1} to 1580 cm^{-1} which is indicative for a coordinated amide group [47]. At the same time there is a red shift of the absorption maximum in the VIS spectrum which is typical for a CuN_4-chromophore when a fifth ligand is axially coordinated [48].

In line with this is also the observation of Wainwright et al. [16] who found that the Ni^{2+} complex of *33* gives a colour change when it is treated with HCl in EtOH. Under these conditions the amino groups are protonated and the Ni^{2+} ion is then coordinated by the four nitrogen atoms of the macrocycle in a square planar geometry.

All these results show that by designing a pendant arm with a chain of two or three carbon atoms it becomes possible to bind the terminal ligating group to an axial position of the metal ion, whereby the properties of the metal ion, such as its coordination geometry and its absorption spectrum are modified. Thus in these cases one has the possibility to control the chemical properties of these complexes by selecting the experimental conditions such as for example the pH of the solution.

4 Kinetical Aspects

The kinetics of metal complex formation with *1* and *5* have been measured by Wilkins et al. [49] using a stopped-flow technique. Metal ions examinated were Mg^{2+}, Ca^{2+}, Sr^{2+}, Ba^{2+}, Zn^{2+}, Ni^{2+} and Cu^{2+} and in all cases the monoprotonated form LH^{3-} of the ligands was the reactive species. The rate constants for the complexation are 10^3–10^4 times larger than those of the unsubstituted macrocycles [42]. This can qualitatively be understood by the different charges of the ligands, the tetraacetates bearing a charge of -3 and the unsubstituted macrocycles one of $+1$. The acetate groups, however, also modify the mechanism since they can rapidly form an adduct with the metal ion. The magnitude of the stability of these adducts indicates that a carboxylate and an amino nitrogen must be involved in the binding. Once the metal ion is coordinated and close to the macrocyclic ring it can be rapidly incorporated. This is similar to the observation by Buckingham et al. [50] that a picked-fence porphyrin with four carboxylate groups reacts much faster in its metal ion complexation than unsubstituted porphyrins. The kinetics were also explained by a rapid preequilibrium in which the metal ion is complexed by at least one carboxylate followed by the rate determining step in which the metal ion is introduced into the ring.

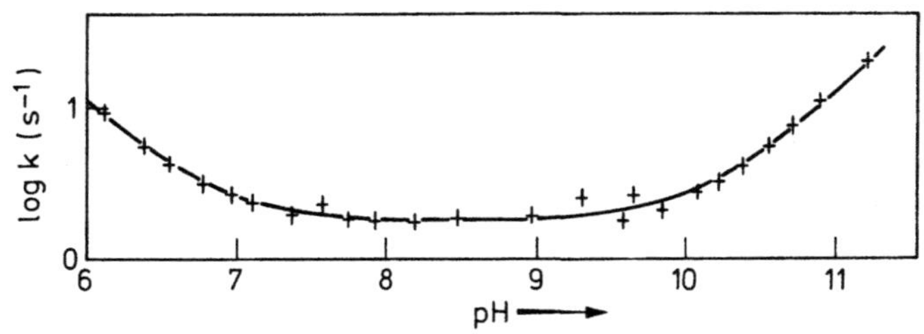

Fig. 15. pH-profile of the forward and backward reaction of the amide coordination in the Cu^{2+} complex *31*

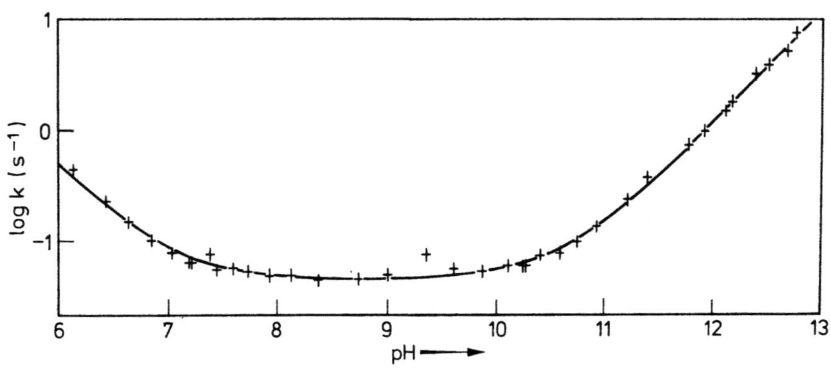

Fig. 16. pH-profile of the forward and backward reaction of the amide coordination in the Cu^{2+} complex *32*

Little is known about the kinetics of the complexation of the ligating group of the side-chain to an axial position of the metal ion (Fig. 13). For the Ni^{2+} complex *18* the reaction is to fast to be followed by stopped-flow or T-jump [51] techniques. The only reaction of this type which was measured is the axial interaction of the amide group in the Cu^{2+} complexes of *31* and *32* [52]. Both the forward and the backward reaction of Fig. 14 were followed by stopped flow kinetics. The results are shown as pH profile in Fig. 15 for *31* and in Fig. 16 for *32*. Similarly to the reactions of open chain amide ligands with the Cu^{2+} ion [53], three terms are necessary to fit the experimental pH dependence (1), k_H describing the proton

$$k_{obs} = k_H[H^+] + k_O + k_{OH}[OH^-] \qquad (1)$$

induced dissociation, k_{OH} the OH^- dependent binding of the deprotonated amide and k_O the pH independent region. The values are given in Table 2, together with rate constants typical for amide complex formation in equatorial position.

As axial ligands change the properties of the metal ion (see sect. 3.2.2) the reciprocal influence would also be expected, i.e. that the metal ion can change the reactivity of the functional group in the side-chain. An example for this was found by Schibler et al. [18,19] when they tried to prepare the Cu^{2+} complex of *34*. Instead a cristalline product with a typical amide band at 1645 cm^{-1} and with an elemental analysis

Table 2. Rate constants for the O- to N-interconversion in different Cu^{2+} amide complexes at 25°

	axial		equatorial [53]
	Cu (*31*)	Cu (*32*)	
k_O (s^{-1})	1.7	$4.4 \cdot 10^{-2}$	$3 \cdot 10^{-2} - 6$
k_H $(M^{-1}\,s^{-1})$	$9.2 \cdot 10^6$	$4.2 \cdot 10^5$	$3 \cdot 10^6 - 4 \cdot 10^7$
k_{OH} $(M^{-1}\,s^{-1})$	$6.9 \cdot 10^3$	81	$9 \cdot 10^3 - 1 \cdot 10^5$

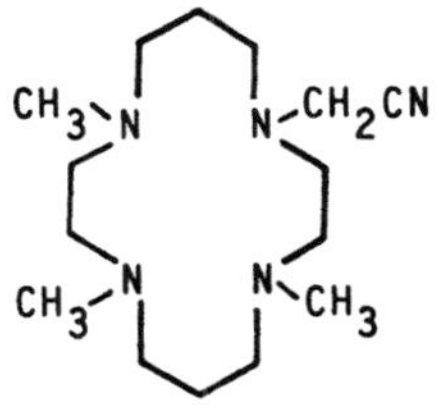

34

which agrees to the Cu^{2+} complex of the amide *32* was obtained. This Cu^{2+} promoted hydrolysis of the nitrile to the amide was studied in more detail. The stopped-flow kinetics show that the reaction is a two phase process and this was interpreted by postulating two consecutive first-order steps. The first and faster reaction shows only a small amplitude change when measured at 643 nm and its rate depends on $[OH^-]$ and can be inhibited by SCN^-. The rate law is given by equation (2) where $c_{complex}$ is the total concentration of the Cu^{2+} in any form.

$$v_1 = k_1K_2c_{complex}[OH^-]/(K_1K_2 + K_2[OH^-] + K_1[SCN^-]) \qquad (2)$$

The constants are $k_1 = 13.0\ s^{-1}$, $K_1 = 9.7 \cdot 10^{-3}$ M and $K_2 = 5.4 \cdot 10^{-3}$ M and their meaning can be taken from Figure 17. The dissociation constant K_2 of the ternary complex with thiocyanate measured separatly by spectrophotometric titration at pH = 7 where no hydrolysis takes place gave $K_2 = 3.9 \cdot 10^{-3}$ M in satisfactory agreement with the value obtained from the kinetics. The first reaction step is attributed to the hydrolysis of the nitrile *34*, to give the amide complex *35*. The second step gives a larger amplitude change and its rate is proportional to $[OH^-]$ (Eq. (3)) with $k_2 = 89\ M^{-1}\,s^{-1}$.

$$v_2 = k_2\,[35] \cdot [OH^-] \qquad (3)$$

The reaction is associated with the rearrangement of the amide group from the O- into the N-coordinated form as shown in Fig. 17. This reaction measured separately gives $k_2 = 81\ M^{-1}\,s^{-1}$ (Table 2), in good agreement with the value obtained from the hydrolysis and thus is a strong support for the assignment of the second step.

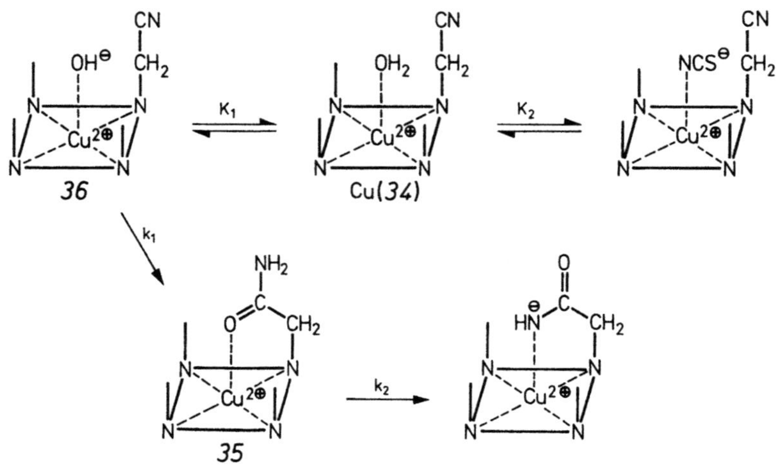

Fig. 17. Mechanisms of the intramolecular hydrolysis of the nitrile complex Cu *34*

For the discussion of the mechanism is was assumed that all Cu^{2+} complexes are pentacoordinated in analogy to the structures found for the complexes with 1,4,8,11-tetramethyl-TA14C4 [54]. The pH profile of the first step (Eq. (2)) with a plateau above pH 12.5 as well as the SCN^- inhibition clearly indicates that the hydroxo complex *36* is the reactive species and the coordinated OH^- is the nucleophile. The fast hydrolysis ($t_{1/2} \simeq 50$ ms at pH 12.5) is mainly due to a favourable proximity of the reactants so that an intramolecular attack can take place (Fig. 18).

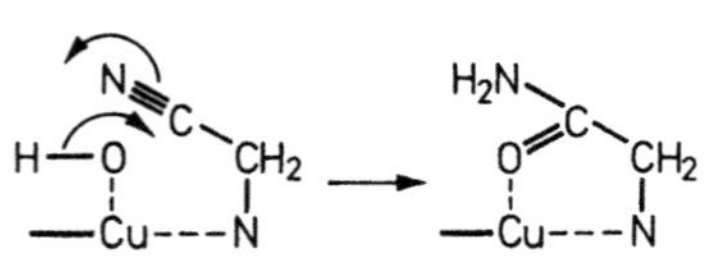

Fig. 18. Transition state of the intramolecular attack of coordinated OH^- onto the nitrile group

The product of the hydrolysis of the Cu^{2+} complex with *12*, which has two nitrile groups, is the Cu^{2+} complex of *37*, which has one amide and one nitrile group. The selective hydrolysis of only one nitrile can be understood by the same type of reaction sequence as for the mononitrile *34*. Once the first nitrile group has been hydrolyzed the amide will coordinate in the axial position with its deprotonated nitrogen. This

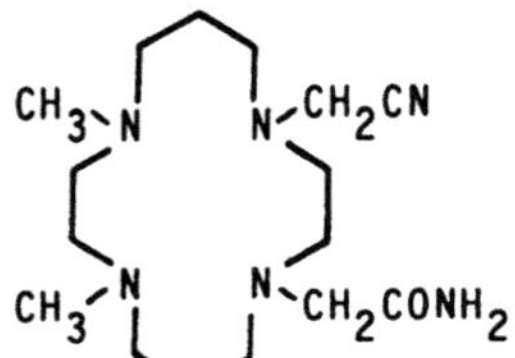

37

will block the axial position for OH^- addition in a similar way as we have described for SCN^-. Thus the reactive species cannot be formed and the further hydrolysis is stopped.

These examples of a very efficient and selective metal ion promoted hydrolysis show how proximity effects and right stereochemical ordering can work even though the strength of the nucleophile is weakened by coordination to the metal ion.

5 Conclusions

In view of new strategies in coordination chemistry macrocycles with functionalized pendant arms represent an interesting class of ligands, which combine both the properties of a macrocyclic structural unit with those of a flexible side-chain to which the functional group is bound. With such systems it is possible to study the effect of axial coordination even for weak ligating groups, since they are brought close to the metal ion, the changes in the properties of the metal ion when the pendant arm becomes coordinated and finally the influence of the metal ion on the reactivity of the functional group.

Since the sterical and geometrical aspects can be kept constant by the macrocycle small effects are expected to be detected. In addition even with a kinetically labile metal ion it is possible to obtain insight into the relation between reactivity and structure.

6 Acknowledgements

This work was supported by the Swiss National Science Foundation which is gratefully acknowledged. This manuscript draws heavely from the work of my research students and coworkers whose names appear in the references. We would also like to thank Dr. J. F. Desreux for communicating structural results prior to publication.

7 References

1. Pedersen, C. J.: J. Am. Chem. Soc. *87*, 7017 (1967). For a review see Pedersen, C. J. in "Synthetic Multidentate Macrocyclic Compounds" (Ed. Izatt, R. H., Christensen, J. J.), Academic Press, New York 1978, p. 1
2. Curtis, N. F.: J. Chem. Soc. 4409 (1960);
   Curtis, N. F., House, D. A.: Chem. Ind. 1708 (1961)
3. Busch, D. H.: Helv. Chim. Acta Fasc. Extraordinarius Alfred Werner, 174 (1967);
   Curtis, N. F.: Coord. Chem. Rev. *3*, 3 (1968);
   Lindoy, L. F., Busch, D. H.: Prep. Inorg. Reactions *6*, 1 (1971);
   Christensen, J., Eatough, D., Izatt, R.: Chem. Rev. *74*, 351 (1974);
   Lindoy, L. F.: Chem. Soc. Rev. *5*, 421 (1975);
   Newkome, G. R., Sauer, J. D., Roger, J. M., Hager, D. C.: Chem. Rev. *77*, 513 (1977);
   Melson, G.: Coordination Chemistry of Macrocyclic Compounds (Ed. Melson, G.), Plenum Press, New York 1979, p. 17
4. Lehn, J. M.: Structure and Bonding *16*, 1 (1973)
5. Stetter, H., Frank, W.: Angew. Chem. *88*, 760 (1976)
6. Stetter, H., Frank, W., Mertens, R.: Tetrahedron *37*, 767 (1981)

7. Delgado, R., Frausto Dasilva, J. J.: Talanta *29*, 815 (1982)
8. Weitl, F., Raymond, K.: J. Am. Chem. Soc. *101*, 2728 (1979)
9. Muller, F. R., Handel, H.: Tetrahedron Lett. *23*, 2769 (1982)
10. Collman, J. P., Halpert, Th. R., Suslick, K. S.: Metal Ion Activation of Dioxygen (Ed. Spiro, Th. G.), Wiley, New York 1980, p. 1
11. Chang, C. K., Traylor, T. G.: Proc. Nat. Acad. Sc. USA *70*, 2647 (1973) and J. Am. Chem. Soc. *95*, 5810, 8475, 8477 (1973)
12. Takahashi, M., Takamoto, S.: Bull. Chem. Soc. Japan *50*, 3413 (1977)
13. Häfliger, H., Kaden, Th. A.: Hev. Chim. Acta *62*, 683 (1979)
14. Buøen, S., Dale, J., Groth, P., Krane, J.: J. Chem. Soc., Chem. Commun. 1172 (1977)
15. Wainwright, K. P.: J. Chem. Soc. Dalton 2117 (1980)
16. Brown, J. K., Wainwright, K. P.: Abstract Tu P47, XII ICCC, Budapest 1982 and Abstract P26, Colloque Int. du CNRS: Composé macrocyclique, Strassbourg 1982
17. Siegfried-Hertli, L., Kaden, Th. A.: unpublished results
18. Schibler, W.: Ph. D. Thesis, Basel 1980
19. Schibler, W., Kaden, Th. A.: J. Chem. Soc., Chem. Commun. 603 (1981)
20. Vitali, P.: Thèse, Strassbourg 1980
21. Louvet, V., Apprion, P., Handel, H.: Tetrahedron Lett. *23*, 2245 (1982)
22. Wagner, F., Barefield, E. K.: Inorg. Chem. *15*, 408 (1976)
23. Stetter, H.: Chem. Ber. *87*, 566 (1954);
Dietrich, B., Lehn, J. M., Sauvage, J. P., Blanzat, J.: Tetrahedron *29*, 1629 (1973)
24. Richman, J. E., Atkins, Th. J.: J. Am. Chem. Soc. *96*, 2268 (1974)
25. Lotz, T. J., Kaden, Th. A.: Helv. Chim. Acta *61*, 1376 (1978)
26. Pierpoint, C., Moore, P.: Communication at the Macrocycles Mini-symp., Polytechnic of North London, 1982
27. Furter, P.: Ph. D. Thesis, Basel 1979
28. Hediger, M., Kaden, Th. A.: J. Chem. Soc., Chem. Commun. *14* (1978) and Helv. Chim. Acta *66*, 861 (1983)
29. Stöcklin, F.: Ph. D. Thesis, Basel 1981
30. Horner, L., Neumann, H.: Chem. Ber. *98*, 3462 (1965)
31. Jäger, E. G.: Z. Chem. *8*, 30, 392, 470 (1968);
Wolf, L., Jäger, E. G.: Z. Anorg. Allgem. Chem. *346*, 76 (1966);
Jäger, E. G.: Z. Anorg. Allgem. Chem. *364*, 177 (1969)
32. Tabushi, J., Taniguchi, Y., Kato, H.: Tetrahedron Lett. 1049 (1977)
33. Henrick, H., Tasker, P.: Inorg. Chim. Acta *47*, 47 (1980)
34. Murase, J., Hamada, K., Kida, S.: Inorg. Chim. Acta *54*, 471 (1981)
35. Barefield, E. K., Chueng, D., van Derveer, D., Wagner, F.: J. Chem. Soc., Chem. Commun. 302 (1981)
36. Buttafava, A., Fabbrizzi, L., Perotti, A., Seghi, B.: J. Chem. Soc., Chem. Commun. 1166 (1982)
37. Wieghardt, K., Bossek, U., Chaudhuri, P., Herrmann, W., Menke, B. C., Weiss, J.: Inorg. Chem. *21*, 4308 (1982)
38. Kaden, Th. A.: Coordination Chemistry 20 (Ed. Banerjea, D.), Pergamon Press, Oxford 1980, p. 77
39. Spirlet, M. R., Rebizant, J., Desreux, J. F., Loucin, M. F.: submitted to Acta cryst.
40. Desreux, J. F.: Inorg. Chem. *19*, 1319 (1980)
41. Desreux, J. F., Mercigny, E., Loucin, M. F.: Inorg. Chem. *20*, 987 (1981)
42. Leugger, A. P., Hertli, L., Kaden, Th. A.: Helv. Chim. Acta *61*, 2296 (1978)
43. Hama, H., Takamoto, S.: Nippon Kagaku Kaishi 1182 (1975)
44. Harris, W. R., Raymond, K. M.: J. Am. Chem. Soc. *101*, 6534 (1979);
Harris, W. R., Raymond, K. M., Weitl, F. C.: J. Am. Chem. Soc. *103*, 2667 (1981)
45. Lotz, T. J., Kaden, Th. A.: J. Chem. Soc., Chem. Commun. 15 (1977)
46. Hediger, M.: Ph. D. Thesis, Basel 1979
47. Kim, M. K., Martell, A. E.: J. Coord. Chem. *4*, 9 (1974)
48. Billo, E. J.: Inorg. Nucl. Chem. Lett. *10*, 613 (1974);
Kurganov, A., Davankov, V.: Inorg. Nucl. Chem. Lett. *12*, 73 (1976)
49. Kasprzyk, S. P., Wilkins, R. G.: Inorg. Chem. *21*, 3349 (1982)
50. Buckingham, D. A., Clark, C. R., Webley, W. S.: J. Chem. Soc., Chem. Commun. 192 (1981)
51. Kaden, Th. A.: unpublished results

52. Hediger, M., Schibler, W., Kaden, Th. A.: to be published
53. Margerum, D. W., Dukes, G. R.: Metal Ions in Biological Systems (Ed. Sigel, H.), Marcel Dekker, New York 1974, Vol. 1, p. 157;
Zuberbühler, A. D., Kaden, Th. A.: Helv. Chim. Acta *55*, 623 (1972); *57*, 1897 (1974); *65*, 753 (1982)
54. Barefield, E. K., Wagner, F.: Inorg. Chem. *12*, 2435 (1973);
Buxtorf, R., Steinmann, W., Kaden, Th. A.: Chimia *28*, 15 (1974);
Buxtorf, R., Kaden, Th. A.: Helv. Chim. Acta *57*, 1035 (1974);
D'Aniello, M. J., Mocella, M. T., Wagner, F., Barefield, E. K., Paul, I. C.: J. Am. Chem. Soc. *97*, 192 (1975)

Bridged, Capped and Fenced Porphyrins

Jack E. Baldwin and Patrick Perlmutter

Dyson Perrings Laboratory, South Parks Road, Oxford OX1 3QX, England

Table of Contents

1 Introduction 182

2 Haemoglobin and Myoglobin 182
2.1 Occurrence and Function 182
2.2 Structure 182
2.3 Dioxygen Binding to Natural and Synthetic Haems 186
2.4 Fenced Porphyrins 187
2.5 Capped Porphyrins 194
2.6 Bridged Porphyrins 198
2.6.1 Singly-Bridged Porphyrins 198
2.6.2 Doubly-Bridged Porphyrins 201

3 Cytochrome P-450 203
3.1 Occurrence and Function 203
3.2 Reactivity 203
3.3 Fenced Porphyrins 205
3.4 Bridged Porphyrins 208

4 Cytochrome *C* Oxidase 210
4.1 Occurrence and Function 210
4.2 Structure and Mechanism 210
4.3 Fenced Porphyrins 211
4.4 Bridged Porphyrins 212
4.5 Dimeric Porphyrins 213

5 A Miscellany 214

6 Symbols and Abbreviations 215

7 References 216

1 Introduction

The object of this review is to bring together many of the diverse metalloporphyrin derivatives, possessing steric protection of at least one face, which have been prepared over the last decade. For it is during this period especially that chemists have used the porphyrin nucleus as the common platform upon which to build models for a wide range of metalloproteins.

The article is divided into sections covering models for haemoglobin and myoglobin, cytochromes P-450, cytochrome *c* oxidase. These account for almost all the bridged, capped and fenced porphyrins so far synthesized. A Section is also included on a few miscellaneous examples. Each section contains a brief description of the natural system, including those properties which should be included in a faithful model, followed by subsections organized according to structural type.

2 Haemoglobin and Myoglobin

2.1 Occurrence and Function

Haemoglobin (Hb) is the red pigment of vertebrate blood. It is also found in a variety of other eukaryotes including invertebrates, some yeasts and even the leguminosae [1]. Its principal function in vertebrates is to transport dioxygen from the lungs to the muscle tissue. Hb also plays an important role in the transport of carbon dioxide from the muscle to the lungs. Myoglobin (Mb) is found in skeletal muscle cells and its function is to store the dioxygen transported to it by Hb. It also enhances the rate of diffusion of dioxygen through the cell to the mitochondria.

2.2 Structure

Both Hb and Mb contain iron(II)protoporphyrin IX (haem) *1* as their prosthetic group. Adult haemoglobin (HbA), a dimer of dimers, consists of two α- and two

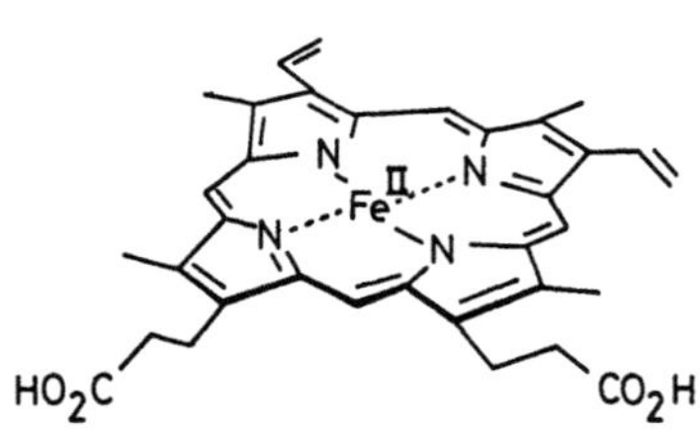

1

β-subunits. Each subunit is made up of one haem group and a polypeptide (globin) chain — 141 and 146 residues for the α and β-globins respectively. Mb, which in many respects is similar to an isolated subunit of Hb, consists of one haem group and a single globin chain of 153 residues.

In the non-ligated (deoxy) forms of both Hb and Mb the iron(II) is pentacoordinate. The four planar ligands are provided by the porphyrin nitrogens and the fifth by the axial imidazole of HisF8).

Very recently, the stereochemistry of the haem-dioxygen-complex in both oxygenated haemoglobin (oxyHb) and oxygenated myoglobin (oxyMb) has been determined. A stereoview of the haem environments of the oxyHb subunits [2] superimposed on that of oxyMb [3] is given in Fig. 1 [2].

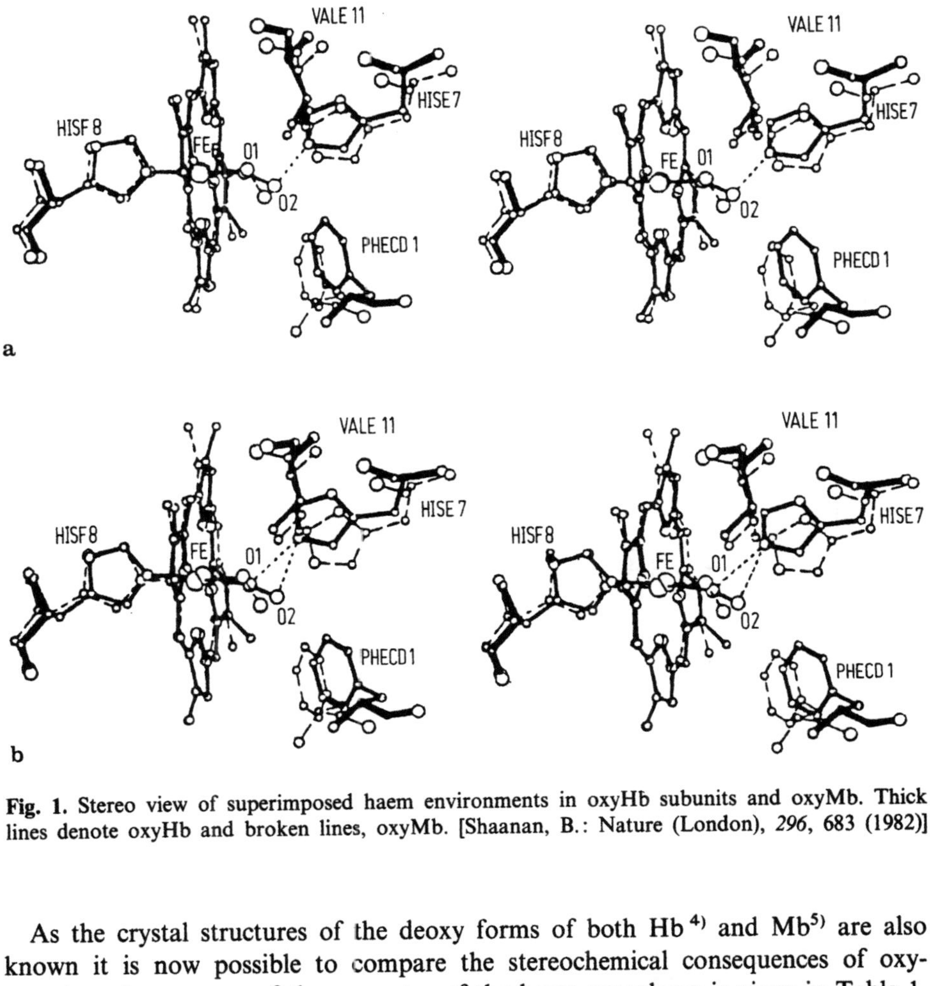

Fig. 1. Stereo view of superimposed haem environments in oxyHb subunits and oxyMb. Thick lines denote oxyHb and broken lines, oxyMb. [Shaanan, B.: Nature (London), *296*, 683 (1982)]

As the crystal structures of the deoxy forms of both Hb [4] and Mb[5] are also known it is now possible to compare the stereochemical consequences of oxygenation. A summary of the geometra of the haem complexes is given in Table 1.

The Fe—N^{ε}_{F8} distance does not alter markedly. The iron, however, which undergoes a spin state change form high spin to low spin, moves toward the porphyrin plane and in so doing pulls the "proximal" imidazole (F8) with it (see Fig. 2a). This movement is a cornerstone of Perutz's mechanical model for the cooperativity shown by Hb in binding dioxygen [9–11] see next Sect.). From Table 1 it is clear that the iron moves an appreciable distance in both pairs of subunits. In fact, in the

Table 1. Geometry of the haem environments in Hb, Mb and model compounds[a,b]

	Fe—Ct[c]	Fe—P_c[d]	Fe—N_P[e]	Fe—N^{ε}_{F8}[f]	Fe—O[1]	O—O	Fe—O—O	Ø[g]	Ref.
Myoglobin	0.42	0.47	2.03	2.22	—	—	—	12	5)
Oxymyoglobin	0.18(3)	0.22(3)	1.95(6)	2.07(6)	1.83(6)	1.22(6)	115(5)	1	3)
Haemoglobin α	0.60	0.60	2.1	2.0	—	—	—	20	4)
Haemoglobin β	0.63	0.63	2.1	2.2	—	—	—	25	4)
Oxyhaemoglobin α	0.13(8)	—	1.98(4)	1.95(10)	1.67(8)	—	156(10)	11	2)
Oxyhaemoglobin β	−0.08(8)	—	1.96(9)	2.06(9)	1.83(13)	—	156(10)	27	2)
Fe(TpivPP) (2-MeIm) · EtOH	0.399	0.43	2.072(5)	2.095(6)	—	—	—	22.8	6)
$Fe(O_2)$ (TpivPP) (2-MeIm) EtOH	0.086	0.119	1.996(4)	2.107(4)	1.898(7)	>1.22(2)	<129(2)	22.2	6)
$Fe(O_2)$ (TpivPP) (1-MeIm)	−0.03	−0.02	1.98(1)	2.07(2)	1.75(2)	>1.16(4)	<131(2)	20	7)
Fe(TPP) (2-MeIm) · EtOH	0.42	0.57	2.086(4)	2.161(5)	—	—	—	7.4	8)

[a] all bond lengths are in Ångströms
[b] all bond angles are in degrees
[c] Ct: centre of the plane formed by the porphyrin nitrogens
[d] P_c: centre of mean plane of the porphyrin
[e] N_P: porphyrin nitrogens
[f] N^{ε}_{F8}: N^{ε} of the proximal imidazole
[g] angle of the proximal imidazole plane with N_1—Fe—N_3

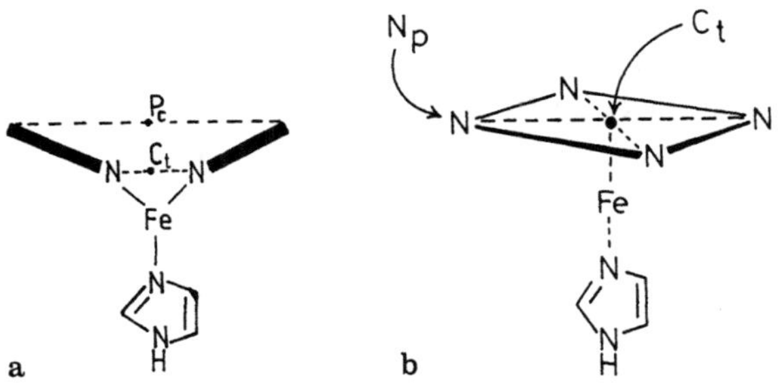

Fig. 2. Side on **a** and perspective **b** view of pentacoordinate iron(II)-porphyrin

β-subunits the iron is now actually displaced *towards* the dioxygen. In the case of oxyMb the movement is not so pronounced and the iron remains slightly further from the porphyrin plane than in the oxyHb α-subunits. The fact that the iron remains outside the porphyrin plane at all may be due to the eclipsed orientation of the imidazole plane with respect to the porphyrin nitrogens. That is to say, if Ø is large (see Fig. 3a) then steric interactions between the imidazole and the porphyrin nitrogens will be minimized. Remembering that the Fe—N^{ε}_{F8} remains essentially constant, this should permit closer approach to the porphyrin plane.

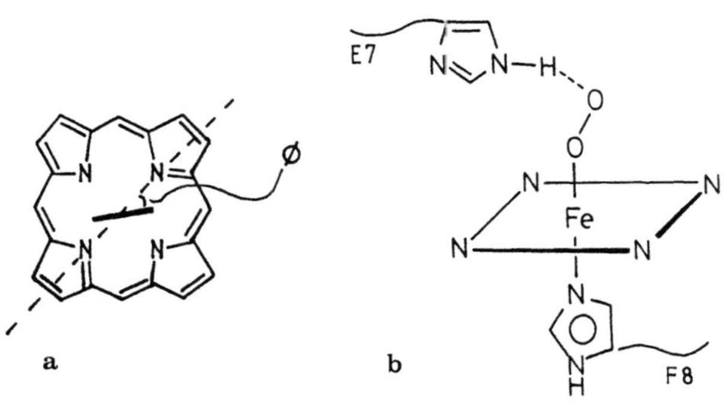

Fig. 3. a Eclipsing angle of proximal imidazole plane and porphyrin axis which includes the Fe—N_P bonds. **b** The hydrogen bond between N^{ε} of HisE7 and a bound dioxygen molecule

The dioxygen molecule is bound in a bent, end-on geometry as predicted by Pauling [12,13]. Furthermore, Phillips and Schoenborn [14], using neutron diffraction analysis, demonstrated that the bound dioxygen forms a medium strength hydrogen bond with the "distal" (E7) imidazole in oxyMb. A similar strength hydrogen bond appears likely to exist in the α-subunits of oxyHb and a weaker one in the β-subunits [2]. The relative strength of these hydrogen bonds could contribute to the control of the affinity for dioxygen of Hb and Mb. However, this does not resolve the controversy surrounding the electronic distribution in the iron-dioxygen bond [12,15–19]. Nevertheless, as Phillips and Schoenborn point out [14], it does seem likely that some development of negative charge does occur at 02. Interestingly, the stronger hydrogen bond in the α-subunits may also explain their greater propensity to autoxidize.

In oxyMb the F—O—O angle' (115 ± 5°) is 15° smaller than that of the "picked-fence" models (see Sect. 2.4). This has been attributed to the HisE7 side chain pressing down on the bound dioxygen [3]. In oxyHb, however, the angle is considerably larger (156 ± 10°). As the haem pocket is more spacious in oxyHb than in oxyMb this may represent the thermodynamically optimum value for the iron-dioxygen bond angle [20,21]. On the other hand, if one regards the "picked-fence" models as providing little steric interference to the bound dioxygen ligand then this large angle may simply reflect a different set of steric constraints.

2.3 Dioxygen Binding to Natural and Synthetic Haems

As several excellent reviews of the binding of dioxygen (and other ligands) to metalloporphyrins have appeared [22-25] only a brief discussion is appropriate here.

One of the most important and fascinating properties of Hb is its ability to bind dioxygen in a cooperative manner. As shown in Fig. 4, this results in a sigmoidal isotherm (a plot of the fraction of binding sites occupied by dioxygen *vs* the pressure of dioxygen) as opposed to the hyperbolic curve expected for a non-interactive system (such as Mb). This enables Hb to offload dioxygen at low oxygen pressure

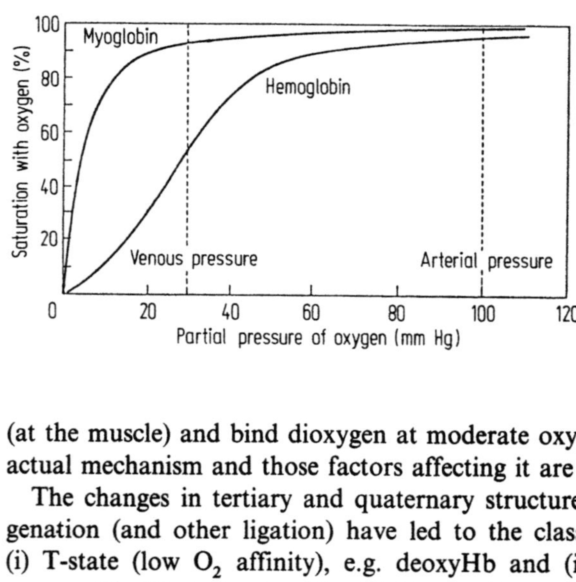

Fig. 4. Dioxygen binding curve for Hb and Mb, from Ref. [82]

(at the muscle) and bind dioxygen at moderate oxygen pressure (at the lungs). The actual mechanism and those factors affecting it are still a matter of debate [9,26-33].

The changes in tertiary and quaternary structure of Hb which accompany oxygenation (and other ligation) have led to the classification of two forms of Hb: (i) T-state (low O_2 affinity), e.g. deoxyHb and (ii) R-state (higher O_2 affinity), e.g. oxyHb. The latter has O_2 affinities similar to those of Mb and isolated α or β chains of Hb (see Table 2). Most modelling studies in this area are concerned with reproducing the properties of one or both of these states.

Simple, planar iron(II) porphyrins react rapidly and irreversibly with dioxygen [34-37]. The oxidation product is usually a μ-oxo-dimer. The mechanism of the reaction, summarized in equations 1 to 6, has been investigated in detail. Many of the postulated intermediates have been characterized by ^1H NMR [38-44], UV-VIS [39-42,44] and IR [45] spectroscopy.

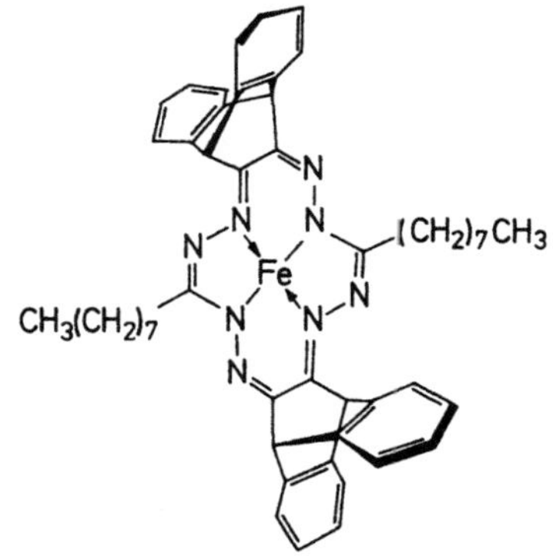

2

$$PFe(II) + L \rightleftharpoons PFe(II)L \tag{1}$$

$$PFe(II)L + L \rightleftharpoons PFe(II)_2 \tag{2}$$

$$PFe(II)L + O_2 \rightleftharpoons PFe(II) \cdot O_2 \tag{3}$$

$$PFe(II)L \cdot O_2 + PFe(II)L \rightleftharpoons LPFe(II){-}O{-}O{-}Fe(II)PL \tag{4}$$

$$LPFe(II){-}O{-}O{-}Fe(II)PL \rightarrow 2[LPFe(IV){=}O] \tag{5}$$

$$LPFe(IV){=}O + LPFe(II) \rightarrow PFe(III){-}O{-}Fe(III)P + 4L \tag{6}$$

The irreversible step in this sequence (5) results from the "decomposition" of the μ-peroxo-dimer formed in step (4). This dimer is in turn formed in a bimolecular process involving two metalloporphyrin nuclei. Inhibition of this reaction will remove one mode of decomposition of the iron(II) porphyrin. Thus one design feature common to all the models discussed in this review is the incorporation of steric protection of at least one face of the metalloporphyrin. In haemoglobin this bimolecular process is prevented because the protein keeps the haems permanently separated (minimum separation is ~25 Å).

The first successful model which employed steric hindrance as a means of retarding μ-oxo dimer formation was, in fact, not a porphyrin. In 1973, Baldwin et al. [46] reported the synthesis of octaazamacrocycle *2* and demonstrated that it was indeed a reversible oxygen carrier (at, albeit, relatively low temperature).

2.4 Fenced Porphyrins

The first example of this type, the "picket fenced" porphyrin *3* was reported by Collman et al. in 1973 [47]. This design has proven to be one of the most versatile and has spawned models for R- and T-state Hb, R- and T-state oxyHb, Mb and oxyMb and cytochrome c oxidase. Several reviews of this work have been published [23,49,50].

The iron(II) derivative, *4*, reversibly binds dioxygen in the presence of any of a number of axial ligands (including 1-alkylimidazoles, 1,2-dimethylimidazole, pyridine, piperidine, tetrahydrothiophen or tetrahydrofuran [37,48]). In benzene solution at 25 °C under an atmosphere of dry dioxygen, and three equivalents of 1-MeIm, has a half-life of approximately two months. Thus the "picket-fence" provides effective protection against irreversible μ-oxo-dimer formation (see Sect. 2.3). As it

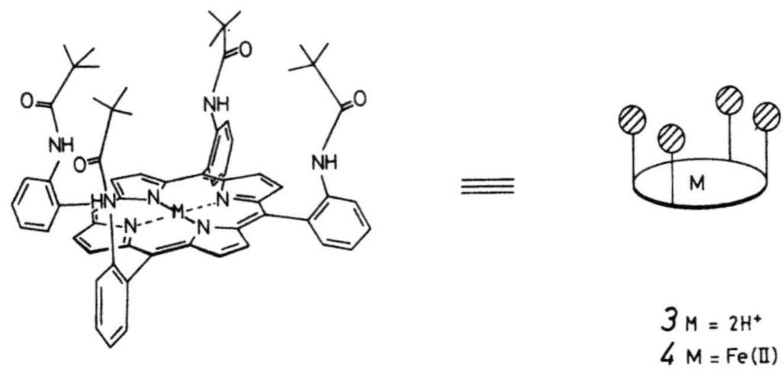

3 M = 2H+
4 M = Fe(II)

is known[48] that *4* can form hexacoordinate complexe with coordinating ligands it is quite likely that this contributes to its remarkable stability. Any hexacoordinate complex of *4* is effectively removed from inner sphere reactions with dioxygen which can lead to irreversible oxidation (see Scheme 1).

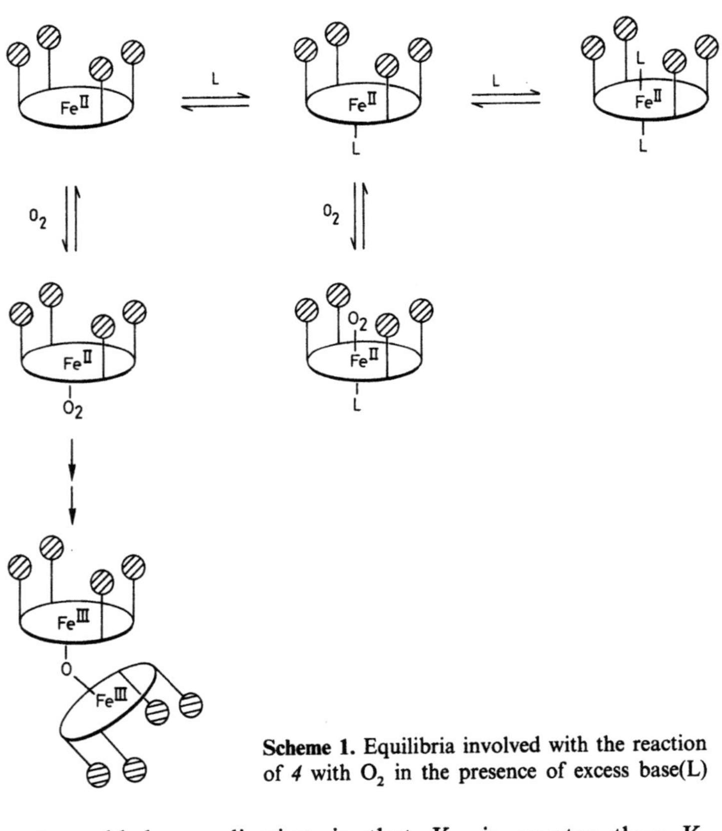

Scheme 1. Equilibria involved with the reaction of *4* with O_2 in the presence of excess base(L)

An added complication is that K_2 is greater than K_1 in the following equilibria:

$$FeTpivPP + L \overset{K_1}{\rightleftharpoons} FeTpivPP.L \tag{7}$$

$$FeTpivPP.L + L \overset{K_2}{\rightleftharpoons} FeTpivPP.L_2 \tag{8}$$

This phenomenon, common to most iron(II) porphyrins [51], has the unfortunate consequence of precluding solution measurements of $K^0 2$ (i.e. dioxygen affinity) for the potential Mb and R-state Hb picket-fenced models (where L = 1-MeIm). This problem, which is not present in the T-state models (L = hindered ligand, e.g. 2-MeIm), was ultimately circumvented by covalent attachment of the axial ligand to the porphyrin periphery [52].

Undoubtedly the most significant achievement in this area has been the successful isolation and crystal structure determination of several hexacoordinate iron picket-fenced *oxy*-complexes [6,7]. The first example was *5* [7,53], which was subsequently used as a model for R-state oxyHb and oxyMb [54]. This also laid to rest the controversy surrounding the geometry of the iron-dioxygen complex in oxy-iron porphyrins [12,15,55] and strongly supported Pauling's prediction of a bent, end-on geometry for oxyHb [12].

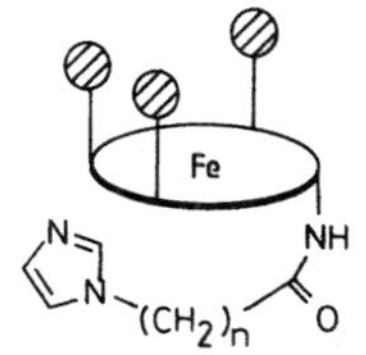

n = 3 *6*
n = 4 *7*

As already mentioned, several "tailed" picket fence porphyrins have been prepared [52]. In dilute solution (~1 mM) at room temperature FePiv$_3$ (5CImP)Por *6* and FePiv$_3$ (4CImP)Por *7* are predominantly pentacoordinate, high spin complexes. At lower temperature, however, dimerization becomes significant. At −25 °C > 50% dimerization occurs leading to the mixed spin system (S = 2 and S = 1):

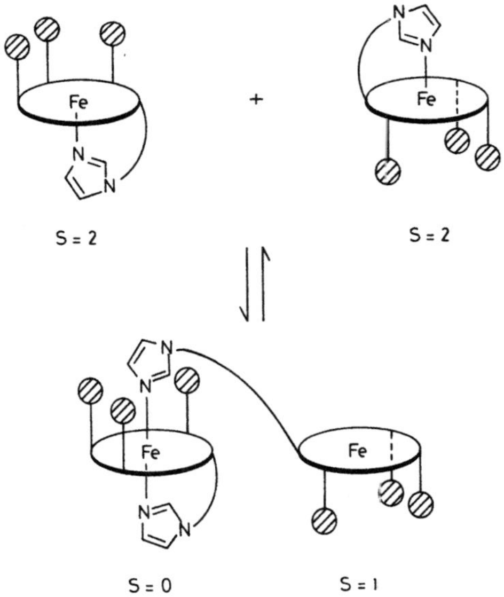

The O_2 affinities of *6* and *7* are very similar to those of R-state Hb. Their half-lives are approximately 33 h, which makes them the most stable *mononuclear* dioxygen carriers reported (see, however, Sect. 2.6).

Attachment to "tailed" ligands was carried out by coupling the appropriate acid chloride with the α,α,α,β-attropisomer (see Scheme 1). The tailed systems are extremely sensitive to light and oxygen (probably due to singlet oxygen formation), and all reactions had to be carried out in an inert atmosphere and/or in the dark. It should be noted that the generation of singlet oxygen is associated with only the free base porphyrin and not its metallated derivative.

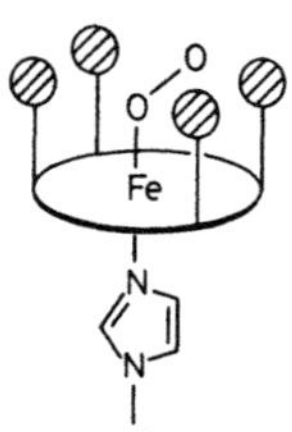

5

Since many picket-fenced models can be prepared with dioxygen affinities very similar to thoise of iron and cobalt Hb and Mb (see Table 2), it was concluded that the protein plays no special role in dioxygen binding apart from preventing irreversible oxidation [57]. At the time that this conclusion was drawn the hydrogen bond between 02 and N^{ε}_{E7}, since established [14], was only conjecture. Nevertheless, the affinity for dioxygen is an important property of these models which is lacking in some other model systems (see next Sect.).

Table 2. Comparison of oxygen and carbon monoxide affinities of iron(II) porphyrins and haemoproteins at 25 °C

	solvent	$P^{O_2}_{1/2}$, torr	$P^{CO}_{1/2}$, torr	*M*	Ref.
Myoglobin (sperm whale)	water (pH ~7)	2.9×10^{-1}	$\sim 2 \times 10^{-2}$	25–40	65–67
Haemoglobin (human "R")	water (pH ~7)	1.7×10^{-1}	$\sim 2.5 \times 10^{-3}$	200–250	67–71
Haemoglobin (human "T")	water (pH ~7)	2.6×10	$\sim 1.5 \times 10^{-1}$	32–1600	67–69
FeC2Cap.1-MeIm	toluene	2.3×10	5.4×10^{-3}	—	58
FeC3Cap.1,5-DCIm	toluene	5.4×10^{a}	4.1×10^{-3}	—	58
Fe(Piv$_3$(5CImP)Por)	toluene	5.8×10^{-1}	2.2×10^{-5}	26600	59, 60
Fe(PocPivP) · 1-MeIm	toluene	3.6×10^{-1}	1.5×10^{-3}	270	61
Fe(PocPivP)1,2-Me$_2$Im	toluene	1.3×10	6.7×10^{-2}	216	61
Fe(TPivPP)1,2-Me$_2$	toluene	3.8×10	8.9×10^{-3}	4280	59, 60
Fe(TPP)1,2-Me$_2$Im	toluene	—	1.4×10^{-1}	—	58
Fe(6,6-cyclophane) · 1,5-DCIm	benzene	6.9×10^{2}	8.4×10^{-2}	—	62–64
Fe(7,7-cyclophane) · 1,5-DCIm	benzene	1.4	9.1×10^{-4}	—	62–64
FeC2Cap.1,2-Me$_2$Im	toluene	4.0×10^{3}	2.0×10^{-1}	—	58

[a] At 0 °C

In line with the idea of "restraint" in Hb [11,49] it was felt that a synthetic haem which provided some impediment to the movement of the iron towards the porphyrin plane upon binding of dioxygen would serve as a useful model for the early stages in the oxygenation of Hb. By increasing the steric demand of the axial ligand, models for both deoxy- and oxy-T-state Hb *8* and *9* respectively, have been prepared

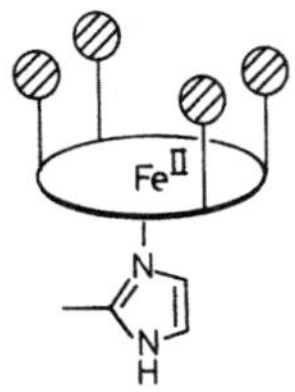

8

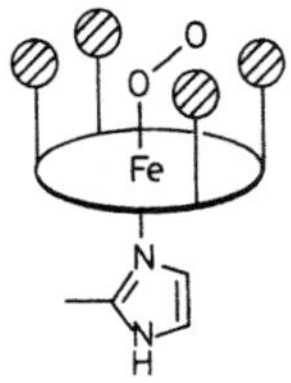

9

and their crystal structures determined [6]. These provided the first opportunity to examine, at a molecular level, the structural changes which occur upon binding of dioxygen to a pentacoordinate high-spin iron(II)porphyrin. A summary of their geometries was given in Table 1. The dioxygen is again bound in a bent, end-on fashion. As expected the iron, which changes to low-spin, moves *toward but not into* the porphyrin plane, pulling the axial 3;MeIm with it.

A comparison of the affinities of these models for CO with those of Hb and Mb is also useful. It is clear from Table 2 that the natural systems have substantially lower affinities. The way in which CO affinity is regulated in haemoproteins has been another matter for conjecture. A number of factors, including distal-side steric effects, electronic influence of peripheral substituents, basicity and protonation state of proximal ligand and polarity of the micro-environment has been suggested [72]. Quite recently the "pocket" porphyrin *10* was synthesized with a view to providing an environment which would permit normal binding of dioxygen whilst hindering the binding of carbon monoxide [61].

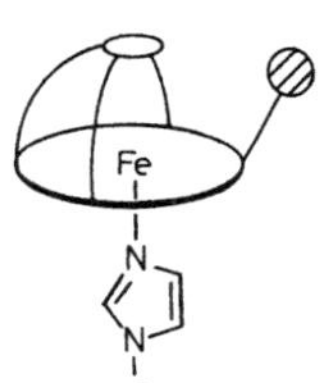

10

The partition coefficients, M (= $P^{O_2}_{1/2}/P^{CO}_{1/2}$), for several system were determined (see Table 2). Comparing the Hb T- and R-state models FeTpivPP(Me_2Im) and FePiv$_3$(5CImP)Por (see below) respectively, with *10* it is evident that whereas the dioxygen affinities are comparable there is a considerable *decrease* in CO affinity

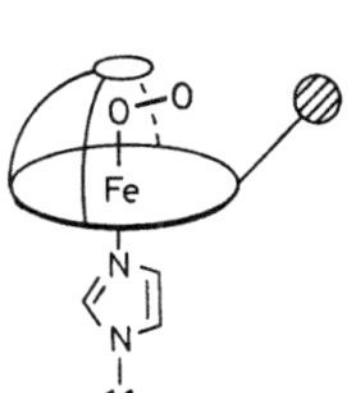

11

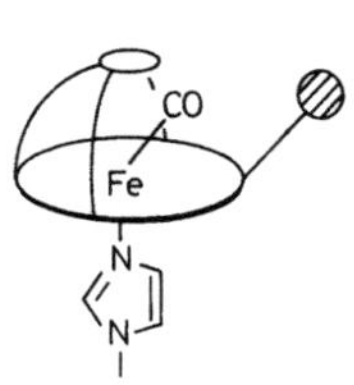

12

(M smaller) for *10*. This difference is interpreted as being due to steric encumbrance close to the binding site preventing the binding of CO in a linear fashion, lowering K^{CO}. The bent geometry (see below) which the Fe—CO unit is thought to be forced to adopt, is that normally observed for Fe—O_2 see *11* and *12*. Thus K^{O_2} is not affected. The value of M for *10* is in fact in reasonable agreement with that of R-state Hb. If this selectivity is due to the steric requirements built into this model then similar factors may contribute to the situation found in natural systems.

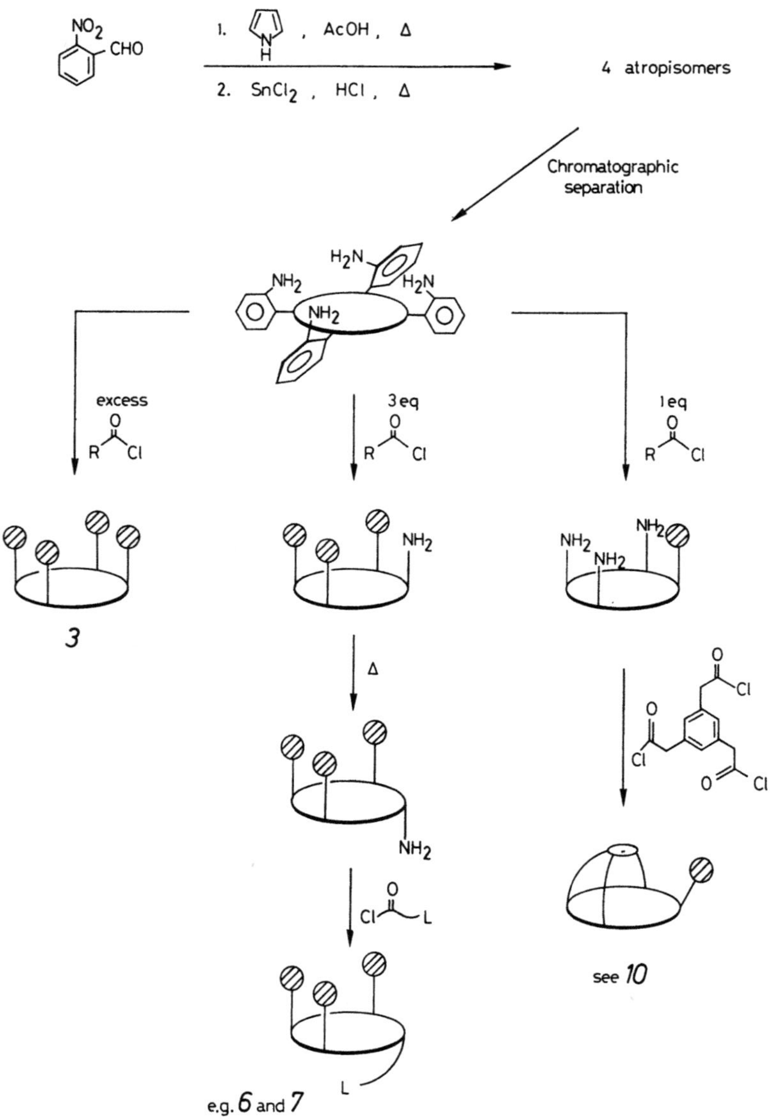

Scheme 2. Synthesis of picket-fenced porphyrin derivatives

It should be pointed out that, on the basis of their studies on the kinetics of O_2 and CO binding to their model systems Traylor [62] and Chang [73] reject the view that CO and O_2 are *differentiated* by distal side steric effects. They also reject the use of *M* as a parameter for comparison of such effects.

A general synthetic scheme for the picket-fenced porphyrins is outlined in Scheme 2. The first reaction in the Scheme, a Rothemund [74] reaction highlights a major limitation in the synthesis of porphyrins with C_4 symmetry. The yield is extremely low. In fact the yield for the α,α,α,α-atropisomer is ~1% after only three reactions. When the substituted benzaldehyde is readily available and cheap this presents no serious problems. As we shall see for some other porphyrin derivatives however, where the benzaldehyde derivative requires several steps in preparation, the target model porphyrin can only be produced in very small quantities.

In the case of the tetra *o*-aminophenylporphyrin, the other atropisomers can be re-equilibrated and more of the required isomer can be isolated by column chromatography. In fact a new simpler method for converting all the atropisomers to predominantly the α,α,α,α-form has been published recently [75]. However many of the reactions are still laborious and tedious with disappointing yields. For the picket-fenced systems, once the correct isomer has been obtained the procedures are generally straightforward and quite versatile.

Metallation can be effected by one of several methods [76]. Usually iron is incorporated as iron(III) and reduced prior to ligation studies. As there are sometimes problems in the reduction step, procedures for the direct introduction of iron(II) are available [52, 76].

Vaska and Amundsen prepared porphyrins *13* and *14*, which are ostensibly protected by "fences" on both faces [77, 78]. The products from exposure of *13* or *14* to O_2 could not be satisfactorily characterized although the authors did conclude that μ-oxodimer formation had not occurred. Very recently a reinvestigation into the properties of these complexes has been undertaken by Balch et al. [42]. Relying principally on ^1H NMR spectroscopy they observed the formation of the *pentacoordinate* dioxygen adducts [T(2,4,6-RO)$_3$PP FeO$_2$] *15* and *16* at low temperature in the absence of any coordinating ligand. Similar adducts were observed for FeTPivPP and Fe[T(3,4,5-MeO)$_3$PP]. FeTPPO$_2$ has also been isolated in an argon matrix and examined by IR spectroscopy.

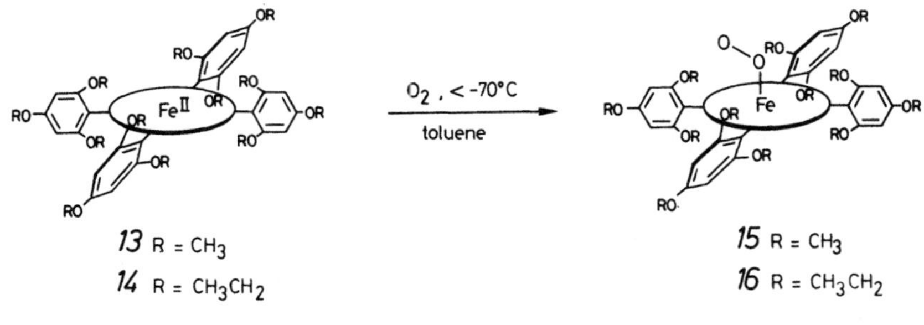

13 R = CH_3

14 R = CH_3CH_2

15 R = CH_3

16 R = CH_3CH_2

Upon warming is solution, *15* or *16* either dissociate back to their deoxy forms or irreversibly oxidize to the iron(III) chloride or hydroxide. No μ-peroxo- or μ-oxodimers were formed. This behaviour is in sharp contrast to that of the other pentacoordinate dioxygen adducts mentioned. This remarkable result confirms Amundsen and Vaska's earlier observations [77]. It seems likely that the strategum of bis-facial protection will be used in the design of new dioxygen carriers. However it should be noted that the Rothemund synthesis employing 2,6-disubstituted benzaldehydes is even less generous in yield than for mono-substituted compounds.

2.5 Capped Porphyrins

The "capping" structure of these compounds was designed to provide a binding site for dioxygen which completely protects the complex from μ-oxo-dimer formation. Their synthesis again relies on an adaptation of the Rothemund synthesis [74]. All four aldehyde groups destined to form the *meso* positions are provided by the pyromellitoyl derivative *17* (available in two steps from salicylaldehyde [80, 81]). This approach has been extended to include a number of naphthaldehyde based systems [81, 82].

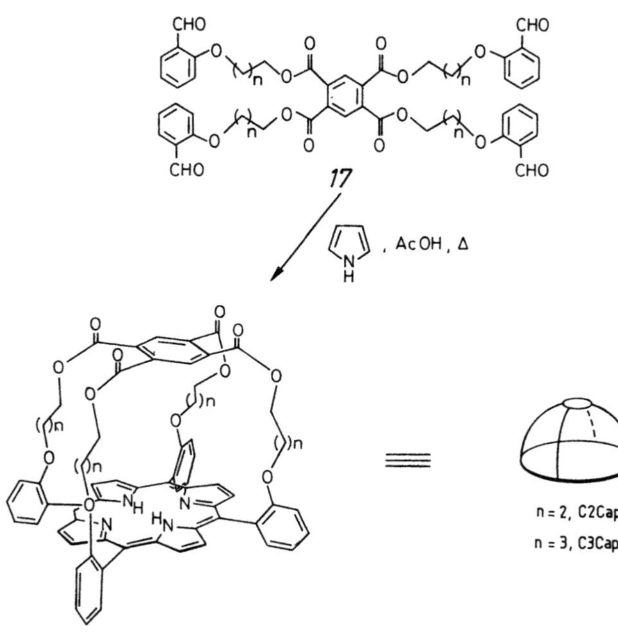

In order to attach "tailed" bases to the uncapped face of the porphyrin, peripheral functionalization was required. This involved nucleophilic nitration of either the free base or zincated capped porphyrin π-cation radical, generated in situ [83, 84]. The mono-nitro derivatives *19* and *21*, obtained in nearly quantitative yields, could then be reduced to the corresponding mono-amino capped porphyrins. This procedure was also successfully applied to a variety of other porphyrin cerivatives. Attempts to attach "tailed" bases have so far met with little success.

In a benzene solution containing excess 1-MeIm, FeC2Cap (n = 2) is pentacoordinate. Haemochrome formation is not observed, presumably because the capp-

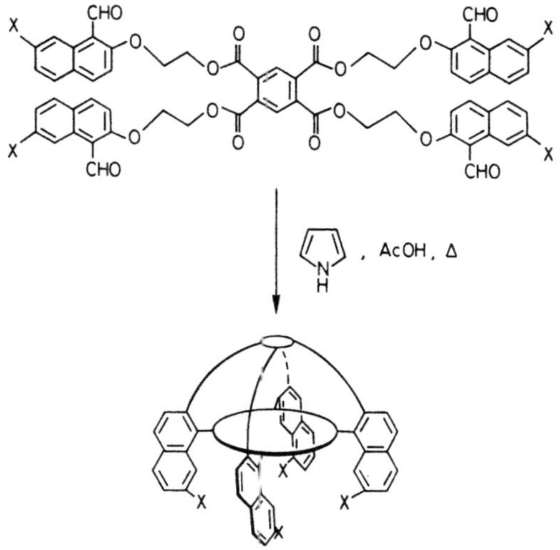

X = H 18
X = OBz
X = OH

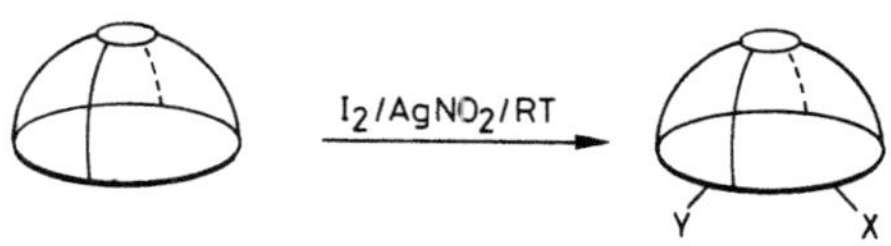

X = H Y = NO_2 19
X = H Y = NH_2 20
X = NO_2 Y = H 21
X = NH_2 Y = H 22

ing cavity is not large enough to accommodate the ligand. This pentacoordinate complex is a reversible dioxygen carrier at room temperature. The affinity of FeC2Cap.1-MeIm for both dioxygen and nitrogenous bases is markedly lower than that of open systems (see Table 3). This, coupled with its inability to form hexacoordinate complexes, results in the relatively short $t_{1/2}$ of 5 h ($t_{1/2}$ in neat pyridine is 20 h [85]).

FeC3Cap (n = 3) [81,86], incorporating a larger capping structure can form hexacoordinate complexes [87,88]. Remarkably, these reversibly bind dioxygen in solution, forming *heptacoordinate* oxy-complexes. The binding of the second base is very weak, behaviour which is inconsistent with axial coordination. Rather it is believed that the second base occupies a non-axial coordination site. The following stereochemistry has been proposed [89,90]:

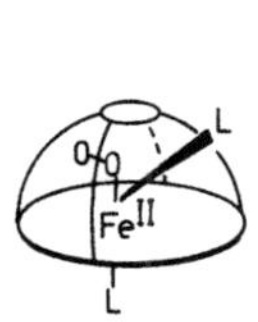

Table 3. Ligation equilibria for capped porphyrins[a, b]

	base	log K_1[d]	log K_2[d]
FeC2Cap	1-MeIm	2.90 ± 0.05	
	1,5-DCIm[c]	3.17 ± 0.05	
	1,2-Me_2Im	3.06 ± 0.05	
FeC3Cap	1-MeIm	3.31 ± 0.05	0.77 ± 0.05
	1,5-DCIm[c]	3.00 ± 0.05	
	1,2-Me_2Im	3.61 ± 0.05	
18	1-MeIm	2.87 ± 0.04	
	1,5-DCIm[c]	3.22 ± 0.07	
	1,2-Me_2Im	2.32 ± 0.04	
19	1-MeIm	3.68 ± 0.05	
	1,2-Me_2Im	3.89 ± 0.05	
FeTPP[e]	1,2-Me_2Im	4.43 ± 0.05	0.39 ± 0.10
Fe(T*p*-OMePP)[e]	1,2-Me_2Im	4.40 ± 0.05	0.31 ± 0.10

[a] taken from reference 58
[b] solvent used was toluene at 23.1 ± 0.1 °C
[c] solvent used was methanol/water (1:1) at 25 °C
[d] see equations (7) and (8)
[e] included for comparison

By using the bulky ligand 1,5-DCI, O_2 and CO affinities for FeC3Cap were determined (see Table 3)[58]. From the data in the Table it is clear that the affinity for CO of both capped porphyrins is very similar to that of open systems. On the other hand, the O_2 affinities are significantly lower. The reasons for this behaviour are not clear. Crystal structure[91, 92] and ^{1}H NMR[93] studies suggest considerable conformational flexibility in the capping structure. Thus it may be that in solution the cap adopts a conformation which disfavours O_2 binding without affecting CO binding. It has also been suggested that capped and bridged porphyrins with connecting chains of seven or more atoms do not interfere with CO binding[94]. This suggestion is somewhat vitiated by Battersby's recent findings[95] (see Sect. 2.7). To shed more light on this matter larger capped porphyrins have recently been prepared[96].

Most recently several "capped-strapped" porphyrins have been prepared[97]. These represent mononuclear oxygen carriers of the type:

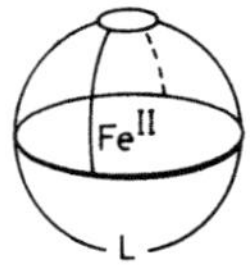

Their synthesis first required the preparation of the centrosymmetric pyromellitoyl diacid/diester *23* and its subsequent elaboration to the α,γ-difunctionalized capped porphyrins *24* and *25* (see Scheme 3). Coupling of the bridging pyridine diacids was

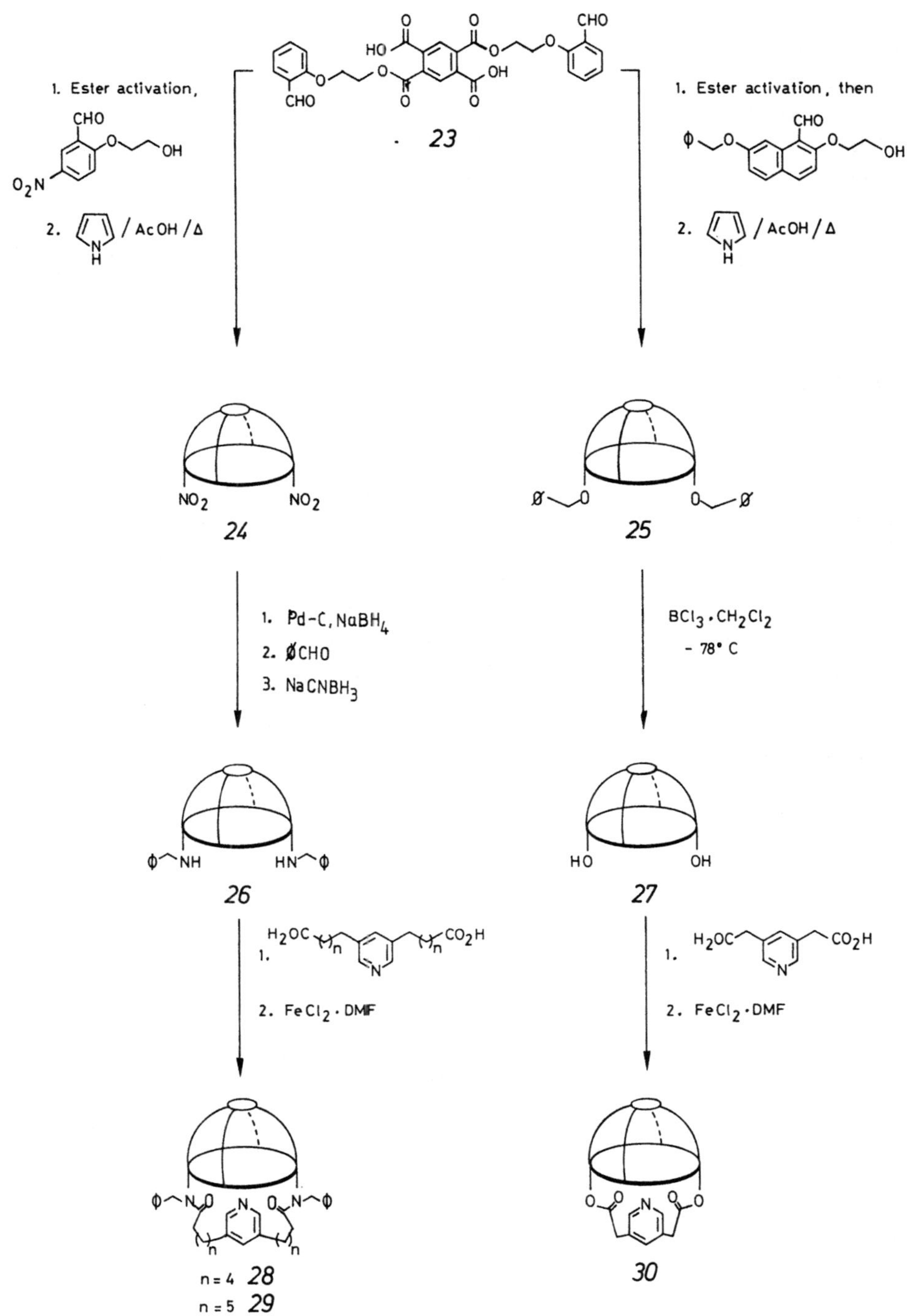

Scheme 3. Synthesis of capped-strapped porphyrins

achieved under high dilution conditions using ester activation with 2-chloro methyl-pyridinium iodide.

Optimum conditions for the bridging reaction were developed using simpler bridges. As with other types of model oxygen carriers [98], the use of sulphur in the bridges yielded extremely light and air sensitive porphyrins. This was most likely due to the porphyrin catalyzed generation of singlet oxygen and subsequent destructive oxidation [56].

At present only preliminary, qualitative information is available on these capped-strapped iron(II)porphyrins. The indications are that they are extremely stable, reversible dioxygen carriers at room temperature in solution.

2.6 Bridged Porphyrins

2.6.1 Singly-Bridged Porphyrins

Since 1971 at least two dozen different singly-bridged porphyrins have been reported (excluding cofacial dimers). Apart from those prepared by Baldwin et al. [99, 100] the bridges are connected to opposite β-positions on the porphyrin ring. The singly-bridged porphyrins fall into two catagories — those containing simple polymethylene bridges and those containing more bulky bridging groups (e.g. biphenyl, anthracene, naphthalene etc). The approaches to the synthesis of these systems also fall into two catagories — those which construct the porphyrin onto the termini of the bridge and those which build a bridge onto a functionalized porphyrin. The last few years have seen the virtually exclusive use of the latter approach. Again with the exception of Baldwin's systems the readily available diesters of mesoporphyrin II were used (*31* R′=H).

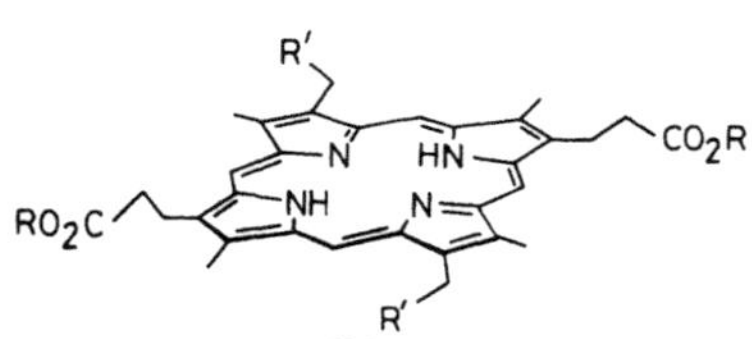

31

A general synthetic scheme is shown below. X and X′ can be used in a number of ways (i) modifying solubility (e.g. X=X′=H to X=X′=*n*-butyl [101]); (ii) attachment of a tailed base (e.g. X=NH$(CH_2)_3$Im, X′=H [102]) or (iii) attachment of a second bridge (see next Sect.). Battersby et al. [103] have also reported bridge syntheses

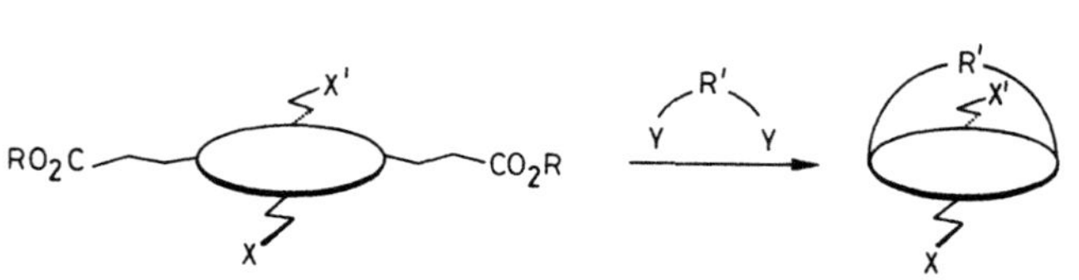

Scheme 4. Synthesis of bridged porphyrins

(Scheme 4) involving the coupling of two fragments attached to the porphyrin by the ester groups (e.g. $R=N(CH_3)(CH_2)_5CO_2CH_3$). The coupling of the bridge to *32* is usually carried out under high dilution conditions with yields of *33* averaging 30%.

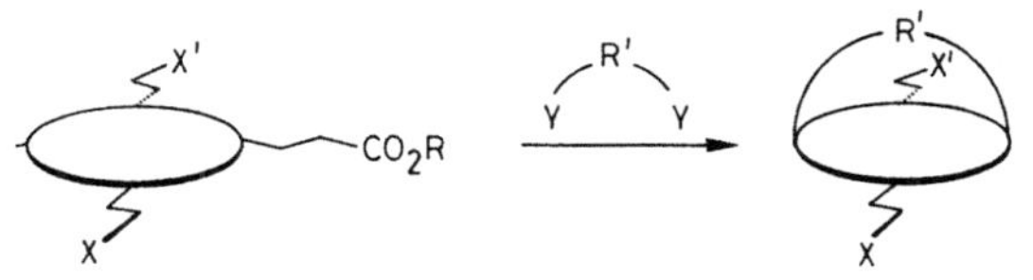

32 *33*

Baldwin's approach [99, 100] involved the stepwise construction of α,γ-diphenylporphyrin derivatives onto the termini of a dialdehyde, an adaptation of Kenner's method [104].

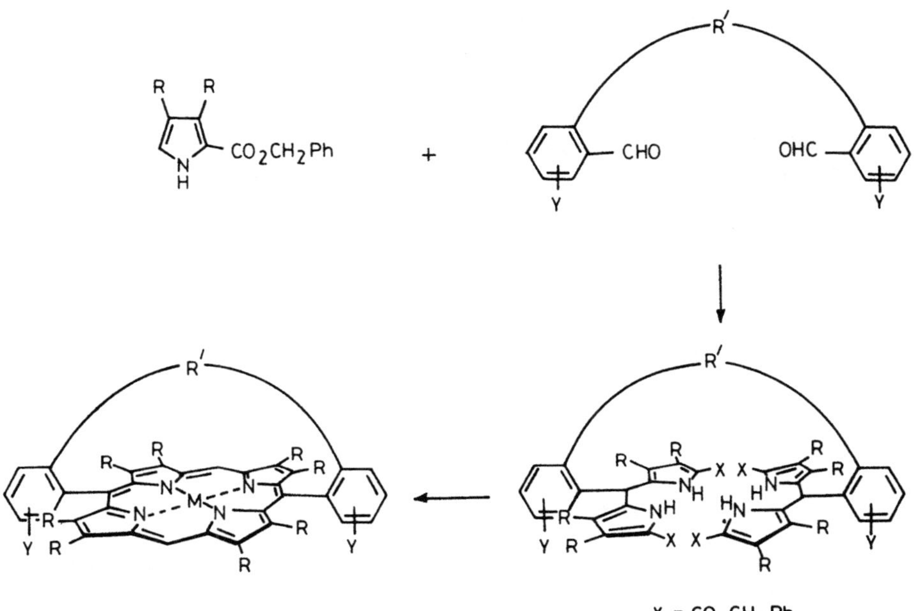

In contrast to model porphyrins whose synthesis depends upon Rothemund reactions, these bridged complexes are available in large quantities.

For several of the singly-bridged iron(II) porphyrins, binding of a second axial ligand is significantly reduced, reflecting the cramped nature of the binding site enclosed by the bridge. By employing bulky ligands, such as 1-tritylimidazole, *pentacoordinate* iron(II)bridged porphyrins can be generated in solution. Hence Chang's "crowned" porphyrin [101], *34*, has a half-life in solution under an atmosphere of oxygen of >1 h at room temperature. (However the majority of singly-bridged iron(II) porphyrins rapidly autoxidize under these conditions.)

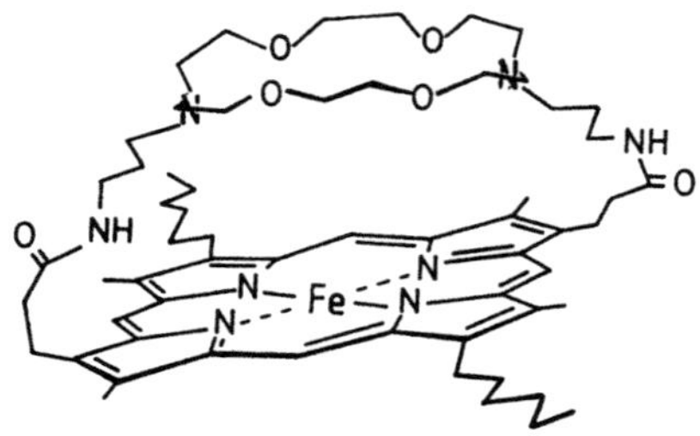

34

Although this is a significant result in itself, most bridged systems have been prepared in order to probe the steric contribution to the differentiation between CO and O_2 binding which occurs in the haemoproteins.

From their studies of the kinetics of binding of various ligands, Traylor et al. [105, 106] have concluded that larger ligands, such as bulky alkyl isonitriles encounter steric effects in all haemoproteins. However, for smaller ligands such as CO, only the more crowded pockets of Mb and the peroxidases provide steric restrictions to binding. They also concluded that R-state Hb does *not* provide distal steric hindrance to CO binding [106]. They have also found that for both the haemoproteins and their models, e.g. (6,6) and (7,7) "cyclophane" porphyrins *35* and *36*, any distal side steric contribution to a reduction in ligand affinity is relfected in a reduction in the *association* rate. This led them to propose that the reduction in CO affinity is due to a pre-equilibrium, prior to binding. That is to say, steric hindrance to binding affects

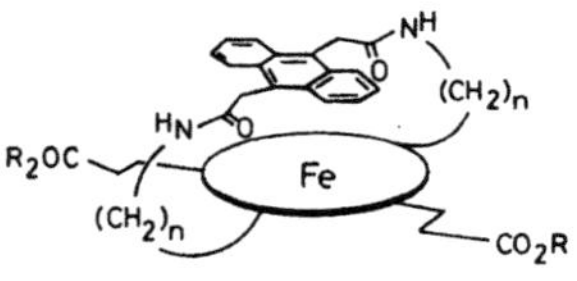

n = 3 *35*
n = 4 *36*

access to the binding site and not the resulting Fe—CO geometry. This latter property would be reflected in an increased dissociation rate (which is not observed). It should also be reflected in ν_{CO} of the Fe—CO complex. Several reports clearly indicate that, contrary to earlier suggestions [107], there is no correlation between ν_{CO} and CO affinity in these systems [73]. Such a correlation should exist if the Fe—CO geometry (and the resulting bond order) are significantly perturbed.

On the matter of distal side steric differentiation of CO and O_2, Traylor et al. [62], using *35* and *36* demonstrated two important points (i) the tighter cavity of *35* reduces the affinity for *both* CO and O_2 when compared to *36* and other R-state Hb models and (ii) the reduction in affinities as the cavity shrinks is quite similar for both gases and is reflected almost entirely in the association rates. Thus it appears that for these models distal side steric effects do not differentiate between O_2 and CO. A similar conclusion was drawn for R-state Hb. Finally, Chang has presented evidence using several singly-bridged systems which supports Traylor's conclusions [73].

2.6.2 Doubly-Bridged Porphyrins

Using the general approach outlined in Scheme 3 Battersby et al. have synthesized the two doubly-bridged systems *37* [102] and *38* [95].

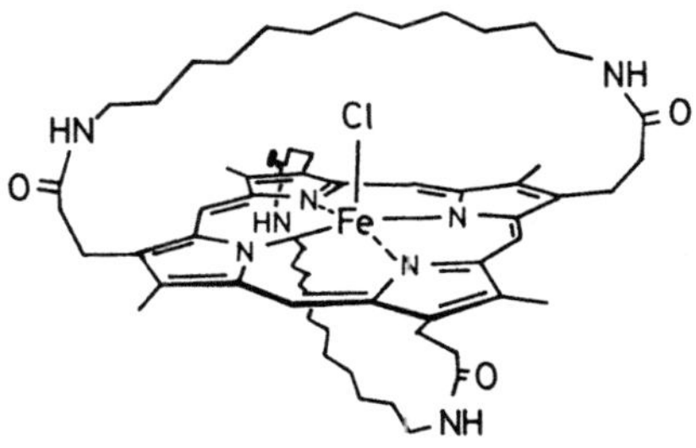

37

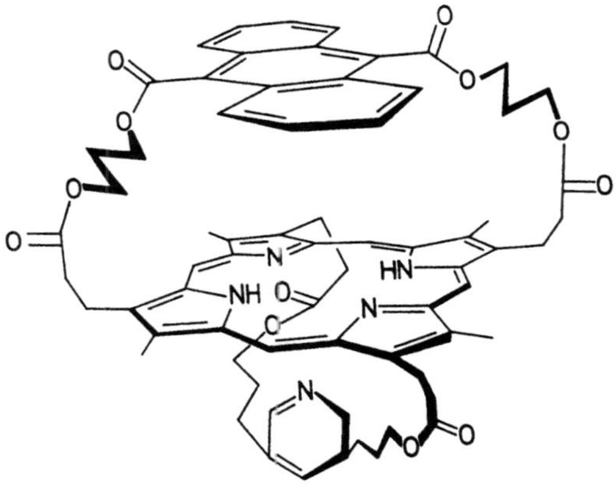

38

Independently, Momenteau et al. have prepared several *meso*-doubly-bridged systems (the so-called "basket handle" porphyrins). Their first synthesis employed the Rothemund condensation of long chain dialdehydes with pyrrole and led to a mixture

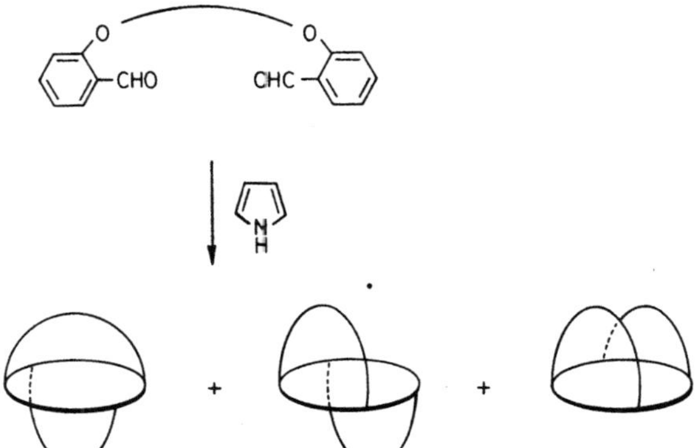

39 *40* *41*

bf all possible isomers [108]. Unfortunately the most important isomer, *39*, was obtained in lowest yield. To circumvent this problem, they first prepared an atropisomeric mixture of the picket-fenced tape porphyrins *42* and reacted it with long chain dibromides [109]. In this case, the major product was the desired isomer *39*. Both

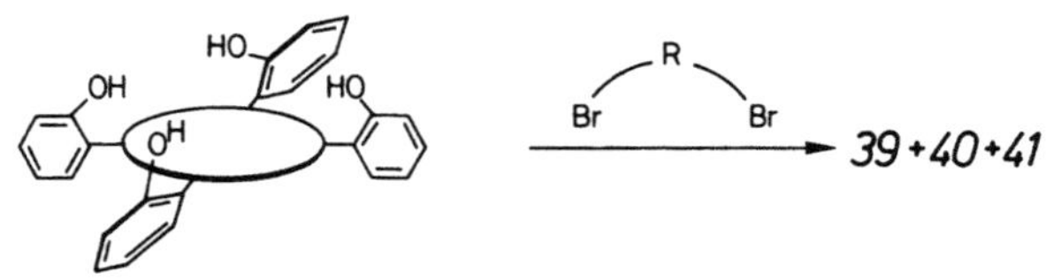

42

More recently Momenteau et al. [110] have prepared the doubly-bridged systems *43* and *44* which include a bridging porphyrin. The synthesis of *44* was simplified in that the desired α, β, α, β-atropisomer can be prepared in pure form, and then elaborated stepwise.

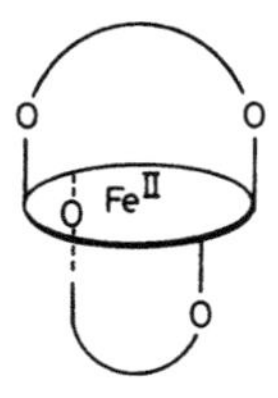

43

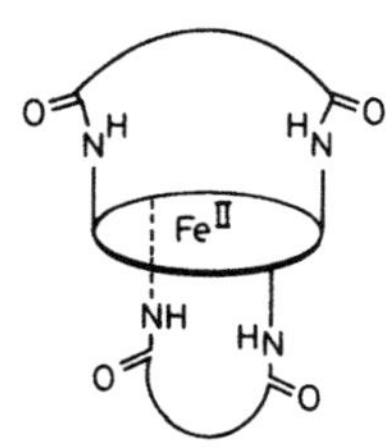

44

The preparation of these model oxygen carriers, bearing *bridging* ligands (see also Sect. 2.6), marks an important advance in design. They are the first examples of a growing family of truly *mononuclear* dioxygen carriers. The permanently bound axial ligand eliminates the problems associated with "tailed" systems [52].

In principle these systems should not require exogenous base. However, $t_{1/2}$ of *38* in the presence of O_2 is only 15 min [95]. By changing the solvent from dichloromethane to dimethylformamide $t_{1/2}$ increased to >2 h, most likely as a result of the solvent's ability to compete with O_2 for the sixth coordination site. Remarkably it was found that the CO of the *38*-CO complex could be displaced by O_2. This is in marked contrast to all other model systems, where the affinity for CO is far greater than that for O_2. It is, more importantly, similar to the behaviour of Mb [111]. Determination of the kinetic parameters is obviously required.

A combination of crystal structure and ^{1}H NMR studies [112] of *45* and *46*, which approximate the two "halves" of *38*, indicates that each bridge can swing across its face of the porphyrin. This mobility in solution helps to explain the low resistance of *38* to autoxidation.

45

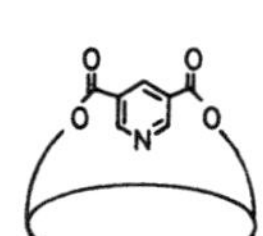

46

In the presence of 1-MeIm and under an atmosphere of oxygen, the trans-trans isomer *39* was the most stable ($t_{1/2} = 25$ min). The oxidation products of *39* and *40* were the iron(III) derivatives. However, isomer *41* which contains steric protection on only one face, rapidly formed μ-oxo-dimer. This, as with Vaska and Amundsen's complexes [77] highlights the effective inhibition of μ-oxo-dimer formation by protection of *both* faces of the porphyrin.

As opposed to the picket-fenced porphyrins, attachment of a "tailed" base *47* increased the stability slightly ($t_{1/2} = 30$ min) compared to its haemochrome derivative [113]. The singly-bridged porphyrin bearing a tailed imidazole *48* [102] has also been prepared however no oxygenation data were provided.

The kinetics of CO and O_2 binding to the "basket handle" derivatives *43* and *44* have been measured [110]. Both bind CO and O_2 more strongly than unhindered iron(II) complexes. Secondly it was found that *44* had an order of magnitude greater affinity for O_2 than did *43*. This was due exclusively to the difference in O_2 *dissociation* rates, which may reflect the more "polar" distal environment provided by the two amide linkages of *44*.

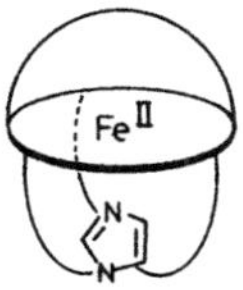

47

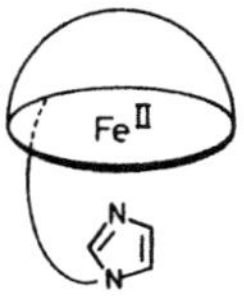

48

3 Cytochrome P-450

3.1 Occurrence and Function

Cytochrome P-450 is the generic term for a large class of monooxygenases which occurs widely in microsomal and mitochondrial fractions of animal tissues as well as in plants and microorganisms. Its name arose from the electronic absorption maximum of the carbonyl adduct which occurs at 450 nm. The term "cytochrome" can be misleading as the function of this enzyme is quite different to that of the cytochromes found in the electron transport chain.

As members of the class of monooxygenases cytochromes P-450 catalyse the hydroxylation of a wide range of substrates including aromatic and aliphatic hydrocarbons. They are responsible for drug detoxification and steroid hydroxylation [82, 114–121].

3.2 Reactivity

The prosthetic group of cytochrome P-450 is protoporphyrin IX. There is strong spectroscopic evidence for a thiolate group (most likely cysteinyl) occupying one of the axial coordination sites [122]. A number of candidates has been suggested for the sixth ligand. These include imidazole [123], tyrosine [124] serine or water [125-127]. This matter has yet to be resolved.

Because of its availability in high purity, cytochrome P-450$_{cam}$ from the bacterium *Pseudomonas Putida* (MW 45,000), has received considerable attention. Many of the steps in this enzyme's complex catalytic cycle are now well understood. A generalized scheme for cytochrome P-450$_{cam}$ is given below.

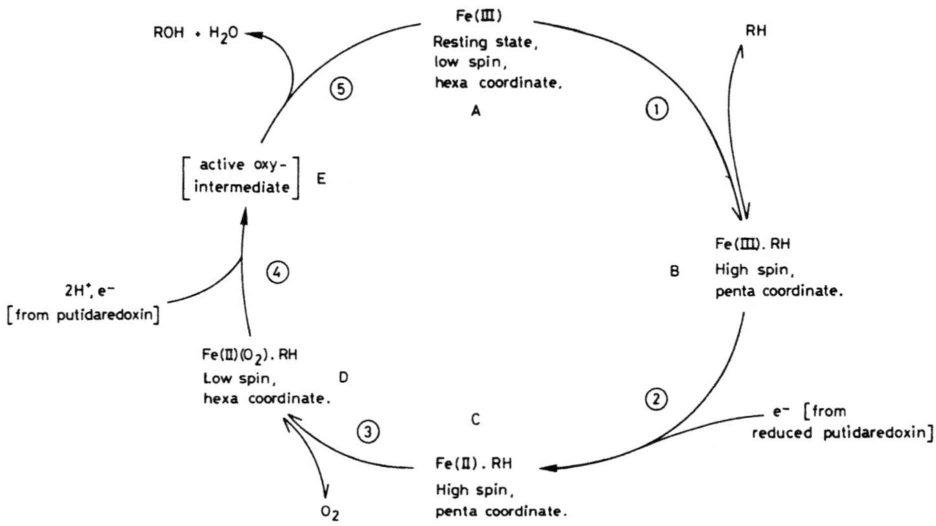

Scheme 5. Catalytic cycle of cytochrome P-450$_{cam}$

Much attention has focussed on steps 4 and 5, i.e. the synthesis of iron porphyrin oxygen models for intermediate E, which can oxygenate substrates including unactivated C—H bonds. The "active oxygen" intermediate has not actually been observed in cytochrome P-450 although similar higher valent iron complexes are observable in peroxidases [128] and catalases [128]. There now exists a large number of models for the chemical properties of this intermediate several of which are discussed in the following sections.

The other area of interest at present is in the synthesis of structural models for various stages in the catalytic cycle, with particular emphasis on reproducing the characteristic "hyperporphyrin" spectrum [129]. As yet, little effort has been devoted to modelling the induction of spin state and coordination number change by substrate binding to resting state A, which may well be allosteric in nature.

Lastly, it is interesting to note the similarity between intermediate D and oxyMb (or Hb). Both contain hexacoordinate, low spin, diamagnetic iron(II)protoporphyrin IX with an oxygen ligand. The other ligand is, of course, different for the two types of systems.

3.3 Fenced Porphyrins

Recently, Groves et al. [130], as part of their continuing interest in models for intermediate E (Scheme 3.1) prepared a higher valent iron-oxygen complex in solution, which they formulated as *49*.

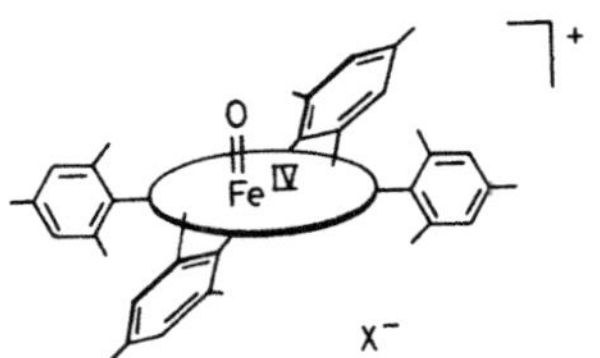

49

When a solution of *49* is treated with norbornene a 78% yield of norbornene oxide is obtained. Treatment of a solution of *49* with $H_2{}^{18}O$ *prior to* reaction with norbornene leads to 99% incorporation of ^{18}O into the norbornene oxide, indicating facile oxygen exchange with water. *49* can also be generated by the base treatment of *50* (prepared by reaction of TMPCl with iodosylbenzene). *50* is also an oxygen transfer reagent and is thought to be at the same formal oxidation state as *49*. No structure for *50* was proposed by the authors.

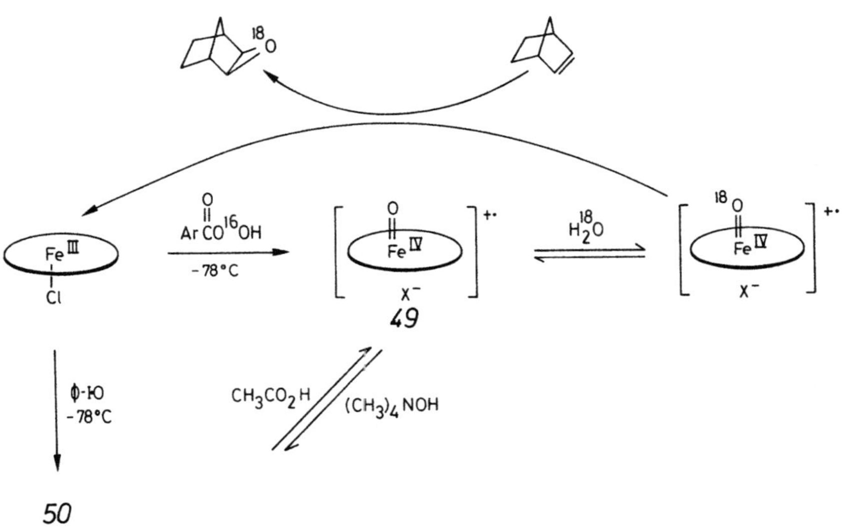

Scheme 6. Groves' higher-valent iron porphyrin complex (*49*), Ref. [130]

The ^1H NMR, electronic, Mössbauer and EPR data all support an iron(IV)porphyrin π-cation radical formulation. This type of intermediate has been proposed several times before from the reaction of iodosylbenzene and iron porphyrins [131]. However this is the first report of direct observation of such an intermediate, a result which may be due in part to the stability conferred on the system by the mesityl "pickets". Together with the spectroscopic evidence the chemical behaviour of *49* indicates that it is a good model for both cytochrome P-450 and peroxidase compounds I.

Collman et al. [98] have prepared cytochrome P-450 models based on TPivPP and TPP derivatives bearing thiols attached by a "tail" (see Scheme 7). To avoid competing thioester formation, the thiols were introduced as their S-trityl or S-acetyl derivatives (for the alkyl thiols) and as the disulphide derivatives (for the aryl thiols). The trityl group was found to be a better protecting group than acetyl as it could be removed in much higher yields.

All the porphyrins bearing a thiol "tail" were found to be extremely sensitive to a combination of light and air. This was attributed to destructive reactions caused by

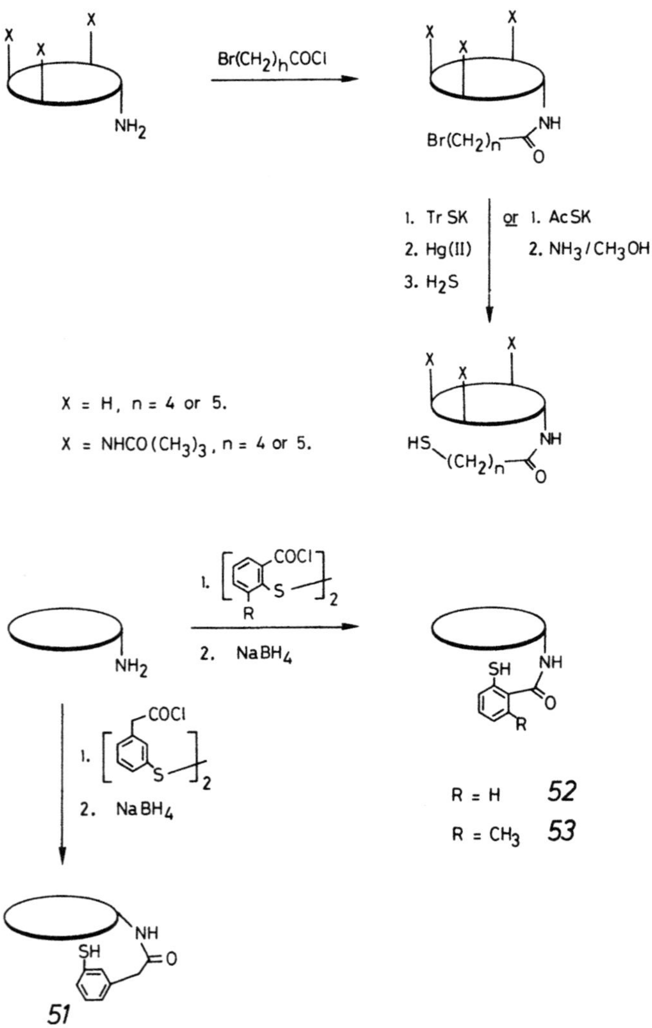

Scheme 7. Synthesis of "tailed" porphyrin models for cytochrome P-450

the porphyrin sensitized production of singlet oxygen *56*. This phenomenon has been observed for the imidazole "tailed" derivatives of TpivPP as well as several capped-strapped porphyrins (see Sects. 2.4 and 2.5). Though inconvenient, this problem can be routinely avoided by manipulating all compounds in an inert atmosphere and using only a red light source.

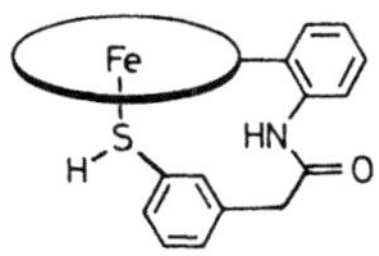

54

Of all the alkyl and aryl thiol iron(II)porphyrin derivatives only *51* appeared to form a significant amount of a pentacoordinate species (*54*). In the presence of CO, however, all the alkyl thiol derivatives formed hexacoordinate, low spin iron(II) complexes as did the aryl derivative *54*. The other two aryl thiols appeared to exist in an equilibrium with the pentacoordinate "tail-of" complex at 25 °C but were completely hexacoordinate at 0 °C:

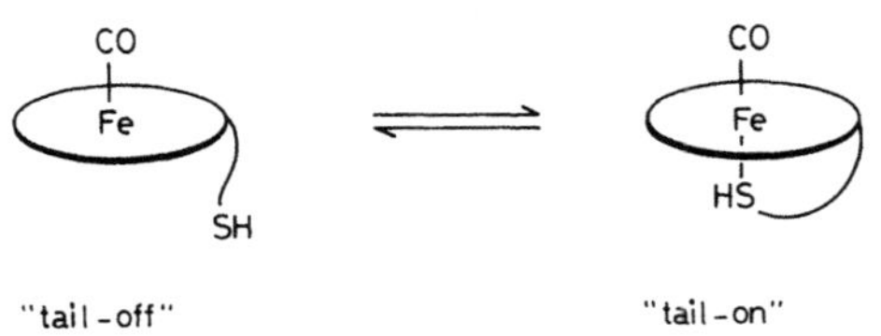

In particular, the CO adduct of *52* displays two IR carbonyl stretching frequencies (2030, 1960) consistent with the mono and bis carbonyl complexes.

Treatment of *51* with the base $KN(C_6H_5)COCH_3$ in the presence of CO gave essentially complete conversion to the mercapt*ide* complex *55*. The UV-VIS and MCD spectra of *55* were similar to those of purified cytochrome P-450 (reduced +CO). As found previously, hexacoordinate iron(II)mercapt*ide* CO adducts give typical "hyperporphyrin" spectra (split Soret bands at ~450 and 380 nm).

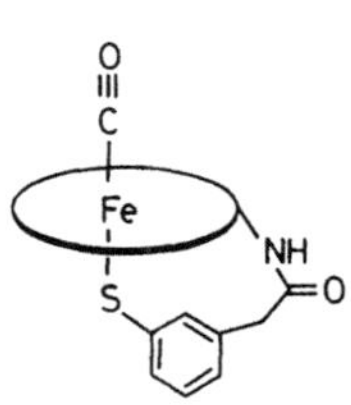

55

It is useful to compare these results to those of Traylor et al. who prepared simple thiol "tailed" protohaems [132]. They found, in agreement with Collman and others, that iron(II) did not form bis thiolate adducts. Hence, *56* and *57* exhibited identical electronic spectra (λ_{max} 384,460 nm). ^1H NMR proved critical in structure

Table 4. Selected parameters of P-450 models[a, b]

	^{1}H NMR	^{13}C NMR	λ_{max}	Ref.
57	$\left.\begin{array}{l}-2.04\\-3.58\end{array}\right\}\alpha$ $\left.\begin{array}{l}-1.49\\-2.17\end{array}\right\}\beta$	197.4	384, 460	[132]
62	0.7 to —0.6 (6 × bridging CH_2)			[133]
61	0.5 to —3.9 (6 × bridging CH_2 α and β CH's)	196.8		[133]
^{13}CO—P-450		200.3	370, 447	[134]

[a] all chemical shifts are in ppm (TMS = 0)
[b] λ_{max} values are in nm

assignment (see Table 4). The chemical shifts of the α and β protons of the "tail" were at very high field ~δ — 3. As this region is unobscured by protein resonances the possibility exists that there may be useful information retrievable from the ^{1}H NMR spectra of cytochrome P-450 in this region. The chemical shift of the carbonyl carbon in the ^{13}CO adduct of *57* is very similar to that of ^{13}CO—P-450 [134] and *61* (see Sect. 3.4).

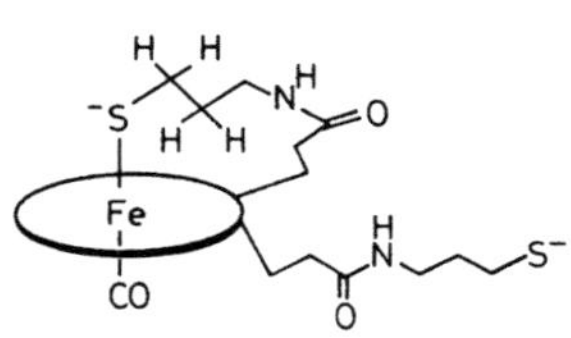

56

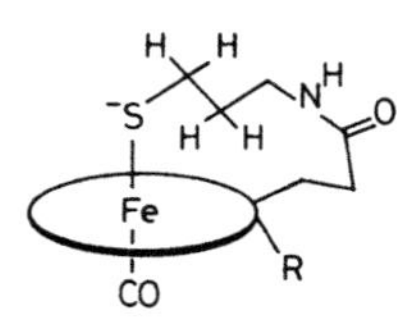

57

Traylor et al. [132] also found that the affinity for CO was *lower* than that of cytochrome P-450 whereas the dissociation rate was *greater*. They also noted that RS^- successfully competed with CO for the iron binding site. Other deprotonated ligands (e.g. Im^-) greatly reduce the affinity for CO of the appropriate iron(II) complex when compared to their protonated forms [135]. As dioxygen affinity increases, and CO affinity decreases, with increasing proximal basicity, it is possible that natural systems "fine-tune" the proximal basicity to suit their needs.

3.4 Bridged Porphyrins

Very recently Battersby et al. [133] reported the synthesis of a bridged iron(II) porphyrin with a thiolate ligand bound to the centre of the bridge (*60*). In order to avoid having iron(III) and RSH present in the molecule together the thiol was carried through the synthesis protected as its S-acetyl derivative. This eliminated the possibility of disulphide formation. (Traylor et al. [132] adopted a similar approach, but used the S-benzyl protected form and released it only after reduction of the iron(III) with sodium dithionite). The sallient features of the synthesis are shown in Scheme 8.

Exposure to dimsyl sodium effected conversion of *59* directly to the pentacoordinate iron(II) derivative *60*. The UV-VIS spectrum was similar to the reduced form of cyto-

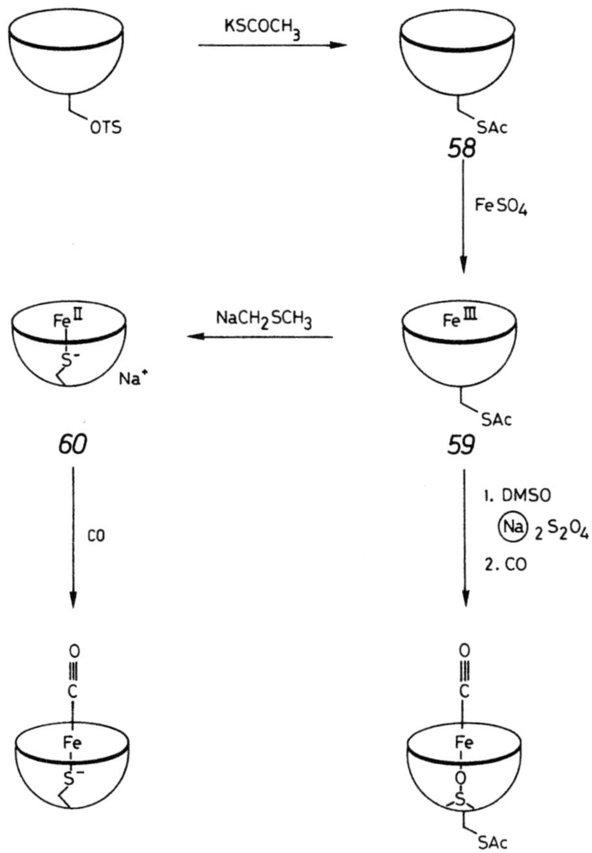

Scheme 8. Synthesis of Battersby's cytochrome P-450 model, Ref. [133]

chrome P-450$_{cam}$. Exposure of *60* to CO yielded hexacoordinate *61*, which displayed the characteristic hyperporphyrin spectrum of P-450. The ^{13}CO complex of *60* showed a single resonance at δ_C 196.8 ppm (see Table 4). This constitutes a significant shift upfield when compared to haems containing X→Fe(II)—^{13}CO (X=O$^-$ or ≡N) which resonate between δ_C 205 and 208 ppm [134]. A similar upfield shift has been observed in ^{13}CO—P-450 [134]. Finally, the methylene protons of the bridge are

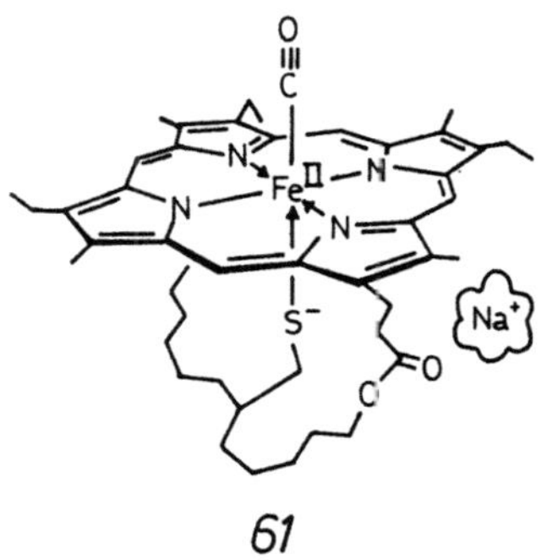

61

considerably shielded (δ_{CH_2} 0.5 to —3.9) in *61* when compared to the chemical shifts in *60* (δ_{CH_2} —1.8 to —0.3). This is in good general agreement with Traylor's findings and almost certainly reflects the greater proximity of the bridge to the porphyrin as a result of the tighter binding of the thiolate to the metal. No reference was made, in any of this work, to attempts to expose the pentacoordinate thiolate complexes to dioxygen.

4 Cytochrome *C* Oxidase

4.1 Occurrence and Function

Cytochrome *c* oxidase occurs in all animals, plants, certain bacteria and yeasts. The enzyme spans the inner mitochondrial membrane. It is responsible for catalysing the four electron reduction of dioxygen to water, the terminal reaction in the mitochondrial electron transport chain [136–143]. This is accomplished by accepting four reducing equivalents from cytochrome *c* and transferring them to the bound dioxygen, forming water:

$$4\,\text{cyt}\,c^{2+} + O_2 + 4\,H^+ \rightarrow 4\,\text{cyt}\,c^{3+} + 2\,H_2O \qquad (9)$$

The physiological importance of this redox process is that free energy is produced which can be stored by the conversion of two equivalents of ADP to ATP.

4.2 Structure and Mechanism

Cytochrome *c* oxidase, which is intensely coloured, can be obtained in reasonably pure form, most commonly from yeast and beef heart. The molecular weight appears to be approximately 140,000 [144]. The components of the enzyme consist of seven peptide subunits, two haems and two copper ions. The structure of both haems is identical (haem *a*). It is well recognized that cytochrome *c* oxidase contains two different cytochromes (*a* and a_3). The variation in the properties of the haem *a* prosthetic groups in each is undoubtedly due to differing protein environments.

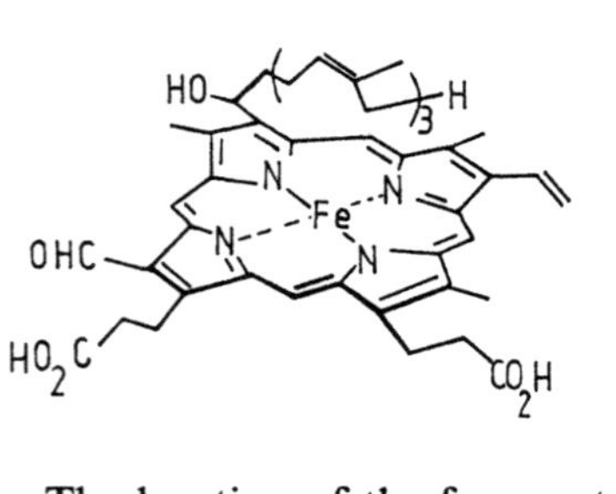

The location of the four metal ions within the enzyme has yet to be established unequivocally although the relative separation between them has been estimated by a number of techniques [145,146]. The following is a rough guide.

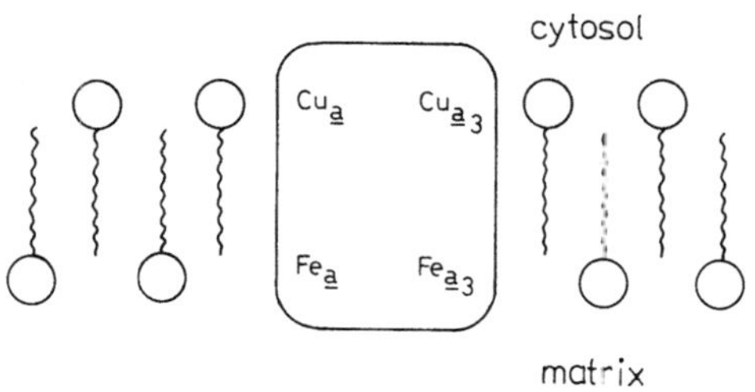

Fig. 4. Simplified representation of cychrome *c* oxidase in the inner mitochondrial membrane

A working model for the enzyme has emerged [138]. It consists of two functional units, cytochromes *a* and a_3 which contain haem environments analogous to those of the cytochromes of the electron transport chain and deoxyHb (or Mb) respectively. There is ample evidence that cytochrome a_3 is associated directly with the binding (and consequent reduction) of the O_2 molecule. Cytochrome *a* serves as an electron shuttle, mediating the transfer of reducing equivalents from cytochrome *c* to the cytochrome a_3. O_2 complex (see Fig. 5).

Only one iron and one copper ion are EPR detectable in the resting fully oxidised state. These are attributed to the cytochrome *a* pair. The fact that the other two metal ions are EPR "silent", whilst most likely being in their oxidized states (i.e. Fe(III) and Cu(II)) has lod to the suggestion that they are strongly antiferromagnetically coupled [147] (exchange integral $J > 200\ \text{cm}^{-1}$). An alternative theory involving an uncoupled high spin iron(IV)porphyrin and a copper(I) complex has also been proposed [148].

Hence some of the features of cytochrome *c* oxidase which invite modelling studies are (i) the catalytic four electron reduction of O_2 to H_2O, (ii) the electron transfer and spectroscopic properties of cytochrome *a* and (iii) and iron(III)—copper(II) complex which exhibits strong antiferromagnetic coupling.

4.3 Fenced Porphyrins

Gunter et al. [149] have prepared a series of models for cytochrome a_3 based on a picket fence design:

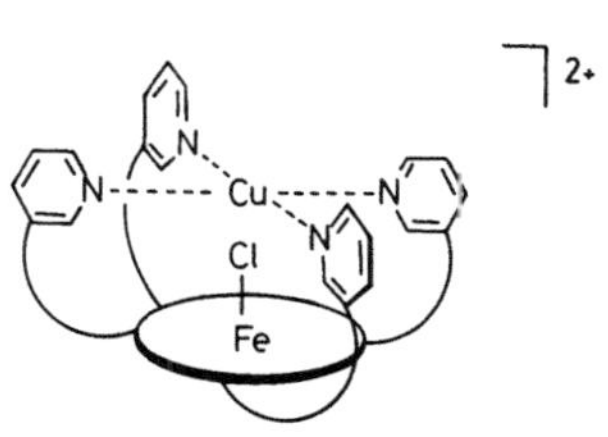

63

Sequential, selective metallation has yielded a number of mixed, bimetallic systems. The Fe(III)—Cu(II) containing complex *63* has been obtained in crystalline form and its structure has been determined [150]. Electronic, EPR, Mössbauer and magnetic susceptibility studies have also been carried out on *63* [150,151]. Although it was demonstrated that inclusion of Cu(II) into the complex significantly perturbs the electronic state of the iron(III), it appears that spin-coupling between the metal ions is close to zero.

4.4 Bridged Porphyrins

Gunter et al. [152] have also prepared a cytochrome a_3 model based on the bridged porphyrin *64*. The synthesis involved preparation of tetramethyldipyrromethane

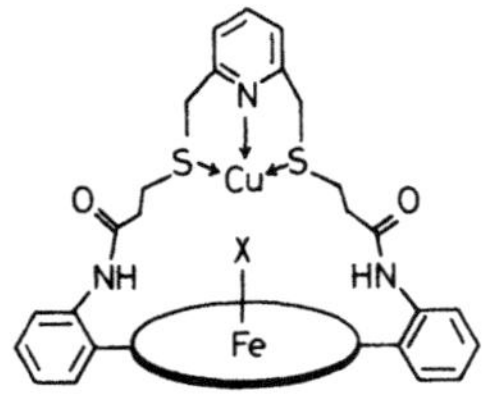

64

65

followed by acid catalysed condensation with *o*-nitrobenzaldehyde to give the α,γ-diarylporphyrin *65* in 60% yield (see Scheme 4.1). The bridging reaction was carried out in excellent yield (70%) under high dilution conditions. (This procedure is essentially complementary to that of Baldwin's [81], which couples two aryl substituted tetramethyldipyrromethanes, which may already be linked by a bridge, in the presence of trimethyl orthoformate.)

Unfortunately no evidence for spin-coupling in *64* was found.

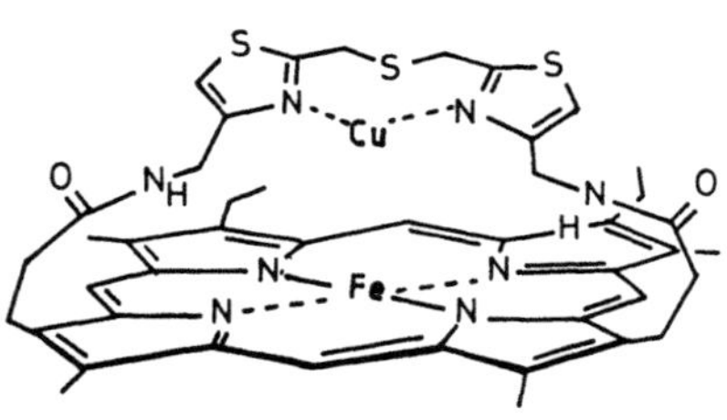

66

Very recently, Chang et al. [154] successfully prepared *66*, a cytochrome a_3 model which shows significant spin-coupling ($-3J = 132 \pm 5\ cm^{-1}$). This system contains several interesting features. It is one of the few examples of a non-symmetrically bridged porphyrin, the points of attachment being at adjacent (rather than opposite) pyrroles. The use of a β- rather than a *meso*-substituted porphyrin may be inportant as high spin iron(III) derivatives of the latter type tend to develop spin mixing with the $S = 3/2$ state [150,153]. The authors also felt it was important to avoid generation of Cu(II) with square planar geometry, as the magnetic orbitals may be mismatched with those of the iron(III).

4.5 Dimeric Porphyrins

The possibility that a bridging ligand may mediate [155] the magnetic exchange proposed to be operating in cytochrome a_3 has led to the preparation of μ-organo-bridged porphyrin dimers. For example, Reed et al. [156] have prepared the iron and manganese imidazolate dimers *67* and *68*. Antiferromagnetic coupling was minimal in these systems.

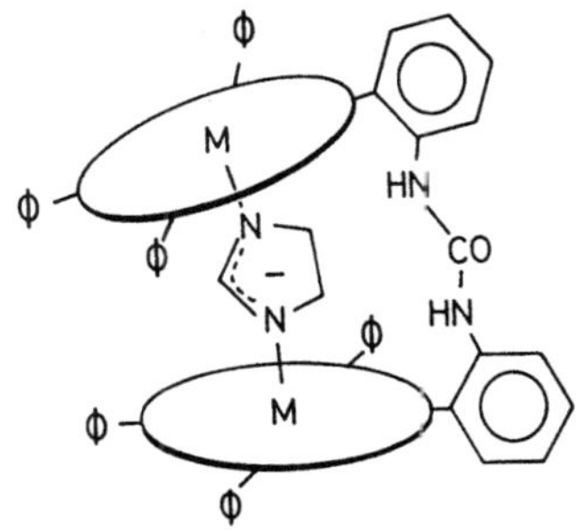

M = Fe *67*
M = Mn *68*

The catalysis of the four electron reduction of O_2 to H_2O has also received considerable attention. Using face-to-face dimeric porphyrins Collman et al. have successfully synthesized highly efficient catalysts for this purpose. (A full account of this work has recently appeared [157]). Reasoning that a μ-peroxo type configuration, involving one molecule of O_2 and two metal ions, should enhance the reduction rate several dimeric porphyrins were prepared. Variation in both the nature of the metal ions and the inter-porphyrin distance resulted in the preparation of the β-linked face-to-face porphyrin dimer *69*.

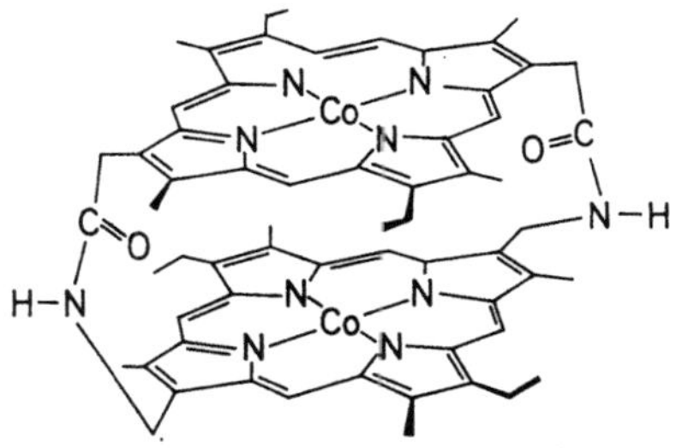

69

This catalyst apparently operates at rates close to those of cytochrome *c* oxidase and considerably greater than those of the previous best catalyst available, platinum. Fortuitously, it was found that substitution of the porphyrin periphery with chlorine enhanced the catalytic properties of the dimer. The synthesis of dimeric porphyrins has been reviewed at length [158] and will not be discussed further here.

Chang et al. [159] have also prepared some bimetallo-face-to-face porphyrin dimers and demonstrated the formation of a 1 : 1 μ-peroxo-bridged complex with O_2. However, no mention of the catalytic properties of these complexes was made.

5 A Miscellany

Several models designed to probe transannular donor-acceptor type interactions, of relevance to the primary reactions in bacterial photosynthesis [160], have been prepared. Kagan et al. [161] prepared the "strati-bis porphyrin" *70* by coupling two tetra *meso*-substituted porphyrins.

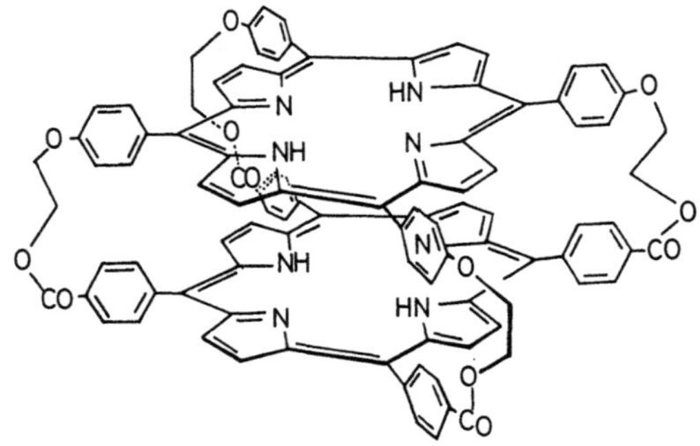

70

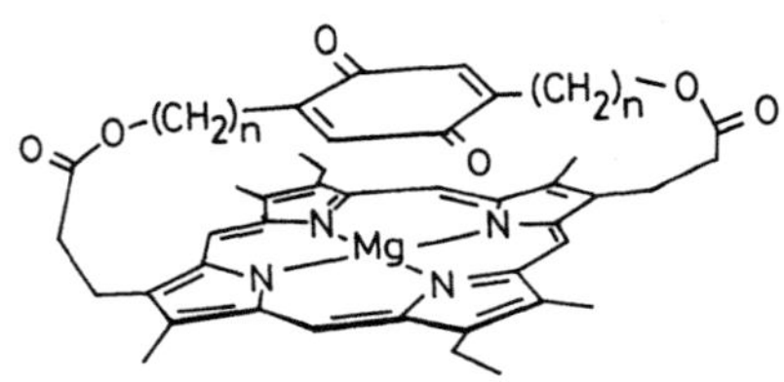

71

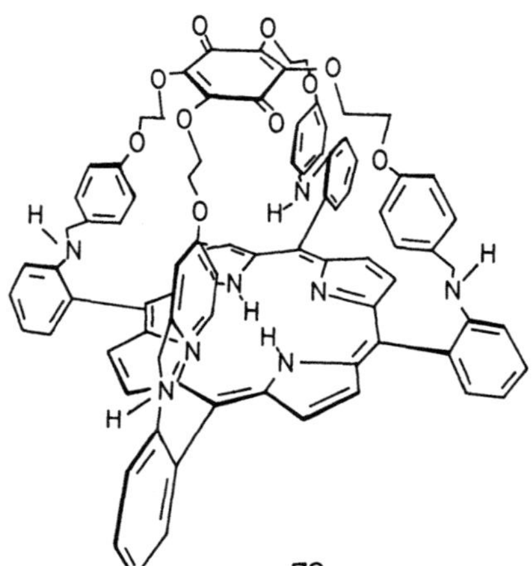

72

Ganesh and Saunders [162] and Lindsey and Mauzerall prepared the quinone-bridged *71* and quinone-capped *72* porphyrins respectively. The synthesis of *72* is especially interesting as the authors claim, and provide some evidence for, considerably enhanced yields for macrocyclisation through the condensation of rigid units which involves *reversible* reactions.

Wollmann and Hendrickson [163] prepared the exotic fenced porphyrin, *meso*-tetraferrocenylporphyrin *73*, as an inseparable mixture of all possible atropisomers.

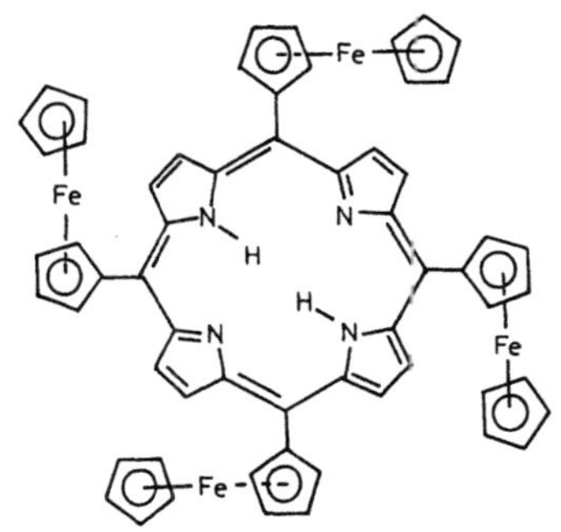

73

The synthesis of the remarkable *meso*-tetracarboranylprophyrins has been reported recently [164]. The crystal structure of one of these was determined and is shown below. Partial degradation of the carboranyl cages to $(-C_2B_9H_{10}R)^-$ groups gave water soluble derivatives. These groups provide four additional metal binding sites.

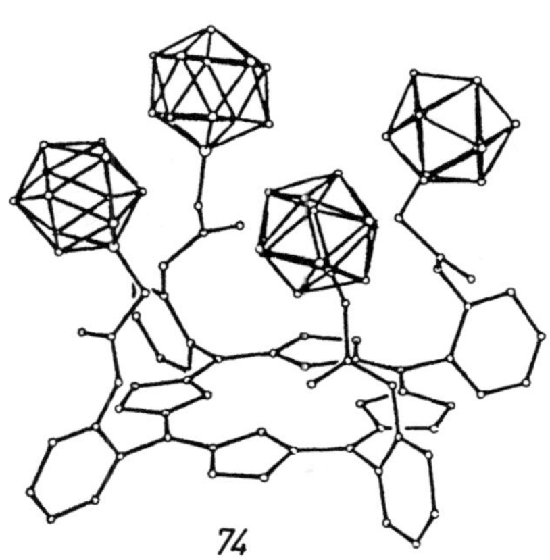

74

6 Symbols and Abbreviations

Por, dianion of porphyrin; C2Cap, dianion of the C2-capped porphyrin 5,10,15,20-[pyromellitoyltetrakis[*o*-(oxyethoxy)phenyl]]porphyrin; C3Cap, dianion of the C3-capped porphyrin 5,10,15,20-[pyromellitoyltetrakis[*o*-(oxypropoxy)phenyl]]porphyrin; TPP, dianion of *meso*-tetraphenylporphine; T(3,4,5-MeO)$_3$PP, dianion of *meso*-tetrakis(3,4,5-trimethoxyphenyl)porphine; TPivPP, dianion of the picket-fenced porphyrin *meso*-α,α,α,α-tetrakis[*o*-(pivalamido)phenyl]porphyrin; Piv$_3$(5CImP)P/ *meso*-α,α,α-tris[*o*-(β-pivalamido)phenyl]-β-[*o*-[5-(*N*-imidazoyl)valeramido]phenyl]-porphine; Piv$_3$(4CImP)P *meso*-α,α,α-tris-[*o*-(β-pivalamido)phenyl]-β-[*o*-[*N*-imidazoyl)valeramido]phenyl]porphine; Hb haemoglobin; Mb myoglobin; haem iron(II)-protoporphyrin IX; His histidine; 1-MeIm, 1-methylimidazole; Me$_2$Im, 1,2-dimethylimidazole; Im, imidazole; 2-MeIm, 2-methylimidazole; 1,5-DCI, 1,5-dicyclohexylimidazole; $P_{1/2}$ 2, the O_2 pressure at half saturation.

7 References

1. Maclean, N.: Studies in Biology, No. 93: Hoemoglobin, London, Edward Arnold 1978
2. Shaanan, B.: Nature (London) *296*, 683 (1982)
3. Phillips, S. E. V.: J. Mol. Biol. *142*, 531 (1980)
4. Fermi, G.: ibid. *97*, 237 (1975)
5. Takano, T.: ibid. *110*, 569 (1977)
6. Jameson, G. B., Molinaro, F. S., Ibers, J. A., Collman, J. P., Brauman, J. I., Rose, E., Suslick, K. S.: J. Am. Chem. Soc. *102*, 3224 (1980)
7. Jameson, G. B., Rodley, G. A., Robinson, W. T., Gagne, R. R., Reed, C. A., Collman, J. P.: Inorg. Chem. *17*, 850 (1978)
8. Jameson, G. B., Molinaro, F. S., Ibers, J. A., Collman, J. P., Brauman, J. I., Rose, E., Suslick, K. S.: J. Am. Chem. Soc. *100*, 6769 (1978)
9. Perutz, M. F.: Nature (London) *228*, 726 (1970)
10. Perutz, M. F.: ibid. *228*, 734 (1970)
11. Perutz, M. F.: Sci. Amer. *239*, 68 (1978)
12. Pauling, L.: Stanford Med. Bull. *6*, 215 (1948)
13. Pauling, L.: Nature (London) *203*, 182 (1964)
14. Phillips, S. E. V., Schoenborn, B. P.: ibid. *292*, 81 (1981)
15. Weiss, J. J.: ibid. *202*, 83 (1964)
16. Reed, C. A., Cheung, S. K.: Proc. Natl. Acad. Sci. USA *74*, 1780 (1977)
17. Olafson, B. D., Goddard, W. A.: ibid. *74*, 1315 (1977)
18. Case, D. A., Huynh, B. H., Karplus, M.: J. Am. Chem. Soc. *101*, 4433 (1979)
19. Drago, R. S., Corden, B. B.: Acc. Chem. Res. *13*, 353 (1980)
20. Hoffmann, R., Chen, M. M. L., Thorn, D. L.: Inorg. Chem. *16*, 503 (1977)
21. Kirchner, R. F., Loew, G. H.: J. Am. Chem. Soc. *99*, 4639 (1977)
22. Jones, R. D., Summerville, D. A. Basolo, F.: Chem. Rev. *79*, 139 (1979)
23. Collman, J. P., Halbert, T. R., Suslick, K. S.: Metal Ion Activation of Dioxygen, (ed. Spiro, T. G.) p. 1, New York, Wiley 1980
24. Basolo, F.: Coord. Chem. — Invited: Lect. Int. Conf. (ed. Banerjea, D.) Oxford, Pergamon 1980
25. Traylor, T. G., Traylor, P. S.: Annu. Rev. Biophys. Bioeng. *11*, 105 (1982)
26. Hopfield, J. J.: J. Mol. Biol. *77*, 207 (1973)
27. Gelin, B. R., Karplus, M.: Proc. Natl. Acad. Sci. USA *74*, 80 (1977)
28. Warshel, A., Weiss, R. M.: J. Am. Chem. Soc. *103*, 446 (1981)
29. Eisenberger, P., Shulman, R. G., Kincaid, B. M., Brown, G. S., Ogawa, S.: Nature (London) *274*, 30 (1978)
30. Perutz, M. F., Hashain, S. S., Duke, P. J., Sessler, J. L., Hahn, J. E.: ibid. *295*, 535 (1982)
31. Perutz, M. F.: Annu. Rev. Biochem. *48*, 327 (1979)
32. Perutz, M. F.: Proc. R. Soc., London, Ser. B. *208*, 135 (1980)
33. Baldwin, J., Chothia, C.: J. Mol. Biol. *129*, 175 (1979)
34. Cohen, I. A., Caughey, W. S.: Biochem *7*, 636 (1968)
35. Hammond, G. S., Wu, C-H., S.: Adv. Chem. Ser., No. 77, 186 (1968)

36. Kao, O. H., Wang, J. H.: Biochem. *4*, 342 (1965)
37. Collman, J. P., Gagne, R. R., Reed, C. A., Halbert, T. R., Lang, G., Robinson, W. T.: J. Am. Chem. Soc. *97*, 1427 (1975) and references cited therein
38. Chin, D.-H., Del Gaudio, J., LaMar, G. N., Balch, A. L.: J. Am. Chem. Soc. *99*, 5486 (1977)
39. Chin, D-H., Balch, A. L., LaMar, G. N.: ibid. *102*, 1446 (1980)
40. Chin, D-H., LaMar, G. N., Balch, A. L.: ibid. *102*, 4344 (1980)
41. Chin, D-H., LaMar, G. N., Balch, A. L.: ibid. *103*, 5945 (1980)
42. Latos-Grazynski, L., Cheng, R-J., LaMar, G. N., Balch, A. L.: ibid. *104*, 5992 (1982)
43. Shirazi, A., Goff, H. M.: ibid. *104*, 6318 (1982)
44. LaMar, G. N., de Ropp, J. S., Latos-Grazynski, L., Balch, A. L., Johnson, R. B., Smith, K. M., Parish, D. W., Cheng, R-J.: ibid. *105*, 782 (1983)
45. Nakamoto, K., Watanabe, T., Arna, T., Urban, M. W.: ibid. *104*, 3744 (1982); see also Kozuka, M., Nakamoto, K.: ibid. *103*, 2162 (1981)
46. Baldwin, J. E., Huff, J.: ibid. *95*, 5757 (1973)
47. Little, R. G., Ibers, J. A., Baldwin, J. E.: ibid. *97*, 7049 (1975)
48. Collman, J. P., Gagne, R. R., Halbert, T. R., Marchon, J.-C., Reed, C. A.: ibid. *95*, 7868 (1973)
49. Collman, J. P., Acc. Chem. Res. *10*, 265 (1977)
50. Collman, J. P., Suslick, K. S.: P. and App. Chem. *50*, 951 (1978)
51. Rougée, M., Brault, D.: Biochem. *14*, 4100 (1975) and references cited therein
52. Collman, J. P., Brauman, J. I., Doxsee, K. M., Halbert, T. R., Bunnenberg, E., Linder, R. E., LaMar, G. N., Del Gaudio, J., Lang, G., Spartalian, K.: J. Am. Chem. Soc. *102*, 4182 (1980)
53. Collman, J. P., Gagne, R. R., Reed, C. A., Robinson, W. T., Rosley, G. A.: Proc. Natl. Acad. Sci, USA *71*, 1326 (1974)
54. Ref. 23, p. 55 et seq.
55. Griffith, J. S.: Proc. R. Soc. Ser. A. *235*, 23 (1956)
56. Maines, M. D., Kappos, A.: J. Biol. Chem. *250*, 2363 (1975)
57. Collman, J. P., Brauman, J. I., Suslick, K. S.: J. Am. Chem. Soc. *97*, 7185 (197)
58. Hashimoto, T.; Dyer, R. L., Crossley, M. J., Baldwin, J. E., Basolo, F.: ibid. *104*, 2101 (1982)
59. Collman, J. P., Brauman, J. I., Doxsee, K. M., Halbert, T. R., Suslick, K. S.: Proc. Natl. Acad. Sci., USA *75*, 564 (1978)
60. Collman, J. P., Brauman, J. I., Doxsee, K. M., Halbert, T. R., Suslick, K. S.: ibid. *76*, 6035 (1979)
61. Collman, J. P., Brauman, J. I., Collins, T. J., Iverson, B., Sessler, J. L.: J. Am. Chem. Soc. *103*, 2450 (1981)
62. Traylor, T. G., Mitchell, M. J., Tsuchiya, S., Campbell, D. H., Stynes, D. V., Koga, N.: ibid. *103*, 5234 (1981)
63. Traylor, T. G., Stynes, D. V.: ibid. *102*, 5938 (1980)
64. Traylor, T. G., Campbell, D., Tsuchiya, S., Mitchell, M., Stynes, D. V.: ibid. *102*, 5939 (1980)
65. Yonetani, T., Yamamoto, H., Woodrow, G. V.: J. Biol. Chem *249*, 682 (1974)
66. Amiconi, G., Antonini, E., Brunori, M., Magnusson, E.: unpublished results, see footnote 29, ref. 58
67. Antonini, E., Brunori, M.: Hemoglobin and Myoglobin in Their Reactions with Ligands, New York, Elsevier 1971
68. MacQuarrie, R., Cubson, Q. H.: J. Biol. Chem. *246*, 5832 (1971)
69. Sharma, V. S., Schmidt, M. R., Ranney, H. M.: ibid. *251*, 4267 (1976)
70. Roughton, F. J. W.: J. Physiol. (London) *126*, 359 (1954)
71. Huestis, W. H., Raftery, H. M.: Biochem. *14*, 1886 (1975)
72. See ref. 25 for a discussion of these effects
73. Ward, B., Wang, C.-B., Chang, C. K.: J. Am. Chem. Soc. *103*, 5236 (1981)
74. Rothemund, P.: ibid. *57*, 2010 (1935)
75. Lindsey, J.: J. Org. Chem. *45*, 5215 (1980)
76. Porphyrins and Metalloporphyrins, (ed. Smith, K. M.) Chapters 5 and 6, Amsterdam, Elsevier 1975
77. Amundsen, A. R., Vaska, L.: Inorg. Chim. *14*, 49 (1975)
78. Vaska, L., Amundsen, A. R., Brady, R., Flynn, B. R., Nakni, H.: Finn. Chem. Lett., 66 (1974)
79. See footnote no. 38 in ref. 42

80. Almog, J., Baldwin, J. E., Dyer, R. L., Peters, M. K.: J. Am. Chem. Soc. *97*, 226 (1975)
81. Almog, J., Baldwin, J. E., Crossley, M. J., DeBernardis, J. F., Dyer, R. L., Huff, J. R., Peters, M. K.: Tetrahedron *37*, 3589 (1981)
82. Haraldsson, G. G.: D. Phil. Thesis, Wolfson College, Oxford 1982
83. Baldwin, J. E., De Bernardis, J. F.: J. Org. Chem. *42*, 396 (1977)
84. Baldwin, J. E., Crossley, M. J., De Bernardis, J. F.: Tetrahedron *38*, 685 (1982)
85. Almog, J., Baldwin, J. E., Huff, J.: J. Am. Chem. Soc. *97*, 227 (1975)
86. Ellis, Jr., P. E., Linard, J. E., Szynamski, T., Jones, R. D., Budge, J. R., Basolo, F.: ibid. *102*, 1889 (1980)
87. Budge, J. R., Ellis Jr., P. E., Jones, R. D., Linard, J. E., Basolo, F., Baldwin, J. E., Dyer, R. L.: ibid. *101*, 4760 (1979)
88. Budge, J. R., Ellis Jr., P. E., Jones, R. D., Linard, J. E., Szymanski, T., Basolo, F., Baldwin, J. E., Dyer, R. L.: ibid. *101*, 4762 (1979)
89. Linard, J. E., Ellis Jr., P. E., Budge, J. R., Jones, R. D., Basolo, F.: ibid. *102*, 1896 (1980)
90. See Rose, E. J., Venkatasubramanian, P. N., Swartz, J. C., Jones, R. D., Basolo, F., Hoffman, B. M.: Proc. Natl. Acad. Sci. USA *79*, 5742 (1982) for a related study of carbon monoxide binding to capped porphyrins
91. Jameson, G. B., Ibers, J. A.: J. Am. Chem. Soc. *102*, 2823 (1980)
92. Sabat, M., Ibers, J. A.: ibid. *104*, 3715 (1982)
93. Clayden, J. J., Moore, G. R., Williams, R. J. P., Baldwin, J. E., Crossley, M. J.: J. Chem. Soc. Perkin. Trans. II, 1693 (1982)
94. Hashimoto, T., Basolo, F.: Comm. Inorg. Chem. *1*, 199 (1981)
95. Battersby, A. R., Hamilton, A. D.: J. Chem. Soc. Chem. Commun., 117 (1980)
96. Vallejos, M.: D. Phil. Thesis, Lineare College, Oxford 1982
97. Baldwin, J. E., Crossley, M. J., Dagley, I. J., Hall, S. R., Haraldsson, G. G., Perlmutter, P., Sutherland, J. M.: unpublished results
98. Collman, J. P., Groh, S. E.: J. Am. Chem. Soc. *104*, 1391 (1982)
99. Baldwin, J. E., Klose, T., Peters, M. K.: J. Chem. Soc. Chem. Commun., 881 (1976)
100. Baldwin, J. E., Crossley, M. J., Klose, T., O'Rear III, E. A., Peters, M. K.: Tetrahedron *38*, 27 (1982)
101. Chang, C. K.: J. Am. Chem. Soc. *99*, 2819 (1977)
102. Battersby, A. R., Hartley, S. G., Turnbull, M. D.: Tetrahedron Lett. 3169 (1978)
103. Battersby, A. R., Buckley, D. G., Hartley, S. G., Turnbull, S. G.: J. Chem. Soc. Chem. Commun., 879 (1976)
104. Jackson, A. H., Kenner, G. W., Wass, J.: J. Chem. Soc. Perkin Trans 1, 1475 (1972)
105. Traylor, T. G., Stynes, D. V.: J. Am. Chem. Soc. *102*, 5938 (1980)
106. Traylor, T. G., Campbell, D., Tsuchiya, S., Mitchell, M., Stynes, D. V.: ibid. *102*, 5939 (1980)
107. Collman, J. P., Brauman, J. I., Halbert, T. R., Suslick, K. S.: Proc. Natl. Acad. Sci. USA *783*, 3333 (1976)
108. Momenteau, M., Loock, B., Mispelter, J., Bisagni, E.: Nouv. J. de Chim. *3*, 77 (1979)
109. Momenteau, M., Mispelter, J., Loock, B., Bisagni, E.: J. Chem. Soc. Perkin Trans. 1, 189 (1983)
110. Momenteau, M., Lavalette, D.: J. Chem. Soc. Chem. Commun., 341 (1982)
111. Heidner, E. J., Ladner, R. C., Perutz, M. F.: J. Mol. Biol. *104*, 707 (1976)
112. Cruse, W. B., Kennard, O., Sheldrick, G. M., Hamilton, A. D., Hartley, S. G., Battersby, A. R.: J. Chem. Soc. Chem. Commun., 700 (1980)
113. Momenteau, M., Loock, B.: J. Mol. Cat. *7*, 315 (1980)
114. Cytochrome P-450, (ed. Sato, R., Omura, T.) New York, Academic Press 1978
115. White, R. E., Coon, M. J.: Annu. Rev. Biochem. *49*, 315 (1980)
116. Coon, M. J., White, R. E.: Metal Ion Activation of Dioxygen, (ed. Spiro, T. G.) New York, Wiley 1980
117. Griffin, B. W., Peterson, J. A., Estabrook, R. W.: the Porphyrins, (ed. D. Dolphin) New York, Academic Press, Vol. VII, 1979
118. Gunsalus, I. C., Sligar, S. G.: Adv. in Enzymol. *47*, 1 (1978)
119. Dawson, J. H., Anderson, L. A., Davis, J. M., Hahn, J. E.: Developments in Biochem., Vol. 13: Biochemistry, Biophysics and Regulation of Cytochrome P-450 (eds. Gustafsson, J. A., Carlstedt-Duke, J., Mode, A., Rafter, J.) Amsterdam, Elsevier 1980

120. Molecular Mechanisms of Oxygen Activation, (ed. Hayaishi, O.) New York, Academic Press 1974
121. Biological Hydroxylation Mechanisms, (eds. Boyd, G. S., Smellie, R. M. S.) New York, Academic Press 1972
122. See Chapter 7 of ref. 82 for a discussion
123. Chevion, M., Peisach, J., Blumberg, W.: J. Biol. Chem. *252*, 3637 (1977)
124. Rein, H., Pirrwitz, J., Friedrich, J., Ristaù, O., Janig, G. R., Ruckpaul, K.: Developments in Biochem., Vol. 13: Biochemistry, Biophysics and Regulation of Cytochrome P-450, (eds. Gustaffson, J.-A., Carlstedt-Duke, J., Mode, A., Refer, J.) Amsterdam, Elsevier 1980
125. Griffin, B. W., Peterson, J. A.: J. Biol. Chem. *250*, 6445 (1975)
126. Ruf, H. H., Wende, P., Ullrich, V.: J. Inorg. Biochem. *11*, 265 (1979)y
127. White, R. E.: Fed. Proc. *38*, 320 (1979)
128. Hewson, W. D., Hager, L. P.: The Porphyrins, (ed. D. Dolphin) New York, Vol. VII, Academic Press 1979
129. Hanson, L. K., Eaton, W. A., Sligan, S. G., Gunsalus, I. C., Gouterman, M., Connel, C. R.: J. Am. Chem. Soc. *98*, 2672 (1976)
130. Groves, J. T., Haushalter, R. C., Nakamura, M., Nemo, T. E., Evans, B. J.: ibid. *103*, 2884 (1981)
131. Footnote no 22 in ref. 130
132. Traylor, T. G., Mincey, T. C., Berzinis, A. P.: J. Am. Chem. Soc. *103*, 7084 (1981)
133. Battersby, A. R., Howson, W., Hamilton, A. D.: J. Chem. Soc. Chem. Commun., 1266 (1982)
134. Berzinis, A. P., Traylor, T. G.: Biochem. Biophys. Res. Commun. *87*, 229 (1979)
135. Mincey, T. C., Traylor, T. G.: J. Am. Chem. Soc. *101*, 765 (1979)
136. Wikström, M., Krah, K., Saraste, M.: Cytochrome Oxidase, A Synthesis, New York, Academic Press 1981
137. Malmström, B. G.: Biochem. Biophys. Acta *549*, 281 (1979)
138. Malmström, B. G.: Metal Ion Activation of Dioxygen, (ed. Spiro, T. G.) New York, Wiley 1980
139. Azzi, A.: Biochem. Biophys. Acta *594*, 231 (1980)
140. Wilson, D. F., Erecińska, M.: The Porphyrins, (ed. D. Dolphin) New York, Academic Press, Vol. VII, 1979
141. Wilson, D. F., Erecińska, M.: Arch. Biochem. Biophys. *188*, 1 (1978)
142. Cytochrome Oxidase (eds. King, T. E., Orii, Y., Chance, B., Okunuki, K.) New York, Elsevier 1979
143. Palmer, G.: Mechanisms of Oxidizing Enzymes, (eds. Singer, T. P., Ondarga, R.) Amsterdam, Elsevier 1978
144. Briggs, M. M., Capaldi, R. A.: Biochem. *16*, 73 (1977)
145. Ohnishi, T., LoBrutto, R., Salerno, J. C., Bruckner, R. C., Frey, T. G.: J. Biol. Chem. *257*, 14821 (1982)
146. Dockter, M. E., Steineman, A., Shatz, G.: ibid. *253*, 311 (1978)
147. Van Gelder, B. F., Beinert, H.: Biochem. Biophys. Acta *189*, 1 (1969)
148. Seiter, C. H., Angelos, A.: Proc. Natl. Acad. Sci. USA *77*, 1806 (1980)
149. Buckingham, D. A., Gunter, M. J., Mander, L. N.: J. Am. Chem. Soc. *100*, 2899 (1978)
150. Gunter, M. J., Mander, L. N., McLaughlin, G. M., Murray, K. S., Berry, K. J., Clark, P. E., Buckingham, D. A.: ibid. *102*, 1470 (1980)
151. Berry, K. J., Clark, P. E., Gunter, M. J., Murray, K. S.: Nouv. J. de Chim. *4*, 581 (1980)
152. Gunter, M. J., Mander, L. N.: J. Org. Chem. *46*, 4792 (1981)
153. Gunter, M. J., Mander, L. N., Murray, K. S., Clark, P. E.: J. A. Chem. Soc. *103*, 6784 (1981)
154. Chang, C. K., Koo, M. S., Ward, B.: J. Chem. Soc. Chem. Commun., 716 (1982)
155. Palmer, G., Babcock, G. T., Vickery, L. E.: Proc. Natl. Acad. Sci. USA, *73*, 2206 (1976)
156. Landrum, J. T., Reed, C. A., Hatano, K., Scheidt, W. R.: J. Am. Chem. Soc. *100*, 3232 (1978)
157. Collman, J. P., Anson, F. C., Bencosme, S., Chong, A., Collins, T., Denisevich, P., Evitt, E., Geiger, T., Ibers, J. A., Jameson, G., Konai, Y., Koval, C., Meier, K., Oakley, P., Pettman, R., Schmittun, E., Sessler, J.: Organic Synthesis Today and Tomorrow, (eds. Trost, B. M., Hutchinson, C. R.) Oxford, Pergamon 1981
158. Dolphin, D., Hiom, J., Paine III, J. B.: Heterocycles *16*, 417 (1981)

159. Chang, C. K.: Inorganic Compounds with Unusual Properties II (ed. King, B. R.) A.C.S. Adv. Chem. Ser. No. 173, 162 (1979)
160. Clayton, R. K.: the Photosynthetic Bacteria (eds. Clayton, R. K., Sistrom, W. R.) New York, Plenum 1978
161. Kagan, N. E., Mauzerall, D., Merrifield, R. B.: J. Am. Chem. Soc. *99*, 5484 (1977)
162. Ganesh, K. M., Sanders, J. K. M.: J. Chem. Soc. Chem. Commun., 1129 (1980)
163. Wollmann, R. G., Hendrickson, D. N.: Inorg. Chem. *16*, 3079 (1977)
164. Haushalter, R. C., Butler, W. M., Rudolph, R. W.: J. Am. Chem. Soc. *103*, 2620 (1981)

Author Index Volumes 101–121

Contents of Vols. 50–100 see Vol. 100
Author and Subject Index Vols. 26–50 see Vol. 50

The volume numbers are printed in italics

Ashe, III, A. J.: The Group 5 Heterobenzenes Arsabenzene, Stibabenzene and Bismabenzene. *105*, 125–156 (1982).
Austel, V.: Features and Problems of Practical Drug Design, *114*, 7–19 (1983).

Balaban, A. T., Motoc, I., Bonchev, D., and Mekenyan, O.: Topilogical Indices for Structure-Activity Correlations, *114*, 21–55 (1983).
Baldwin, J. E., and Perlmutter, P.: Bridged, Capped and Fenced Porphyrins. *121*, 181–220 (1984).
Barkhash, V. A.: Contemporary Problems in Carbonium Ion Chemistry I. *116/117*, 1–265 (1984).
Barthel, J., Gores, H.-J., Schmeer, G., and Wachter, R.: Non-Aqueous Electrolyte Solutions in Chemistry and Modern Technology. *111*, 33–144 (1983).
Bestmann, H. J., Vostrowsky, O.: Selected Topics of the Wittig Reaction in the Synthesis of Natural Products. *109*, 85–163 (1983).
Beyer, A., Karpfen, A., and Schuster, P.: Energy Surfaces of Hydrogen-Bonded Complexes in the Vapor Phase. *120*, 1–40 (1984).
Boekelheide, V.: Syntheses and Properties of the $[2_n]$ Cyclophanes, *113*, 87–143 (1983).
Bonchev, D., see Balaban, A. T., *114*, 21–55 (1983).
Bourdin, E., see Fauchais, P.: *107*, 59–183 (1983).

Charton, M., and Motoc, I.: Introduction, *114*, 1–6 (1983).
Charton, M.: The Upsilon Steric Parameter Definition and Determination, *114*, 57–91 (1983).
Charton, M.: Volume and Bulk Parameters, *114*, 107–118 (1983).
Chivers, T., and Oakley, R. T.: Sulfur-Nitrogen Anions and Related Compounds. *102*, 117–147 (1982).
Consiglio, G., and Pino, P.: Asymmetrie Hydroformylation. *105*, 77–124 (1982).
Coudert, J. F., see Fauchais, P.: *107*, 59–183 (1983).

Dyke, Th. R.: Microwave and Radiofrequency Spectra of Hydrogen Bonded Complexes in the Vapor Phase. *120*, 85–113 (1984).

Edmondson, D. E., and Tollin, G.: Semiquinone Formation in Flavo- and Metalloflavoproteins. *108*, 109–138 (1983).
Eliel, E. L.: Prostereoisomerism (Prochirality). *105*, 1–76 (1982).

Fauchais, P., Bordin, E., Coudert, F., and MacPherson, R.: High Pressure Plasmas and Their Application to Ceramic Technology. *107*, 59–183 (1983).
Fujita, T., and Iwamura, H.: Applications of Various Steric Constants to Quantitative Analysis of Structure-Activity Relationshipf, *114*, 119–157 (1983).

Gerson, F.: Radical Ions of Phanes as Studied by ESR and ENDOR Spectroscopy. *115*, 57–105 (1983).
Gielen, M.: Chirality, Static and Dynamic Stereochemistry of Organotin Compounds. *104*, 57–105 (1982).
Gores, H.-J., see Barthel, J.: *111*, 33–144 (1983).
Groeseneken, D. R., see Lontie, D. R.: *108*, 1–33 (1983).
Gurel, O., and Gurel, D.: Types of Oscillations in Chemical Reactions. *118*, 1–73 (1983).
Gurel, D., and Gurel, O.: Recent Developments in Chemical Oscillations. *118*, 75–117 (1983).

Heilbronner, E., and Yang, Z.: The Electronic Structure of Cyclophanes as Suggested by their Photoelectron Spectra. *115*, 1–55 (1983).
Hellwinkel, D.: Penta- and Hexaorganyl Derivatives of the Main Group Elements. *109*, 1–63 (1983).
Hess, P.: Resonant Photoacoustic Spectroscopy. *111*, 1–32 (1983).
Hilgenfeld, R., and Saenger, W.: Structural Chemistry of Natural and Synthetic Ionophores and their Complexes with Cations. *101*, 3–82 (1982).

Iwamura, H., see Fujita, T., *114*, 119–157 (1983).

Kaden, Th. A.: Syntheses and Metal Complexes of Aza-Macrocycles with Pendant Arms having Additional Ligating Groups. *121*, 157–179 (1984).
Karpfen, A., see Beyer, A.: *120*, 1–40 (1984).
Káš,J., Rauch, P.: Labeled Proteins, Their Preparation and Application. *112*, 163–230 (1983).
Keat, R.: Phosphorus(III)-Nitrogen Ring Compounds. *102*, 89–116 (1982).
Kellogg, R. M.: Bioorganic Modelling — Stereoselective Reactions with Chiral Neutral Ligand Complexes as Model Systems for Enzyme Catalysis. *101*, 111–145 (1982).
Kniep, R., and Rabenau, A.: Subhalides of Tellurium. *111*, 145–192 (1983).
Krebs, S., Wilke, J.: Angle Strained Cycloalkynes. *109*, 189–233 (1983).
Kosower, E. M.: Stable Pyridinyl Radicals, *112*, 117–162 (1983).

Labarre, J.-F.: Up to-date Improvements in Inorganic Ring Systems as Anticancer Agents. *102*, 1–87 (1982).
Laitinen, R., see Steudel, R.: *102*, 177–197 (1982).
Landini, S., see Montanari, F.: *101*, 111–145 (1982).
Lavrent'yev, V. I., see Voronkov, M. G.: *102*, 199–236 (1982).
Lontie, R. A., and Groeseneken, D. R.: Recent Developments with Copper Proteins. *108*, 1–33 (1983).
Lynch, R. E.: The Metabolism of Superoxide Anion and Its Progeny in Blood Cells. *108*, 35–70 (1983).

McPherson, R., see Fauchais, P.: *107*, 59–183 (1983).
Majestic, V. K., see Newkome, G. R.: *106*, 79–118 (1982).
Manabe, O., see Shinkai, S.: *121*, 67–104 (1984).
Margaretha, P.: Preparative Organic Photochemistry. *103*, 1–89 (1982).
Mekenyan, O., see Balaban, A. T., *114*, 21–55 (1983).
Montanari, F., Landini, D., and Rolla, F.: Phase-Transfer Catalyzed Reactions. *101*, 149–200 (1982).
Motoc, I., see Charton, M.: *114*, 1–6 (1983).
Motoc, I., see Balaban, A. T.: *114*, 21–55 (1983).
Motoc, I.: Molecular Shape Descriptors, *114*, 93–105 (1983).
Müller, F.: The Flavin Redox-System and Its Biological Function. *108*, 71–107 (1983).
Murakami, Y.: Functionalited Cyclophanes as Catalysts and Enzyme Models. *115*, 103–151 (1983).
Mutter, M., and Pillai, V. N. R.: New Perspectives in Polymer-Supported Peptide Synthesis. *106*, 119–175 (1982).

Newkome, G. R., and Majestic, V. K.: Pyridinophanes, Pyridinocrowns, and Pyridinycryptands. *106*, 79–118 (1982).

Oakley, R. T., see Chivers, T.: *102*, 117–147 (1982).

Painter, R., and Pressman, B. C.: Dynamics Aspects of Ionophore Mediated Membrane Transport. *101*, 84–110 (1982).
Paquette, L. A.: Recent Synthetic Developments in Polyquinane Chemistry. *119*, 1–158 (1984)
Perlmutter, P., see Baldwin, J. E.: *121*, 181–220 (1984).
Pillai, V. N. R., see Mutter, M.: *106*, 119–175 (1982).
Pino, P., see Consiglio, G.: *105*, 77–124 (1982).
Pommer, H., Thieme, P. C.: Industrial Applications of the Wittig Reaction. *109*, 165–188 (1983).
Pressman, B. C., see Painter, R.: *101*, 84–110 (1982).

Rabenau, A., see Kniep, R.: *111*, 145–192 (1983).
Rauch, P., see Káš, J.: *112*, 163–230 (1983).
Recktenwald, O., see Veith, M.: *104*, 1–55 (1982).
Reetz, M. T.: Organotitanium Reagents in Organic Synthesis. A Simple Means to Adjust Reactivity and Selectivity of Carbanions. *106*, 1–53 (1982).
Rolla, R., see Montanari, F.: *101*, 111–145 (1982).
Rossa, L., Vögtle, F.: Synthesis of Medio- and Macrocyclic Compounds by High Dilution Principle Techniques, *113*, 1–86 (1983).
Rzaev, Z. M. O.: Coordination Effects in Formation and Cross-Linking Reactions of Organotin Macromolecules. *104*, 107–136 (1982).

Saenger, W., see Hilgenfeld, R.: *101*, 3–82 (1982).
Sandorfy, C.: Vibrational Spectra of Hydrogen Bonded Systems in the Gas Phase. *120*, 41–84 (1984).
Schmeer, G., see Barthel, J.: *111*, 33–144 (1983).
Schöllkopf, U.: Enantioselective Synthesis of Nonproteinogenic Amino Acids. *109*, 65–84 (1983).
Schuster, P., see Beyer, A., see *120*, 1–40 (1984).
Shibata, M.: Modern Syntheses of Cobalt(III) Complexes. *110*, 1–120 (1983).
Shinkai, S., and Manabe, O.: Photocontrol of Ion Extraction and Ion Transport by Photofunctional Crown Ethers. *121*, 67–104 (1984).
Shubin, V. G.: Contemporary Problems in Carbonium Ion Chemistry II. *116/117*, 267–341 (1984).
Siegel, H.: Lithium Halocarbenoids Carbanions of High Synthetic Versatility. *106*, 55–78 (1982).
Sinta, R., see Smid, J.: *121*, 105–156 (1984).
Smid, J., and Sinta, R.: Macroheterocyclic Ligands on Polymers. *121*, 105–156 (1984).
Steudel, R.: Homocyclic Sulfur Molecules. *102*, 149–176 (1982).
Steudel, R., and Laitinen, R.: Cyclic Selenium Sulfides. *102*, 177–197 (1982).
Suzuki, A.: Some Aspects of Organic Synthesis Using Organoboranes. *112*, 67–115 (1983).
Szele, J., Zollinger, H.: Azo Coupling Reactions Structures and Mechanisms. *112*, 1–66 (1983).

Tabushi, I., Yamamura, K.: Water Soluble Cyclophanes as Hosts and Catalysts, *113*, 145–182 (1983).
Takagi, M., and Ueno, K.: Crown Compounds as Alkali and Alkaline Earth Metal Ion Selective Chromogenic Reagents. *121*, 39–65 (1984).
Takeda, Y.: The Solvent Extraction of Metal Ions by Crown Compounds. *121*, 1–38 (1984).
Thieme, P. C., see Pommer, H.: *109*, 165–188 (1983).
Tollin, G., see Edmondson, D. E.: *108*, 109–138 (1983).

Ueno, K. see Takagi, M.: *121*, 39–65 (1984).

Veith, M., and Recktenwald, O.: Structure and Reactivity of Monomeric, Molecular Tin(II) Compounds. *104*, 1–55 (1982).
Venugopalan, M., and Veprek, S.: Kinetics and Catalysis in Plasma Chemistry. *107*, 1–58 (1982).

Vepřek, S., see Venugopalan, M.: *107*, 1–58 (1983).
Vögtle, F., see Rossa, L.: *113*, 1–86 (1983).
Vögtle, F.: Concluding Remarks. *115*, 153–155 (1983).
Vostrowsky, O., see Bestmann, H. J.: *109*, 85–163 (1983).
Voronkov. M. G., and Lavrent'yev, V. I.: Polyhedral Oligosilsequioxanes and Their Homo Derivatives. *102*, 199–236 (1982).

Wachter, R., see Barthel, J.: *111*, 33–144 (1983).
Wilke, J., see Krebs, S.: *109*, 189–233 (1983).

Yamamura, K., see Tabushi, I.: *113*, 145–182 (1983).
Yang, Z., see Heilbronner, E.: *115*, 1–55 (1983).

Zollinger, H., see Szele, I.: *112*, 1–66 (1983).

GPSR Compliance

The European Union's (EU) General Product Safety Regulation (GPSR) is a set of rules that requires consumer products to be safe and our obligations to ensure this.

If you have any concerns about our products, you can contact us on ProductSafety@springernature.com

In case Publisher is established outside the EU, the EU authorized representative is:

Springer Nature Customer Service Center GmbH
Europaplatz 3
69115 Heidelberg, Germany

Batch number: 10620881

Printed by Printforce, the Netherlands